Advanced Machining Processes

Manufacturing Design and Technology Series

Series Editor
J. Paulo Davim

PUBLISHED

Advanced Machining Processes: Innovative Modeling Techniques
Angelos P. Markopoulos and J. Paulo Davim

Additive Manufacturing and Optimization: Fundamentals and Applications
V. Vijayan, Suresh B. Kumar, and J. Paulo Davim

Technological Challenges and Management: Matching Human and Business Needs
Carolina Machado and J. Paulo Davim

Drills: Science and Technology of Advanced Operations
Viktor P. Astakhov

Advanced Machining Processes

Innovative Modeling Techniques

Edited by
Angelos P. Markopoulos
J. Paulo Davim

CRC Press
Taylor & Francis Group
Boca Raton London New York

CRC Press is an imprint of the
Taylor & Francis Group, an **informa** business

CRC Press
Taylor & Francis Group
6000 Broken Sound Parkway NW, Suite 300
Boca Raton, FL 33487-2742

First issued in paperback 2019

© 2018 by Taylor & Francis Group, LLC
CRC Press is an imprint of Taylor & Francis Group, an Informa business

No claim to original U.S. Government works

ISBN-13: 978-1-138-03362-7 (hbk)
ISBN-13: 978-0-367-88598-4 (pbk)

Library of Congress Cataloging-in-Publication Data

Names: Markopoulos, Angelos P., 1976- author. | Davim, J. Paulo, author.
Title: Advanced machining processes : innovative modeling techniques / Angelos P. Markopoulos, J. Paulo Davim.
Description: Boca Raton : Taylor & Francis, a CRC title, part of the Taylor & Francis imprint, a member of the Taylor & Francis Group, the academic division of T&F Informa, plc, [2017] | Series: Manufacturing design & technology | Includes bibliographical references.
Identifiers: LCCN 2017026510| ISBN 9781138033627 (hardback : acid-free paper) | ISBN 9781315305271 (ebook)
Subjects: LCSH: Machine-tools--Numerical control | Machining--Data processing.
Classification: LCC TJ1189 .M289 2017 | DDC 671.3/5011--dc23
LC record available at https://lccn.loc.gov/2017026510

Visit the Taylor & Francis Web site at
http://www.taylorandfrancis.com

and the CRC Press Web site at
http://www.crcpress.com

Contents

List of figures.. vii
List of tables... xvii
Preface... xix
Editors... xxi
Contributors.. xxiii

Chapter 1   **A particle finite element method applied to modeling and simulation of machining processes** 1
Juan Manuel Rodríguez, Pär Jonsén, and Ales Svoboda

Chapter 2   **Smoothed particle hydrodynamics for modeling metal cutting** 25
Mohamed N.A. Nasr

Chapter 3   **Failure analysis of carbon fiber reinforced polymer multilayer composites during machining process**............. 51
Sofiane Zenia and Mohammed Nouari

Chapter 4   **Numerical modeling of sinker electrodischarge machining processes** 81
Carlos Mascaraque-Ramírez and Patricio Franco

Chapter 5   **Modeling of interaction between precision machining process and machine tools** 107
Wanqun Chen and Dehong Huo

Chapter 6   **Large-scale molecular dynamics simulations of nanomachining** 141
Stefan J. Eder, Ulrike Cihak-Bayr, and Davide Bianchi

Chapter 7   **Multiobjective optimization of support vector regression parameters by teaching-learning-based optimization for modeling of electric discharge machining responses**...179

Ushasta Aich and Simul Banerjee

Chapter 8   **Modeling of grind-hardening**..211

Angelos P. Markopoulos, Emmanouil L. Papazoglou, Nikolaos E. Karkalos, and Dimitrios E. Manolakos

Chapter 9   **Finite element modeling of mechanical micromachining**..245

Samad Nadimi Bavil Oliaei and Murat Demiral

Chapter 10   **Modeling of materials behavior in finite element analysis and simulation of machining processes: Identification techniques and challenges**.......................281

Walid Jomaa, Augustin Gakwaya, and Philippe Bocher

Index 319

List of figures

Figure 1.1 Remeshing steps in a standard PFEM machining numerical simulation.. 6

Figure 1.2 2D plane strain PFEM model of orthogonal cutting: (a) initial set of particles and (b) initiation of the chip 19

Figure 1.3 Intermediate stages of the chip formation: (a) time 8.04×10^{-4} s and (b) time 1.6×10^{-3} s....................................... 19

Figure 1.4 Cutting force and feed force for test case no. 4...................... 20

Figure 1.5 Effective plastic strain rate 21

Figure 1.6 Temperature distribution.. 22

Figure 1.7 Von Mises stress field 22

Figure 2.1 Deformation zones in metal cutting, with the shear plane angle (φ) shown... 27

Figure 2.2 Concept of FEM. (a) Cantilever beam (physical case) and (b) finite element of a cantilever beam.............................. 28

Figure 2.3 Lagrangian versus Eulerian meshes—material under shear loading.. 29

Figure 2.4 Orthogonal (2D) cutting models, using different FE formulations. (a) Eulerian model, (b) Lagrangian model, and (c) ALE model 31

Figure 2.5 SPH versus FEM (linear elements)—geometrical representation.. 33

Figure 2.6 Smoothing/support domain 34

Figure 3.1   Boundary condition and geometry of the
                   tool–workpiece couple 54

Figure 3.2   Progressive failure analysis of the chip formation
                   with 3D model for 45° fiber orientation. (a) Primary
                   rupture. (b) Secondary rupture and complete chip
                   formation. (c) Experimental result of Iliescu et al. 63

Figure 3.3   Progressive failure analysis of chip formation with 3D
                   model for 90° fiber orientation. (a) Primary rupture.
                   (b) Secondary rupture and complete chip formation.
                   (c) Schematization of the experimental chip formation
                   process by Teti 64

Figure 3.4   Progressive failure analysis of chip formation with 3D
                   model for −45° fiber orientation. (a) Primary rupture.
                   (b) Secondary rupture and complete chip formation.
                   (c) Schematization of the experimental chip formation
                   process 65

Figure 3.5   Cutting force F_c obtained during FE simulation
                   for different fiber orientations with unidirectional
                   composite compared with experimental results
                   ($V_c = 60$ m/min, $a_p = 0.2$ mm, $\alpha = 0°$) 65

Figure 3.6   Depth of damage dm obtained during FE simulation
                   for different fiber orientations with unidirectional
                   composite ($V_c = 60$ m/min, $a_p = 0.2$ mm, $\alpha = 0°$) 66

Figure 3.7   Effect of tool rake angle on machining forces,
                   $V = 60$ m/min, $a_p = 200$ µm, $R = 15$ µm, $\gamma = 11°$ 67

Figure 3.8   Effect of tool rake angle on the chip formation process
                   during cutting of CFRP composites and for fiber
                   orientation at 45°: (a) by shear $\alpha = 10°$, and (b) by
                   buckling $\alpha = -5°$ 68

Figure 3.9   Illustration of the bouncing-back phenomenon 69

Figure 3.10   The effect of clearance angle on machining forces,
                   $V = 60$ m/min, $a_p = 200$ µm, $\alpha = 10°$, $r_\varepsilon = 15$ µm 69

Figure 3.11   The effect of tool edge radius on machining forces,
                   $V = 60$ m/min, $a_p = 200$ µm, $\alpha = 10°$, $\gamma = 11°$ 70

Figure 3.12   Cutting depth effect on machining forces,
                   $V = 60$ m/min, $r_\varepsilon = 15$ µm, $\alpha = 10°$, $\gamma = 11°$ 71

Figure 3.13   Cutting depth effect on chip size, $V = 60$ m/min,
                   $r_\varepsilon = 15$ µm, $\alpha = 10°$, $\gamma = 11°$ 72

Figure 3.14 Size chip measurement: fiber orientation 45°,
$V = 60$ m/min, $r_\varepsilon = 15$ µm, $\alpha = 10°$, $\gamma = 11°$ 72

Figure 3.15 Cutting depth effect on the damage depth,
$V = 60$ m/min, $r_\varepsilon = 15$ µm, $\alpha = 10°$, $\gamma = 11°$ 73

Figure 3.16 Velocity effect on cutting forces for fiber orientation
at 45°: $a_p = 200$ µm, $\alpha = 10°$ 73

Figure 3.17 Two adjacent layers with interlaminar interface 74

Figure 3.18 Damage of the interface between two adjacent
layers, showing the delamination process for four
configurations: (a) 45°/0°, (b) 45°/45°, (c) 45°/−45°, and
(d) −45°/90° 75

Figure 3.19 Steps of hole drilling (a) contact between the tool
and the workpiece, (b) material removal, and (c) hole
completely drilled 76

Figure 3.20 Comparison between experimental and 3D
simulation thrust forces... 77

Figure 3.21 Drill entry delamination: (a) simulation result and
(a′) experimental result. Drill exit delamination:
(b) simulation result and (b′) experimental result 78

Figure 4.1 Schematic representation of the sinker EDM process......... 84

Figure 4.2 Different states of plasma channel during the EDM
process.. 85

Figure 4.3 Examples of scanning electron microscope (SEM)
images for workpiece and electrode in the EDM
process: (a) Stainless steel workpiece and (b) copper
electrode.. 85

Figure 4.4 Different phases of the EDM processes: (a) Voltage
diagram and (b) current intensity diagram 87

Figure 4.5 Heat input distribution on the workpiece surface
during the EDM process... 88

Figure 4.6 Basic diagram of conduction heat transfer 90

Figure 4.7 Dielectric fluid turbulence around the workpiece
surface 91

Figure 4.8 Convection heat transfer throughout the dielectric
fluid in EDM processes... 92

Figure 4.9 Example of simulation mesh for an EDM process 93

Figure 4.10 Diagram with the concept of equivalent temperature 95

Figure 4.11 Variation of equivalent temperature at the node of study from heat transfer with adjacent nodes 95

Figure 4.12 Examples of heat transfer at different nodes of the simulation mesh 96

Figure 4.13 Example of end points in the workpiece meshing 97

Figure 4.14 Variable duration of the discharge and cooling cycles 99

Figure 4.15 Planes defined for a 2D/3D simulation 103

Figure 4.16 Example of equivalent temperature matrix to define the progressive mesh 104

Figure 5.1 Flowchart of the integrated method 110

Figure 5.2 Establishment of state space model based on the FE model 111

Figure 5.3 The topography requirements of the KDP crystal 116

Figure 5.4 Nano-indentation experiment. (a) Nano-indentation experiment system and (b) the curve of load-displacement sampled on the KDP crystal surface 117

Figure 5.5 The FE cutting simulation model 120

Figure 5.6 The simulated cutting force. (a) Cutting force in *y* direction and (b) cutting force in *z* direction 121

Figure 5.7 Fly-cutting machining. (a) Schematic diagram of the fly-cutting machining process, (b) the fly-cutting machining path, and (c) cutting force profile 122

Figure 5.8 Cutting force of the three typical parts (a) A′, (b) B′, and (c) C′ 123

Figure 5.9 The configuration of the fly-cutting machine tool 124

Figure 5.10 The FE model of air spindle 125

Figure 5.11 Outline of the dynamic modeling approach for the aerostatic bearing 126

Figure 5.12 Triangular element 127

Figure 5.13 Meshing principle for the modeling method based on the pressure distribution 130

Figure 5.14 Finite element distribution of the bearing.
(a) Finite element distribution of the axial bearing.
(b) Finite element distribution of the radial bearing......... 130

Figure 5.15 Generation of the spring element group. (a) The
pressure distributions of the gas film. (b) The spring
element group 131

Figure 5.16 The FE model of the fly-cutting machine tool 132

Figure 5.17 Dynamic modes of the machine tool: (a–h) 1st to 8th
order modes vibration of the machine tool 133

Figure 5.18 Tool tip response comparison between the FE method
and the integration method 133

Figure 5.19 Flow chart of the IMPMTS of the KDP crystal
fly-cutting machining 134

Figure 5.20 Typical cutting force response of (a) part A, (b) part B,
and (c) part C 135

Figure 5.21 The surface generation by the proposed simulation
method 136

Figure 5.22 The tested result of the machined surface 137

Figure 6.1 The Lennard–Jones potential for $\varepsilon = 1$ and $\sigma = 1$ 144

Figure 6.2 (a) The initial 3D Voronoi construction that serves
as the basis for the isotropic polycrystalline MD
model of the workpiece. (b) Top view of the random,
fractal, Gaussian surface, with topographic shading
(dark = low/high, light = mid) 147

Figure 6.3 Six examples of abrasive particle geometries
obtained by cleaving bcc crystals along {1 0 0} and
{1 1 1} planes. The large particle in the top left is the
plate-shaped type used in the examples throughout
this chapter. The other types (counterclockwise
from left) are cubic, octahedral, rod-shaped,
cubo-octahedral, and truncated octahedral 150

Figure 6.4 Gaussian size distribution (a) and random lateral
placement and orientation (b) of 60 plate-shaped,
abrasive particles 150

Figure 6.5 Fully assembled system consisting of a rough,
polycrystalline workpiece about to be machined by

60 plate-shaped, hard, abrasive particles. Shading
is according to a grayscale version of the hybrid
scheme proposed in Eder et al., where the surface
has topographic (dark = low/high, light = mid) and
the bulk crystallographic (dark = grains and white =
grain boundaries) shading. The abrasives are shown
in mid-gray.. 151

Figure 6.6   How to determine which atoms are currently
considered removed material (dark, attached to
abrasives), substrate (dark, at bottom), or within the
shear zone (light, in between), depending on the
atomic advection velocity v. The abrasives move at a
constant speed of $v^{(abr)}$.. 154

Figure 6.7   Affiliating the chips of removed matter with the
abrasives that caused them at normal pressures
of 0.1 GPa (a) and 0.4 GPa (b) using a partly
knowledge-based clustering algorithm. Different
shades represent different abrasives.................................... 156

Figure 6.8   Substrate tomographs with EBSD-IPF grain
orientation shading of the initial system configuration
(a). Abrasives are mid-gray. In the IPF triangle legend
in (b), the individual grain orientations within the
workpiece are superimposed as black clusters 160

Figure 6.9   Exemplary atomic displacement tomograph with
normalized vector lengths. The shading corresponds
to atomic drift velocities ranging from 0 m/s to
8 m/s to resolve the *slow* displacements within the
workpiece (lightest shading = 4 m/s). Removed
matter and shear zone have saturated to dark
shading. Abrasives are mid-gray... 161

Figure 6.10   After 1 ns of nanomachining: (a) $\sigma_z = 0.1$ GPa,
(b) $\sigma_z = 0.4$ GPa, and (c) $\sigma_z = 0.7$ GPa. Shading scheme
identical to Figure 6.5.. 163

Figure 6.11   Substrate tomographs after 5 ns of grinding at 0.1 GPa
(a and b), 0.4 GPa (c and d), and 0.7 GPa (e and f).
Abrasives are mid-gray. (a,c,e) Shading according
to grain orientation (EBSD-IPF standard, see legend
below). (b,d,f) Shading according to temperature (see
bar below, the removed matter in (f) is the hottest) 164

Figure 6.12 Mean wear depth h_w (a), mean shear zone thickness h_{shear} (b), arithmetic mean height z_{subst} (c), and root-mean-square roughness S_q (d) over time.................... 165

Figure 6.13 Mean shear stress σ_x (a), final wear depth h_w (b), mean normalized real contact area A_c/A_{nom} (c), final arithmetic mean height z_{subst} (d), final root-mean-square roughness S_q (e), mean contact temperature T_c (f), and final mean shear zone thickness h_{shear} (g) over normal pressure σ_z 167

Figure 6.14 Detail tomographs of slice no. 9 located at $y = 28.5$ nm after 5 ns of machining at 0.5 GPa. Abrasives are mid-gray. (a) EBSD-IPF grain orientation shading (see SST legend in Figure 6.11), and (b) temperature shading (dark $= 300$ K/450 K and light $= 375$ K) 169

Figure 6.15 Detail tomographs of slice no. 15 located at $y = 46.5$ nm after 5 ns of machining. Abrasives are mid-gray. Left: 0.4 GPa, center: 0.5 GPa, and right: 0.6 GPa. (a–c) EBSD-IPF grain orientation shading (see SST legend in Figure 6.11), (d–f) advection velocity shading (dark: $\langle v_x \rangle = 0$ m/s or 80 m/s, light: $\langle v_x \rangle = 40$ m/s), (g–i) atomic displacement vector plots (arrow shading according to equivalent velocities ranging from 0 m/s to 8 m/s), and (j–l) temperature shading (dark $= 300$ K/450 K and light $= 375$ K).............. 170

Figure 6.16 (a) Shear stress σ_x and (b) final wear depth $h_w^{(end)}$ over the normalized contact area A_c/A_{nom} with $A_{nom} = 3595$ nm^2 171

Figure 7.1 Schematic of electrical discharge machining process...... 183

Figure 7.2 Nonlinear SVM regression model................................. 185

Figure 7.3 ε-Insensitive loss function 186

Figure 7.4 Sequence diagram of modified TLBO to search optimum unique set of C, ε, and σ by simultaneous minimization of MATE$_1$ and MATE$_2$ 200

Figure 7.5 Changes in MATE in the estimation of MRR (MATE$_1$).... 201

Figure 7.6 Changes in MATE in the estimation of ASR (MATE$_2$) 201

Figure 7.7 Change of SR ratio along C, ε, and σ during simultaneous minimization of MATE$_1$ and MATE$_2$ 202

Figure 7.8 Steps for concurrent estimation of MRR and ASR from unified structure of SVM regression learning system 202

Figure 7.9 Effect of current and pulse-on time on MRR at pulse-off time 125 μs.. 204

Figure 7.10 Effect of current and pulse-off time on MRR at pulse-on time 125 μs 204

Figure 7.11 Effect of pulse-on time and pulse-off time on MRR at current 10.5 A 204

Figure 7.12 Effect of current and pulse-on time on ASR at pulse-off time 125 μs.. 205

Figure 7.13 Effect of current and pulse-off time on ASR at pulse-on time 125 μs 205

Figure 7.14 Effect of pulse-on time and pulse-off time on ASR at current 10.5 A 205

Figure 8.1 AISI D2 and AISI O1 temperature-dependent material properties.. 227

Figure 8.2 Specific heat capacity of steel 228

Figure 8.3 Workpiece temperature field of xz plane for cutting parameters $u_w = 0.195$ m/sec and $a_e = 0.3$ mm 231

Figure 8.4 Workpiece with the adjusted mesh....................................... 232

Figure 8.5 Temperature field for AISI O1 workpiece, when the 20th node is activated for depth of cut $a_e = 0.3$ mm and feed speed (a) 0.195 m/s, (b) 0.2815 m/s, and (c) 0.3765 m/s.. 233

Figure 8.6 Temperature field for AISI O1 workpiece, when the 90th node is activated for depth of cut $a_e = 0.3$ mm and feed speed (a) 0.195 m/s, (b) 0.2815 m/s, and (c) 0.3765 m/s.... 234

Figure 8.7 Temperature time variation 236

Figure 8.8 Comparison of maximum temperature by using or not using grinding fluid 241

Figure 8.9 Comparison of HPD by using or not using grinding fluid 241

Figure 9.1 Different cutting scenarios based on undeformed chip thickness value (a) $t_u < h_{min}$, (b) $t_u \cong h_{min}$, and (c) $t_u > h_{min}$......249

Figure 9.2 Cutting and thrust forces for different edge radii 257

Figure 9.3 Effective stresses and chip morphology at different edge radii 258

Figure 9.4 Micromachining-induced stress distributions of effective stresses with respect to depth beneath the machined layer at different edge radii 259

Figure 9.5 Temperature distributions at two different edge radii 260

Figure 9.6 Average cutting and thrust forces at different cutting speeds 260

Figure 9.7 Chip morphology at different cutting speeds and edge radii 261

Figure 9.8 Comparison of measured and predicted micromachining forces for different frictional conditions 262

Figure 9.9 Velocity field in front of the cutting edge at two different edge radii 263

Figure 9.10 Laser scanning microscope image of BUE 264

Figure 9.11 Modified geometry of the cutting edge including BUE 264

Figure 9.12 Finite element simulations of chip morphology and effective stresses for different frictional conditions 265

Figure 9.13 Micromachining force predictions at different frictional conditions (uncut chip thickness of 1 μm and cutting speed of 62 m/min) 267

Figure 9.14 Dimensions and orientations for orthogonal machining of single-crystal workpiece material 271

Figure 9.15 Evolution of cutting forces for different cutting directions of (1 1 0) plane 272

Figure 9.16 Chip morphologies at cutting length of 0.5 μm for different rotation angles of (1 1 0) plane 273

Figure 10.1 Finite element model: geometry and mesh 302

Figure 10.2 Predicted chip curling for material models: (a) JCP1, (b) JCP2, (c) JCP3, (d) JCP4-1, (e) JCP4-2, and (f) JCP4-3 after 0.012 s machining time, and (g) experimental result 305

Figure 10.3 Comparison of chip characteristics 305

Figure 10.4 Comparison of cutting force signals 306

Figure 10.5   Comparison of thrust force signals....................................307

Figure 10.6   Comparison of experimental and predicted average
                   force values 307

Figure 10.7   Predicted temperature distribution for material
                   models: (a) JCP1, (b) JCP2, (c) JCP3, (d) JCP4-1,
                   (e) JCP4-2, and (f) JCP4-3 after 0.012 s machining time.....308

Figure 10.8   Comparison of predicted cutting temperature
                   history at 10 μm beneath the machined surface..............309

Figure 10.9   Comparison of predicted stress–strain curves at
                   10 μm beneath the machined surface...............................309

Figure 10.10   Comparison of predicted and experimental
                   stress–strain curves under (a) quasi-static and
                   (b) dynamic conditions 310

List of tables

Table 1.1 Material properties for the process simulations 14

Table 1.2 Cutting data in simulations 18

Table 1.3 Measured and simulated cutting forces................................ 20

Table 2.1 Al 6061-T6 chip thickness and cutting force component 38

Table 2.2 AISI 4340 serrated chip tooth thickness (μm) 38

Table 3.1 Mechanical properties of the aeronautical CFRP
composite T300/914.. 55

Table 3.2 Plastic and damage parameters of UD-CFRP T300/914 56

Table 3.3 Material parameters used to model interface cohesive
elements.. 57

Table 4.1 Material properties for EDM numerical modeling 100

Table 4.2 Process parameters for EDM numerical modeling 101

Table 4.3 Parameters for process simulation 101

Table 5.1 Measured Young's modulus E and
micro-hardness H of type II KDP crystal................................ 118

Table 7.1 Process parameters and their levels 183

Table 7.2 Searching ranges of SVM internal structural
parameters—C, ε, and σ.. 192

Table 7.3 Results of tuning internal structural parameters
(C, ε, and σ) of SVM for unified learning 198

Table 7.4 Testing of estimated MRR.. 203

Table 7.5 Testing of estimated ASR 203

Table A.1   Initial learner population for searching optimum
unique set of C, ε, and σ by modified TLBO............................ 207

Table A.2   Difference of Lagrange multipliers (α_i, α_i^*)
for normalized MRR and normalized ASR............................ 207

Table 8.1   Workpiece material properties 225

Table 8.2   Air and grinding wheel properties..228

Table 8.3   Process parameters... 228

Table 8.4   Results using the analytical model..229

Table 8.5   Results using ANSYS software..235

Table 8.6   Grinding conditions...236

Table 8.7   Comparison of experimental and simulation results.......... 237

Table 8.8   Water properties..238

Table 8.9   Results for using grinding fluid..239

Table 9.1   The coefficients of the material model...................................... 253

Table 9.2   The coefficients of the material model...................................... 254

Table 9.3   BUE parameters obtained at a cutting speed of
62 m/min..265

Table 9.4   Material parameters of single-crystal copper...................... 271

Table 9.5   Cutting direction setup ($[d\,e\,f]$) for (1 1 0) crystal plane
(Figure 9.15) 271

Table 9.6   Average cutting energies obtained with CP theory
for different rotation angles of (1 1 0) plane 272

Table 10.1   Identification techniques and validity domains of
selected JCP models for aged Inconel 718 alloy................... 303

Table 10.2   JCP parameters for aged Inconel 718 alloy............................ 303

Table 10.3   JCP4 parameters for aged Inconel 718 alloy.......................... 303

Table 10.4   Physical properties of the superalloy Inconel 718 and
cutting insert.. 304

Preface

Machining is one of the most important manufacturing methods used for the production of mechanical components worldwide. Conventional and nonconventional machining processes are vital for the production of high-quality components from many different material categories. Automotive, aerospace, and medical industries are only some of the sectors in which machined components of high-dimensional accuracy, exceptional properties, complex sizes, and usually from difficult-to-machine materials are employed. The research in the refinement of machining or the introduction of new features is ongoing and fast growing.

Modeling and simulation are powerful tools in many applications, including manufacturing and especially machining. Modeling can provide valuable information on the fundamental understanding of the material removal process but more importantly it provides the means to predict many machining parameters in a reliable manner. The value of modeling in machining can be recognized by the number of publications appearing in the contemporary relevant literature, which can be described as vast.

As more sophisticated machining methods are used, more elaborate models need to be proposed. This includes the use of innovative techniques in well-known methods, the proposal of novel methods, and the introduction of modeling procedures that are used in other scientific sectors in machining modeling practice. Furthermore, advances in computers, with an increase in computational power and data storage, provide opportunities for more complicated and detailed simulations. This research book aims at providing all the related data that are needed to employ innovative modeling techniques in advanced machining processes, with a thorough presentation of methods and techniques completed by case studies from experts, who have contributed their work in this book.

Chapter 1 covers the topic of particle finite element modeling and its application in machining. Chapter 2 pertains to smoothed particle hydrodynamics method as applied in cutting processes. Chapter 3 presents the application of modeling in the case of machining of carbon fiber-reinforced polymer. Chapter 4 pertains to the numerical modeling of a nonconventional material removal process, that is, sinker electrodischarge machining. Chapter 5 covers the modeling of interaction between process and machine tool for precision machining. Chapter 6 is dedicated to large-scale molecular dynamics simulations for nanomachining. Chapter 7 presents the teaching learning-based optimization of electro-discharge machining. Chapter 8 contains information on the analytical and numerical modeling of abrasive processes and grind-hardening procedure. Chapter 9 is dedicated to the finite element modeling of micromachining. The final chapter, Chapter 10, is dedicated to material modeling in finite element simulation of machining. Color versions of all figures will be hosted on a companion website. Visit the book's CRC Press website for further details: www.crcpress.com/9781138033627.

This book is intended for both the academia and the industry. The former pertains to pre- and postgraduate students, PhD students, and researchers from universities and institutes who are involved in machining and modeling and interested in exploring more aspects of these subjects. The latter pertains to industries that have R&D departments interested in machining modeling for the improvement of their products, mainly in the sectors of automotive and aerospace industry, medical, medical tools, and devices developers, metalworking industry, machine tools and cutting tools manufacturers, and modeling software companies.

The editors acknowledge the aid from CRC Press and express their gratitude for this opportunity and for their professional support. The help of Alexandra Micha from National Technical University of Athens (NTUA) is also acknowledged and is greatly appreciated. The editors also express their gratitude to all the chapter authors for their availability and for delivering the high-quality research material at hand.

Angelos P. Markopoulos
National Technical University of Athens

J. Paulo Davim
University of Aveiro

MATLAB® is a registered trademark of The MathWorks, Inc. For product information, please contact:

The MathWorks, Inc.
3 Apple Hill Drive
Natick, MA 01760-2098 USA
Tel: 508-647-7000
Fax: 508-647-7001
E-mail: info@mathworks.com
Web: www.mathworks.com

Editors

Angelos P. Markopoulos earned his PhD in mechanical engineering from the National Technical University of Athens, Greece in 2006. He is currently an assistant professor in the same university. He is the author or coauthor of more than 60 papers in international journals and conferences: 1 book and more than 10 book chapters in edited books. His research includes topics such as precision and ultraprecision machining processes with special interest in high-speed hard machining, grinding, micromachining, and advanced modeling methods and techniques. He is member of the international editorial review board of two journals and a regular reviewer of several journals in the above-mentioned areas. He is also member of the Technical Chamber of Greece and the Hellenic Association of Mechanical and Electrical Engineers.

J. Paulo Davim earned his PhD in mechanical engineering in 1997; MSc in mechanical engineering (materials and manufacturing processes) in 1991; Dipl.-Ing Engineer's degree (5 years) in mechanical engineering in 1986 from the University of Porto (FEUP), Porto, Portugal; the Aggregate title (Full Habilitation) from the University of Coimbra, Coimbra, Portugal in 2005; and a DSc from London Metropolitan University, London, United Kingdom in 2013. He is a Eur Ing by Fédération Européenne d'Associations Nationales d'Ingénieurs (FEANI)–Brussels and senior chartered engineer by the Portuguese Institution of Engineers with an MBA and specialist title in engineering and industrial management. Currently, he is professor at the Department of Mechanical Engineering, University of Aveiro, Portugal. He has more than 30 years of teaching and research experience in manufacturing, materials and mechanical engineering with special emphasis in machining and tribology. He also has interest in management and industrial engineering and higher education for sustainability and engineering education. He has received several scientific awards. He has worked as an evaluator of projects for international research agencies as well as an examiner of PhD thesis for many universities. He is the editor-in-chief of several international journals, guest editor of journals, books editor, book series editor, and scientific advisory for many international journals and conferences. Presently, he is an editorial board member of 30 international journals and acts as a reviewer for more than 80 prestigious Web of Science journals. In addition, he has also published as an editor (and coeditor) of more than 100 books and as an author (and coauthor) of more than 10 books, 60 book chapters, and 400 articles in journals and conferences (more than 200 articles in journals indexed in Web of Science/ h-index 39+ and SCOPUS/h-index 48+).

Contributors

Ushasta Aich
Department of Mechanical
  Engineering
Jadavpur University
Kolkata, India

Simul Banerjee
Department of Mechanical
  Engineering
Jadavpur University
Kolkata, India

Davide Bianchi
AC2T research Gmbh
Wiener Neustadt, Austria

Philippe Bocher
Department of Mechanical
  Engineering
École de Technologie Supérieure
Montréal, Québec, Canada

Wanqun Chen
Center for Precision Engineering
Harbin Institute of Technology
Harbin, China

and

School of Mechanical and Systems
  Engineering
Newcastle University
Newcastle, United Kingdom

Ulrike Cihak-Bayr
AC2T research Gmbh
Wiener Neustadt, Austria

Murat Demiral
Department of Mechanical
  Engineering
Çankaya University
Ankara, Turkey

Stefan J. Eder
AC2T research Gmbh
Wiener Neustadt, Austria

Patricio Franco
Departamento de Ingeniería de
  Materiales y Fabricación
Universidad Politécnica de
  Cartagena
Cartagena, Spain

Augustin Gakwaya
Department of Mechanical
  Engineering
Université Laval
Québec City, Québec, Canada

Dehong Huo
School of Mechanical and Systems
  Engineering
Newcastle University
Newcastle, United Kingdom

Walid Jomaa
Department of Mechanical
 Engineering
Université Laval
Québec City, Québec, Canada

Pär Jonsén
Division of Mechanics of Solid
 Materials
Department of Engineering
 Sciences and Mathematics
Lulea University of Technology
Luleå, Sweden

Nikolaos E. Karkalos
Section of Manufacturing
 Technology
School of Mechanical Engineering
National Technical University of
 Athens
Athens, Greece

Dimitrios E. Manolakos
Section of Manufacturing
 Technology
School of Mechanical Engineering
National Technical University of
 Athens
Athens, Greece

Angelos P. Markopoulos
Section of Manufacturing
 Technology
School of Mechanical Engineering
National Technical University of
 Athens
Athens, Greece

Carlos Mascaraque-Ramírez
Departamento de Tecnología
 Naval
Universidad Politécnica de
 Cartagena
Cartagena, Spain

Samad Nadimi Bavil Oliaei
Department of Mechanical
 Engineering
ATILIM University
Ankara, Turkey

Mohamed N.A. Nasr
Department of Mechanical
 Engineering
Faculty of Engineering
Alexandria University
Alexandria, Egypt

Mohammed Nouari
GIP-InSIC
Université de Lorraine
Saint-Dié-des-Vosges, France

Emmanouil L. Papazoglou
Section of Manufacturing
 Technology
School of Mechanical Engineering
National Technical University
 of Athens
Athens, Greece

Juan Manuel Rodríguez
Division of Mechanics of Solid
 Materials
Department of Engineering
 Sciences and Mathematics
Lulea University of Technology
Luleå, Sweden

Ales Svoboda
Division of Mechanics of Solid
 Materials
Department of Engineering
 Sciences and Mathematics
Lulea University of Technology
Luleå, Sweden

Sofiane Zenia
GIP-InSIC
Université de Lorraine
Saint-Dié-des-Vosges, France

chapter one

A particle finite element method applied to modeling and simulation of machining processes

Juan Manuel Rodríguez, Pär Jonsén, and Ales Svoboda

Contents

1.1 Introduction 2
1.2 The particle finite element method 3
    1.2.1 The particle finite element method in solid mechanics 4
    1.2.2 The particle finite element method in the numerical
        simulation of metal cutting processes 4
        1.2.2.1 Data transfer of internal variables 5
1.3 Governing equations for a Lagrangian continuum 5
    1.3.1 Momentum equation 6
    1.3.2 Thermal balance 7
    1.3.3 Mass balance 7
    1.3.4 Boundary conditions 8
        1.3.4.1 Mechanical problem 8
        1.3.4.2 Thermal problem 8
1.4 Variational formulation 8
    1.4.1 Momentum equations 8
    1.4.2 Mass conservation equation 9
    1.4.3 Thermal balance equation 9
1.5 Finite element discretization 10
1.6 The constitutive model 12
    1.6.1 Thermo-elastoplasticity model at finite strains 12
        1.6.1.1 Elastic response 12
        1.6.1.2 Yield condition 13
        1.6.1.3 Flow rule 13
        1.6.1.4 The Johnson–Cook constitutive model 13
1.7 Stress update algorithm 14
    1.7.1 Thermo-elastoplasticity model at finite strains 14
    1.7.2 Transient solution of the discretized equations 14

        1.7.3   Mechanical problem 16
        1.7.4   Thermal problem 17
1.8   Example, result, and discussion 18
        1.8.1   Cutting and feed forces... 20
        1.8.2   Material response... 21
1.9   Conclusion 22
References.. 23

Metal cutting process is a nonlinear dynamic problem that includes geometrical, material, and contact nonlinearities. In this work, a Lagrangian finite element approach for the simulation of metal cutting process is presented based on the so-called particle finite element method (PFEM). The governing equations for the deformable bodies are discretized with the finite element method (FEM) via a mixed formulation using simplicial elements with equal linear interpolation for displacements, pressure, and temperature. The use of PFEM for modeling of metal cutting processes includes the use of a remeshing process, α-shape concepts for detecting domain boundaries, contact mechanics laws, and material constitutive models. In this chapter, a 2D PFEM-based numerical modeling of metal cutting processes has been studied to investigate the effects of cutting velocity on tool forces, temperatures, and stresses in machining of Ti–6Al–4V. The Johnson–Cook plasticity model is used to describe the work material behavior. Numerical simulations are in agreement with experimental results.

1.1   Introduction

Numerical modeling of machining processes is continuously attracting researchers for better understanding of chip formation mechanisms, heat generation in cutting zone, tool–chip interfacial frictional characteristics, and quality and integrity of the machined surfaces. In predictive process engineering for machining processes, prediction of physics-related process field variables such as temperature and stress fields becomes highly important.

The numerical simulation of metal cutting processes is complicated mainly by two factors: First, the constitutive model of the piece material. It must adequately represent deformation behavior during high strain rate loading and low strain rate loading under a range of temperatures, and account for hardening and softening processes. The second challenge is concerned with the modeling and numerical realization of large configuration changes. Numerical simulations of machining process involve large strains and angular distortions, multiple contacts and self-contact, generation of new boundaries, and fracture with multiple cracks and defragmentation. All of them are difficult to handle with standard FEMs. Different numerical techniques have been developed to deal with the second challenge in the numerical simulation of metal cutting processes, the material point method [1], the smooth particles hydrodynamics method [2], the constrained natural element method (C-NEM) [3], the discrete element method (DEM) [4], and more recently the PFEM [5–7] are among them. The wide variety of numerical techniques developed till now to model machining process demonstrates not only that the modeling of machining has been a subject of intensive research, but also that this field still requires further attention. The aforementioned considerations constitute solid and compelling reasons to continue the research in numerical techniques to model metal cutting processes.

The purpose of this work is to apply the PFEM [1–3] to solve the problems associated with the large configuration changes and to use the Johnson–Cook constitutive law to model the complex material behavior.

The remaining of the chapter is organized as follows: Section 1.2 is devoted to the description of PFEM and the modifications introduced here concerning insertion and removal of particles. In Section 1.3, the basic equations for conservation of linear momentum, mass conservation, and heat transfer for a continuum using a generalized Lagrangian framework are presented. Section 1.4 deals with the variational formulation of the continuous problem that is presented in Section 1.3. Then, the mixed finite element discretization using simplicial element with equal order approximation for the displacement, the pressure, and the temperature is presented, and the relevant matrices and vectors of the discretized problem are given in Section 1.5. Section 1.6 presents the J_2 plasticity model at finite deformation and a description of the Johnson–Cook constitutive model. Details of the implicit solution of the Lagrangian FEM equations in time using an updated Lagrangian approach and a Newton–Raphson-type iterative scheme are presented in Section 1.7. Section 1.8 focuses on a set of representative numerical simulations of metal cutting processes using PFEM. Finally, some concluding remarks are presented in Section 1.9.

1.2   The particle finite element method

The PFEM is a FEM-based particle method [8], initially proposed for the solution of the continuous fluid mechanics equations. The main objectives are, on the one hand, to develop a method to eliminate the convective terms in the governing equations. On the other hand, the introduction of a technology, based on the α-shape method used in other areas, that is able to deal with free boundary surfaces is the second objective. The interpretation of the method has evolved from a meshless method, in which the nodes are supposed to be the particles that move according to simple rules of motion, to a sort of updated Lagrangian approach in which the advantages of the standard FEM formulation for the solution of the incremental problem are used.

1.2.1 The particle finite element method in solid mechanics

The PFEM is a set of numerical strategies that are combined for the solution of large deformation problems. The standard algorithm of the PFEM for the solution of solid mechanics problems contains the following steps:

1. Definition of the domain(s) Ω_n in the last converged configuration, $t = {}^n t$, keeping existing spatial discretization $\bar{\Omega}_n$.
2. Transference of variables by a smoothing process from Gauss points to nodes.
3. Discretization of the given domain(s) in a set of particles of infinitesimal size elimination of existing connectivities $\bar{\Omega}_n$.
4. Reconstruction of the mesh through a triangulation of the domain's convex hull and the definition of the boundary applying the α-shape method [9,10], defining a new spatial discretization $\bar{\Omega}_n$.
5. A contact method to recognize the multibody interaction.
6. Transference of information, interpolating nodal variables into the Gauss points.
7. Solution of the system of equations for ${}^{n+1}t = {}^n t + \Delta t$.
8. Go back to step 1 and repeat the solution process for the next time step.

1.2.2 The particle finite element method in the numerical simulation of metal cutting processes

The standard PFEM presents some weaknesses when applied in orthogonal cutting simulation. For example, the external surface generated using α-shape may affect the mass conservation, the chip shape, absence of equilibrium on the boundary due to the introduction of artificial perturbations, and generation of unphysical welding of the workpiece material and the chip.

To deal with this problem, in this work the use of a constrained Delaunay algorithm is proposed. Furthermore, addition and removal of particles are the principal tools, which are employed for sidestepping the difficulties that are associated with deformation-induced element distortion and for resolving the different scales of the solution.

In the numerical simulation of metal cutting process, despite the continuous Delaunay triangulation, elements arise with unacceptable aspect ratios; for this reason, the mesh is also subjected to a Laplacian smoothing algorithm to smooth the mesh. For each node in a mesh, a new position based on the position of neighbors is obtained. In the case that a mesh is topologically a rectangular grid, then this operation produces the Laplacian of the mesh. These procedures are applied locally and not in every time step. Specific size metrics control node insertion and removal, whereas the Laplacian smoothing algorithm drives the repositioning of nodes.

In summary, the enhancement of the PFEM takes place in three main areas: the dynamic process for the discretization of the domain into particles, varying the number of them depending on the deformation of the body; transference of the internal variables, from a nodal smoothing through a variable projection; and the boundary recognition, eliminating the geometric criterion of the α-shape method.

1.2.2.1   Data transfer of internal variables

The transference of internal variables or element information between evolving meshes within the field of PFEM is critical in the numerical simulation of process such as machining as shown in References 7 and 11. In this work, the authors show that the nodal scheme presented in Reference 12 generates unphysical springback of the machined surface and subestimation of the cutting and feed forces.

Due to the insertion, removal, and relocation of particles through Laplacian smoothing, the transference of Gauss point variables is set directly through a mesh projection instead of traditional nodal smoothing [5]. The projection is carried out by a direct search of the position of the integration point of the new connectivity, over the former mesh. The use of this transference scheme gives an improved computational efficiency. The use of the projector operator to transfer internal variables guarantees the preservation of the stress-free state for the portions of the domains that do not yield. In this zone, there is no insertion, removal, or relocation of particles; most of the finite elements of the stress-free region remain the same before and after the Delaunay triangulation, resulting in no diffusion of the internal variables in the stress-free zone. More details about the data transfer of internal variables in the numerical simulations are found in References 5, 7, and 11.

A summary about the PFEM that is applied in metal cutting processes is shown in Figure 1.1.

1.3   Governing equations for a Lagrangian continuum

Consider a domain containing a deformable material that evolves in time due to the external and internal forces and prescribed displacements and thermal conditions from an initial configuration at time $t = 0$ to a current

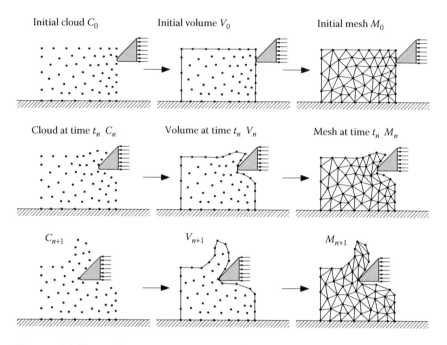

Figure 1.1 Remeshing steps in a standard PFEM machining numerical simulation.

configuration at time $t = t_n$. The volume V and its boundaries Γ at the initial and current configurations are denoted as $(^0V, {}^0\Gamma)$ and $(^nV, {}^n\Gamma)$, respectively. The aim is to find a spatial domain that the material occupies and at same time to obtain velocities, strain rates, stresses, and temperature in the updated configuration at time $^{n+1}t = {}^nt + \Delta t$. In the following lines, a left super index denotes the configuration where the variable is computed.

1.3.1 Momentum equation

The equation of conservation of linear momentum for a deformable continuum is written in a Lagrangian description as

$$\rho \frac{Dv_i}{Dt} - \frac{\partial^{n+1}\sigma_{ij}}{\partial^{n+1}x_j} - {}^{n+1}b_i = 0, \quad i,j = 1,\ldots,n_s \text{ in } {}^{n+1}V \tag{1.1}$$

where:

^{n+1}V is the analysis domain in the updated configuration at time ^{n+1}t with boundary $^{n+1}\Gamma$

v_i and b_i are the velocity and the body force components along the Cartesian axis

ρ is the density

n_s is the number of space dimensions

$^{n+1}x_j$ is the position of the material point at time ^{n+1}t

$^{n+1}\sigma_{ij}$ is the Cauchy stress in ^{n+1}V

Dv_i/Dt is the material derivative of the velocity field

The Cauchy stress is split into the deviatoric s_{ij} and pressure p components as

$$\sigma_{ij} = s_{ij} + p\delta_{ij} \tag{1.2}$$

where δ_{ij} is the Kronecker delta.

The pressure is assumed to be positive for a tension state.

1.3.2   Thermal balance

The thermal balance equation in the current configuration is written in a Lagrangian framework as

$$\rho c \frac{DT}{Dt} - \frac{\partial}{\partial^{n+1}x_i}\left(k\frac{\partial T}{\partial^{n+1}x_i}\right) + {}^{n+1}Q = 0, \quad i,j = 1,\ldots,n_s \text{ in } {}^{n+1}V \tag{1.3}$$

where:

T is the temperature

c is the thermal capacity

k is the thermal conductivity

Q is the heat source

1.3.3   Mass balance

The mass conservation equation can be written for solids domain as

$$-\frac{1}{k}\frac{Dp}{Dt} + \varepsilon_V = 0 \tag{1.4}$$

where:

k is bulk elastic moduli of the solid material

Dp/Dt is the material derivative of the pressure field

ε_V is the volumetric strain rate defined as the trace of the rate of deformation tensor, which is defined as

$$d_{ij} = \frac{1}{2}\left(\frac{\partial v_i}{\partial \chi_j} + \frac{\partial v_j}{\partial \chi_i}\right) \tag{1.5}$$

For a general time interval $[{}^n t, {}^{n+1} t]$, Equation 1.4 is discretized as

$$-\frac{1}{k}\left({}^{n+1}p - {}^n p\right) + \varepsilon_V \Delta t = 0 \tag{1.6}$$

Equations 1.1, 1.3, and 1.4 are completed by the standard boundary conditions.

1.3.4 Boundary conditions

1.3.4.1 Mechanical problem
The boundary conditions at the Dirichlet Γ_v and Neumann Γ_t boundaries with $\Gamma = \Gamma_v \cup \Gamma_t$ are

$$\upsilon_i - \upsilon_i^P = 0 \text{ on } \Gamma_v \tag{1.7}$$

$$\sigma_{ij} n_j - t_i^P = 0 \text{ on } \Gamma_t, \quad i = 1, \dots n_s \tag{1.8}$$

where υ_i^P and t_i^P are the prescribed velocities and the prescribed tractions, respectively.

1.3.4.2 Thermal problem

$$^{n+t}T - {}^{n+t}T^P = 0 \text{ on } {}^{n+t}\Gamma_T \tag{1.9}$$

$$k\frac{\partial T}{\partial n} + {}^{n+1}q_n^P = 0 \text{ on } {}^{n+1}\Gamma_q \tag{1.10}$$

where:
T^P and q_n^P are the prescribed temperature and the prescribed normal heat flux at the boundaries Γ_T and Γ_q, respectively
n is a vector in the direction normal to the boundary

1.4 Variational formulation

1.4.1 Momentum equations

Multiplying Equation 1.1 by arbitrary test function w_i with dimensions of velocity and integrating over the updated domain ^{n+1}V gives the weighted residual form of the momentum equations as [13,14]

$$\int_{^{n+1}V} \left(-\frac{\partial^{n+1}\sigma_{ij}}{\partial^{n+1}x_j} - {}^{n+1}b_i \right) w_i = 0 \tag{1.11}$$

In Equation 1.11, the inertial term $\rho(Dv_i/Dt)$ is neglected because in the problems of interest in this work, this term is much smaller than the other terms appearing in Equation 1.11.

Integrating by parts the terms involving σ_{ij} and using the traction boundary conditions (1.8) yield the weak form of the momentum equation as

$$\int_{^{n+1}V} \delta\varepsilon_{ij}^{n+1}\sigma_{ij}\,dV - \int_{^{n+1}V} w_i^{n+1}b_i\,dV - \int_{^{n+1}\Gamma_t} {}^{n+1}w_i^{n+1}t_i^p\,dV = 0 \qquad (1.12)$$

where $\delta\varepsilon_{ij} = (1/2)[(\partial w_i/\partial^{n+1}x_j) + (\partial w_i/\partial^{n+1}x_i)]$ is a virtual strain rate field

Equation 1.12 is the standard form of the principle of virtual power [13,14].

Using Equation 1.2, Equation 1.12 gives the following expression:

$$\int_{^{n+1}V} \delta\varepsilon_{ij}^{n+1}s_{ij}\,dV + \int_{^{n+1}V} \delta\varepsilon_{ij}^{n+1}p\delta_{ij}\,dV - \int_{^{n+1}V} w_i^{n+1}b_i\,dV - \int_{^{n+1}\Gamma_t} w_i^{n+1}t_i^p\,dV = 0 \quad (1.13)$$

Introducing, **w, s**, and $\delta\varepsilon$, the vectors of test function, deviatoric stresses, and virtual strain rates, respectively; **b** and **t**^{p}, the body forces and traction vectors, respectively; and **m**, an auxiliary vector, in Equation 1.13 yields

$$\int_{^{n+1}V} \delta\varepsilon^{T\,n+1}s\,dV + \int_{^{n+1}V} \delta\varepsilon^{T}\mathbf{m}^{n+1}p\,dV - \int_{^{n+1}V} \mathbf{w}^{T\,n+1}\mathbf{b}\,dV - \int_{^{n+1}\Gamma_t} \delta\varepsilon^{T}\mathbf{w}^{n+1}\mathbf{t}^p\,dV = 0 \quad (1.14)$$

where the matrices introduced in Equation 1.14 are defined in References 15 and 16.

1.4.2 Mass conservation equation

To obtain the mass conservation equation, Equation 1.6 is multiplied by an arbitrary test function q, defined over the analysis domain. Integrating over ^{n+1}V yields

$$\int_{^{n+1}V} -\frac{q}{k}\left({}^{n+1}p - {}^{n}p\right)dV + \int_{^{n+1}V} q\varepsilon_V \Delta t\,dV = 0 \qquad (1.15)$$

1.4.3 Thermal balance equation

Application of the standard weighted residual methods to the thermal balance equations (1.3), (1.9), and (1.10) leads, after standard operations, to [14]

$$\int\limits_{n+1_V} \hat{w}\rho c \frac{DT}{Dt} dV + \int\limits_{n+1_V} \frac{\partial \hat{w}}{\partial^{n+1} x_i}\left(k \frac{\partial T}{\partial^{n+1} x_i}\right) dV + \int\limits_{n+1_V} \hat{w}^{n+1} Q dV$$

$$+ \int\limits_{n+1_V} \hat{w}^{n+1} q_n^p d\Gamma = 0 \qquad (1.16)$$

where $\hat{w}$ is the space weighting function for the temperature.

1.5 Finite element discretization

The analysis domain is discretized into finite elements with n nodes in the standard manner, leading to a mesh with a total number of N_e elements and N nodes. In the present work, a simple three-noded linear triangle ($n = 3$) with local linear shape functions N_i^e defined for each node n of the element e is used. The displacement, the velocity, the pressure, and the temperature are interpolated over the mesh in terms of their nodal values in the same manner using the global linear shape function N_j that is spanning over the nodes sharing node j. In matrix format and 2D problems, we have

$$\mathbf{u} = N_u \bar{u}, \; \mathbf{v} = N_v \bar{v}, \; \mathbf{p} = N_p \bar{p}, \; \mathbf{T} = N_T \bar{T} \qquad (1.17)$$

where:

$$\bar{u} = \begin{Bmatrix} \bar{u}^1 \\ \bar{u}^2 \\ \vdots \\ \bar{u}^N \end{Bmatrix} \text{with } \bar{u}^i = \begin{Bmatrix} u_1^i \\ u_2^i \end{Bmatrix}, \quad \bar{v} = \begin{Bmatrix} \bar{v}^1 \\ \bar{v}^2 \\ \vdots \\ \bar{v}^N \end{Bmatrix} \text{with } \bar{v}^i = \begin{Bmatrix} v_1^i \\ v_2^i \end{Bmatrix}, \quad \bar{p} = \begin{Bmatrix} \bar{p}^1 \\ \bar{p}^2 \\ \vdots \\ \bar{p}^N \end{Bmatrix}, \bar{T} = \begin{Bmatrix} \bar{T}^1 \\ \bar{T}^2 \\ \vdots \\ \bar{T}^N \end{Bmatrix}$$

$$N_u = N_v = [N_1 N_2 ... N_N]$$
$$N_p = N_T = [N_1 N_2 ... N_N] \qquad (1.18)$$

with $N_j = N_j \mathbf{I}_2$ where $\mathbf{I}_2$ is the 2×2 identity matrix.

Substituting Equation 1.17 into Equations 1.13 and 1.16 while choosing a Galerkin formulation with $w_i = q = \hat{w}_i = N_i$ leads to the following system of algebraic equations:

$$F_{res,mech} = F_{u,int}\left(\bar{u}, \bar{p}\right) - F_{u,ext} = 0$$

$$F_{res,mass} = F_{p,press}\left(\bar{p}\right) - F_{p,vol}\left(\bar{u}\right) + F_{p,stab}\left(\bar{p}\right) - F_{p,press,n}\left(^n\bar{p}\right) = 0 \qquad (1.19)$$

$$F_{res,therm} = F_{0,dyn}\left(\dot{\bar{T}}\right) - F_{0,int}\left(\bar{T}\right) + F_{0,ext}\left(\bar{T}\right) = 0$$

where:

$$F_{u,\text{int}}\left(\bar{u},\bar{p}\right) = \int_{^{n+1}V} B_u^{T\ n+1}\sigma dV \tag{1.20}$$

$$F_{u,\text{ext}} = \int_{^{n+1}V} N^{T\ n+1}b dV - \int_{^{n+1}\Gamma_t} N^{T\ n+1}t^p d\Gamma \tag{1.21}$$

$$F_{p,\text{press}}\left(\bar{p}\right) = \int_{^{n+1}V} \frac{1}{k} N^T N \bar{p} dV \tag{1.22}$$

The term $F_{p,\text{press}}, n(^n\bar{p})$ is exactly the same term as in Equation 1.22, but the nodal pressure is evaluated at time $^n t$:

$$F_{p,\text{vol}}\left(\bar{u}\right) = Q^{Tn+1}\bar{u} \tag{1.23}$$

where the element form of the Q matrix is given by

$$Q^{(e)} = \int_{^{n+1}V^{(e)}} B_{ui}^{(e)T} m N_j^{(e)} dV \tag{1.24}$$

$$F_{p,\text{stab}}\left(\bar{p}\right) = \overset{N_e}{\underset{e=1}{A}} \int_{^{n+1}V^{(e)}} \frac{\alpha}{\mu}\left(N^{(e)}N^{T(e)} - \tilde{N}^{(e)}\tilde{N}^{T(e)}\right)p^{(e)} dV \tag{1.25}$$

$$F_{T,\text{dyn}}\left(\dot{\bar{T}}\right) = \int_{^{n+1}V} \rho c N N^T \dot{T} dV \tag{1.26}$$

$$F_{q,\text{int}}\left(\bar{T}\right) = \int_{^{n+1}V} k B_0^T B_0 \bar{T} dV - \int_{^{n+1}V} N^T Q dV \tag{1.27}$$

$$F_{q,\text{ext}}\left(\bar{T}\right) = \int_{^{n+1}\Gamma_q} N^T q_n^p d\Gamma \tag{1.28}$$

In Equation 1.19, $\dot{T}$ denotes the material time derivative of the nodal temperature.

In finite element computations, the aforementioned force vectors are obtained as the assemblies of element vectors. Given a nodal point, each component of the global force associated with a particular global node is obtained as the sum of the corresponding contributions from the element force vectors of all elements that share the node. In this work, the element force vectors are evaluated using Gaussian quadrature.

Note that Equation 1.19 involves the geometry at the updated configuration (^{n+1}V). This geometry is unknown; hence the solution of Equation 1.19 has to be found iteratively. The iterative solution scheme proposed in this work is presented in the next section.

1.6   The constitutive model

1.6.1   Thermo-elastoplasticity model at finite strains

In metal forming processes such as machining, elastic strains are in the order of 10^{-4}, whereas plastic strains can be in the order of 10^{-1} to 10 [17]. In case elastic strains are neglected, the model is not able to predict the residual stresses and the springback of the machined surface. For this reason, the constitutive model is developed for small elastic and large plastic deformation, instead of a more complex model that uses large elastic and large plastic deformation [5,7,11]. An example of modeling of machining processes that uses a fluid mechanics approach is presented in Reference 18. A valuable implication of the small elastic strains is that the rate of deformation tensor $d_{ij} = \{(1/2)[(\partial v_i / \partial x_j) + (\partial v_j / \partial x_i)]\}$ inherits the additive structure of classical small-strain elastoplasticity:

$$d = d^e + d^p \tag{1.29}$$

where d^e and d^p are the elastic and plastic parts of the rate of deformation tensor, respectively.

1.6.1.1   Elastic response

Let a material with a hypoelastic constitutive equation such as

$$L_v(\tau) = c : (d - d^p) \tag{1.30}$$

where:

$L_v(\bullet)$ denotes the Lie objective stress rate
τ denotes the Kirchhoff stress tensor

It can be assumed that the special elasticity **c** tensor is given by

$$c = 2\mu\left(\mathbf{I} - \frac{1}{3}\mathbf{1}\otimes\mathbf{1}\right) + \kappa\mathbf{1}\otimes\mathbf{1} \tag{1.31}$$

where:

I and **1**, with components $I_{abcd} = [(\delta_{ac}\delta_{bd} + \delta_{ad}\delta_{cd})/2]$ and $1_{ab} = \delta_{ab}$, are the fourth and second-order symmetric unit tensor, respectively
The parameters μ and κ represent the shear and the bulk elastic modulus.

1.6.1.2   Yield condition

The yield condition used in this work is the von Misses–Huber yield criterion. This criterion is formulated in terms of the second invariant of the Kirchhoff stress tensor $J_2 = (1/2)\text{dev}\tau : \text{dev}\tau$. Hence, the von Mises yield criterion (henceforth simply called the Mises criterion) can be stated as

$$f(\tau) = \sqrt{2J_2} - \sqrt{\frac{2}{3}}\sigma_y = \|\text{dev}\tau\| - \sqrt{\frac{2}{3}}\sigma_y\left(\bar{\varepsilon}^p, \dot{\bar{\varepsilon}}^p\right) = 0 \qquad (1.32)$$

where:

σ_y denotes the flow stress

$\bar{\varepsilon}^p$ is the hardening parameter or plastic strain

1.6.1.3   Flow rule

As is customary in the framework of incremental plasticity, the concept of flow rule is applied to obtain the plastic rate of deformation tensor $\mathbf{d}^p$ in terms of the plastic flow direction tensor $n = \left(\text{dev}\tau/\|\text{dev}\tau\|\right)$ associated with the yield surface:

$$d^p = \dot{\lambda}n = \dot{\lambda}\frac{\text{dev}\tau}{\|\text{dev}\tau\|} \qquad (1.33)$$

and the evolution equation for the accumulated effective plastic strain $\bar{\varepsilon}^p$ is governed by

$$\dot{\bar{\varepsilon}}^p = \sqrt{\frac{2}{3}}\dot{\lambda} \qquad (1.34)$$

where $\dot{\lambda}$ is the consistency parameter or plastic multiplier subject to the standard Kuhn–Tucker loading/unloading conditions:

$$\dot{\lambda} \geq 0, f(\tau) \leq 0, \dot{\lambda}f(\tau) = 0 \qquad (1.35)$$

Along with the consistency condition [19], complete the formulation of the model:

$$\dot{\lambda}\dot{f}(\tau) = 0$$

1.6.1.4   The Johnson–Cook constitutive model

The titanium alloy Ti–6Al–4V used in this study is a commonly used material in aerospace and biomedical industries for its superior properties. The isotropic constitutive model proposed by Johnson and Cook has provided a description of the material behavior when subjected to large strains, high-strain rates, and thermal softening. This model has been widely used in machining simulation [20–22].

Table 1.1 Material properties for
the process simulations

A(MPa)	860
B(MPa)	612
n	0.78
C	0.08
m	0.66

$$\sigma_y = \left(A + B\left(\bar{\varepsilon}^p\right)^n\right)\left(1 + C\ln\left(\frac{\dot{\bar{\varepsilon}}^p}{\dot{\bar{\varepsilon}}_0}\right)\right)\left(1 - \left(\frac{T - T_0}{T_m - T_0}\right)\right) \qquad (1.36)$$

In the Johnson–Cook model, $\bar{\varepsilon}^p$ is the plastic strain; $\dot{\bar{\varepsilon}}^p$ is the plastic strain rate; $\dot{\bar{\varepsilon}}_0$ is the reference plastic strain rate (s^{-1}); T is the temperature of the workpiece; T_m is the melting temperature of the workpiece material; and $T_0 = 293.15$ is the room temperature. Material constant A is the yield strength; B is the hardening modulus; C is the strain rate sensitivity; n is the strain-hardening exponent; and m is the thermal softening exponent. Although a more realistic simulation model for the machining process should also take the state of the work material into account due to a previous machining pass or manufacturing process, the material enters the workpiece without any strain or stress history in our model. Table 1.1 gives the Johnson–Cook properties used in our numerical simulations.

1.7   Stress update algorithm

1.7.1   Thermo-elastoplasticity model at finite strains

An implicit integration of the constitutive model presented in Section 1.6.1.4 is summarized in Box 1.1.

1.7.2   Transient solution of the discretized equations

Equations 1.19 are solved in time with an uncoupled (mechanical–thermal) implicit Newton–Raphson-type iterative scheme. The basic steps within a time increment $[n.n+1]$ are as follows:

- Initialize variables

$$\left({}^{n+1}x^1, {}^{n+1}\bar{u}^1, {}^{n+1}\tau^1, {}^{n+1}\bar{p}^1, {}^{n+1}\bar{T}^1, {}^{n+1}\bar{\varepsilon}^p\right) \leftarrow \left({}^{n}x, {}^{n}\bar{u}, {}^{n}\tau, {}^{n}\bar{p}^1, {}^{n}\bar{T}^1, {}^{n}\bar{\varepsilon}^p\right)$$

- In the following lines, $(\cdot)^i$ denotes a value computed at the ith iteration.
- Iteration loop: $i = 1, ..., N_{\text{iter}}$ for each iteration

**BOX 1.1   IMPLICIT INTEGRATION SCHEME
OF THE THERMO-ELASTOPLASTICITY
MODEL AT FINITE STRAINS**

Given $^{n+1}u, {}^{n}\tau, {}^{n}\bar{\varepsilon}^p, {}^{n}\rho_i, {}^{n}c_v, \Delta t, \mu$

$$f = 1 + \nabla u_{n+1}$$

$$^{n+1}e = \frac{1}{2} 1 - {}^{n+1}f^{-T} \cdot {}^{n+1}f^{-1}$$

$$\tau_{n+1}^{\text{trial}} = {}^{n+1}f \cdot {}^{n}\tau \cdot {}^{n+1}f^T + c : {}^{n+1}e$$

Check for plastic loading

$$\text{dev}\left({}^{n+1}\tau^{\text{trial}}\right) = {}^{n+1}\tau^{\text{trial}} - \frac{1}{3}tr\left({}^{n+1}\tau^{\text{trial}}\right) 1 \otimes 1$$

$$f\left({}^{n+1}\tau^{\text{trial}}\right) = \left\|\text{dev}\left({}^{n+1}\tau^{\text{trial}}\right)\right\| - \sqrt{\frac{2}{3}}\sigma_y\left({}^{n+1}\bar{\varepsilon}^p, {}^{n}\rho_i, {}^{n}c_v\right)$$

IF $f\left({}^{n+1}\tau^{\text{trial}}\right) \prec 0$

$$^{n+1}\tau = {}^{n+1}\tau^{\text{trial}}$$

$$^{n+1}\bar{\varepsilon}_p = {}^{n+1}\bar{\varepsilon}_p$$

ELSE
Go to return mapping
END IF
The return mapping

$$^{n+1}n = \frac{\text{dev}\left({}^{n+1}\tau^{\text{trial}}\right)}{\left\|\text{dev}\left({}^{n+1}\tau^{\text{trial}}\right)\right\|}$$

FIND $\Delta\lambda$ from the solution of the yielding equation using
Newton–Raphson

$$\text{dev}\left({}^{n+1}\tau\right) = \text{dev}\left({}^{n+1}\tau^{\text{trial}}\right) - 2\mu\Delta\lambda\,{}^{n+1}n$$

$$^{n+1}\bar{\varepsilon}_p = {}^{n}\bar{\varepsilon}_p + \sqrt{\frac{2}{3}}\Delta\lambda$$

1.7.3   Mechanical problem

Step 1: Compute the nodal displacement increments and the nodal pressure from Equation 1.19

$$K \begin{bmatrix} \Delta u \\ \Delta \bar{p} \end{bmatrix} = - \begin{bmatrix} {}^{n+1}F^i_{\text{res,mech}} \\ {}^{n+1}F^i_{\text{res,mass}} \end{bmatrix} \qquad (1.37)$$

The iteration matrix **K** is given by

$$\mathbf{K} = \begin{bmatrix} K_{uu} & K_{up} \\ K_{pu} & K_{pp} \end{bmatrix} \qquad (1.38)$$

where K_{uu}, K_{up}, K_{pu}, and K_{pp} are given by the following expressions:

$$K_{uu,ij} = \int_{{}^{n+1}V} B_i^{eT} \left(C_i^{\text{dev}} - 2pI \right) B_j^e dV + \int_{{}^{n+1}V} G_i^T \sigma G_j dV$$

$$K_{up,ij} = \int_{{}^{n+1}V} B_i^{eT} m N_j^e dV \qquad (1.39)$$

$$K_{pu} = K_{up}$$

$$K_{pp} = \int_{{}^{n+1}V} \frac{1}{\kappa} N^{eT} N^e dV + \int_{{}^{n+1}V} \frac{\alpha}{\mu} \left(N^e N^{eT} - \tilde{N}^e \tilde{N}^{eT} \right) dV$$

where C_T^{dev} is the deviatoric part of the consistent algorithmic matrix emanating from the linearization of Equation 1.19 with respect to the nodal displacements [13].

Step 2: Update the nodal displacements and nodal pressure

$$ {}^{n+1}\bar{u}^{i+1} = {}^{n+1}\bar{u}^i + \Delta u$$

$$ {}^{n+1}\bar{p}^{i+1} = {}^{n+1}\bar{p}^i + \Delta\bar{p} \qquad (1.40)$$

Step 3: Update the nodal coordinates and the incremental deformation gradient

$$ {}^{n+1}x^{i+1} = {}^{n+1}x^i + \Delta u$$

$$ {}^{n+1}F_{ij}^{i+1} = \frac{\partial {}^{n+1}x_i^{i+1}}{\partial {}^n x_j} \qquad (1.41)$$

Step 4: Compute the deviatoric Cauchy stresses from Box 1.1.
Step 5: Compute the stresses

$$^{n+1}\sigma_{ij} = \text{dev}\left(^{n+1}\tau\right) + \bar{p}\delta_{ij} \tag{1.42}$$

Step 6: Check convergence
    Verify the following conditions:

$$\left\|^{n+1}\bar{u}^{i+1} - {}^{n+1}\bar{u}^i\right\| \le e_u \left\|^n\bar{u}\right\|$$
$$\left\|^{n+1}\bar{p}^{i+1} - {}^{n+1}\bar{p}^i\right\| \le e_p \left\|^n\bar{p}\right\| \tag{1.43}$$

where e_u and e_p are the prescribed error norms. In the examples presented in this chapter, the error norms are set to $e_u = e_p = 10^{-3}$. If conditions (1.43) are satisfied, the solution of the thermal problem in the updated configuration ^{n+1}x is accepted. Otherwise, make the iteration counter $i \leftarrow i+1$ and repeat Steps 1–6.

1.7.4 Thermal problem

- Iteration loop: $i = 1,..., N_{\text{iter}}$ for each iteration

Step 7: Compute the nodal temperatures

$$\left[\int_{^{n+1}V} \frac{\rho c}{\Delta t} NN^T dV + \int_{^{n+1}V} kB_0^T B_0 \right] \Delta\bar{T} = -^{n+1}F^i_{\text{res,therm}} \tag{1.44}$$

Step 8: Update the nodal temperatures

$$^{n+1}\bar{T}^{i+1} = {}^{n+1}\bar{T}^i + \Delta\bar{T} \tag{1.45}$$

Step 9: Check convergence

$$\left\|^{n+1}\bar{T}^{i+1} - {}^{n+1}\bar{T}^i\right\| \le e_T \left\|^n\bar{T}\right\| \tag{1.46}$$

where e_T is the error norm in the balance of energy. In the examples presented in this chapter, the error norm is set to $e_T = 10^{-5}$. If condition (1.46) is satisfied, then make $n \leftarrow n+1$ and proceed to the next time step. Otherwise, make the iteration counter $i \leftarrow i+1$ and repeat Steps 7–9.

1.8   Example, result, and discussion

The ability of PFEM to adaptively insert and remove particles and to improve mesh quality is crucial in the problems presented from here on. Then, it is possible to maintain a reasonable shape of elements and also to capture gradients of strain, strain rate, and temperature. The PFEM strategy does not require a criterion for modeling of the chip separation from the workpiece.

The friction condition is an important factor that influences chip formation. Friction on the tool–chip interface is a nonconstant function that is dependent of normal and shear stress distribution. Normal stresses are largest in the sticking contact region near the tool tip. The stress in the sliding zone along the contact interface from the tool tip to the point where the chip separates from the tool rake face is controlled by frictional shear stress. A variety of complex friction models exist; however, the lack of input data to these models is a limiting factor. The model for tool–chip interface employed in this study is the Coulomb friction model. The friction coefficient $\mu = 0.5$ was selected following the value used in References 23–25.

The heat generated in metal cutting has a significant effect on the chip formation. The heat generation mechanisms are the plastic work done in the primary and secondary shear zones and the sliding friction in the tool–chip contact interface. Generated heat does not have sufficient time to diffuse away, and the rise in temperature in the work material is mainly due to localized adiabatic conditions. A standard practice in the numerical simulations of mechanical cutting is to assume the fraction of plastic work that is transformed into heat equal to 0.9 [23,25,26].

An orthogonal cutting operation was employed to mimic 2D plain strain conditions. The depth of cut, used for all the test cases, was equal to 3 mm. The dimension of the workpiece was 8×1.6 mm. A horizontal velocity corresponding to the cutting speed was applied to the particles at the right side of the tool as given in Table 1.2. The particles along the bottom and the left sides of the workpiece were fixed. Material properties for the workpiece material are shown in Table 1.1. Material properties of the tool were assumed as thermoelastic.

The workpiece was discretized with 105 particles (Figure 1.2a). The tool geometry was discretized by 2232 tree-node thermomechanical elements.

Table 1.2 Cutting data in simulations

Test no	Cutting speed v_c (m/min)	Feed (mm/rev)	Cutting depth (mm)
1	30	0.05	3.0
2	30	0.15	3.0
3	60	0.05	3.0
4	60	0.15	3.0
5	120	0.05	3.0
6	120	0.15	3.0

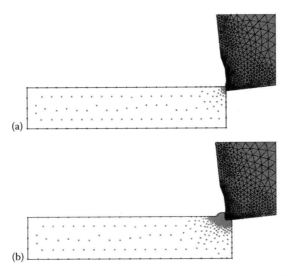

Figure 1.2 2D plane strain PFEM model of orthogonal cutting: (a) initial set of particles and (b) initiation of the chip.

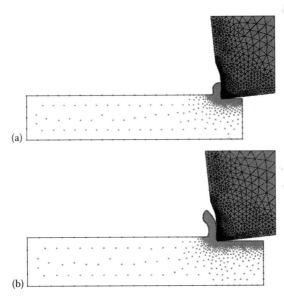

Figure 1.3 Intermediate stages of the chip formation: (a) time 8.04×10^{-4} s and (b) time 1.6×10^{-3} s.

Due to adaptive insertion and removal of particles, the average number of particles increased up to 6329. The effect of insertion of particles near the tool tip is illustrated in Figures 1.2b and 1.3a, b for the test case no. 4. The insertion of particles was controlled by the equidistribution of plastic power.

1.8.1 Cutting and feed forces

The loading histories of simulated forces for test no. 3 (Figure 1.4) were evaluated at the chip–tool interface. Average values of the computed forces in the steady-state region are compared with the experimental results in Table 1.3. The error used for the evaluation of the computed results is computed as

$$\text{Error} = \frac{\text{Computed} - \text{measured}}{\text{Measured}} \times 100\% \qquad (1.47)$$

Table 1.3 shows that the cutting force was overestimated in all tests (in average) in more than 46%. Meanwhile, the feed force was overestimated

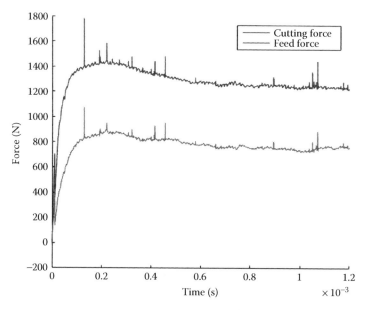

Figure 1.4 Cutting force and feed force for test case no. 4.

Table 1.3 Measured and simulated cutting forces

	Measured		Simulated PFEM			
	$F_c(N)$	$F_f(N)$	$F_c(N)$	Error (%)	$F_f(N)$	Error (%)
1	405	491	672.6	66.0	485	−1.22
2	922	735	1349	46.31	764	3.95
3	396	454	588.8	48.6	456	0.44
4	868	701	1264	45.6	771	9.99
5	424	478	551.3	30.02	449	−6.07
6	838	746	1194	42.4	744	−0.27

by about 1%. The errors in Table 1.3 must be related to the context in which they will be used, namely the cutting tool manufacturing industry. Literature overview [4] shows that in the industrial production of nominally identical cutting tool as well as variations in material properties of nominally the same material can cause variations around 10% in forces. As the average error we get in cutting forces is of the order of 46%, we recommend the use of better constitutive models to model the titanium Ti–6Al–4V, for example, the dislocation density constitutive models that are developed and applied by the authors of the present chapter in References 6 and 23.

1.8.2 Material response

All figures presented in this section correspond to the steady-state conditions. The results shown are for the cutting velocity of 60 m min^{-1} and feed of 0.15 mm. Figure 1.5 illustrates the distribution of plastic strain rates in the primary and the secondary shear zones. Figure 1.5 presents a maximum plastic strain rate value of 40,377 s^{-1}.

Temperature fields are presented in Figure 1.6. Maximum temperature was generated in the contact between the chip and rake face of the tool.

The von Mises stress fields are presented in Figure 1.7. The maximum value of the stress takes place in the tool in which it looses the contact with the machined surface.

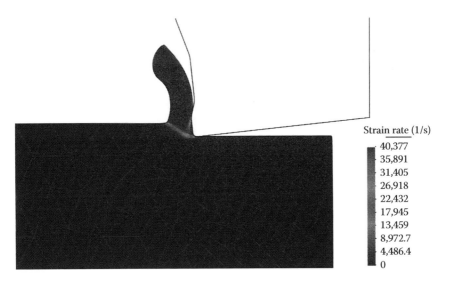

Strain rate (1/s)

40,377
35,891
31,405
26,918
22,432
17,945
13,459
8,972.7
4,486.4
0

Figure 1.5 Effective plastic strain rate.

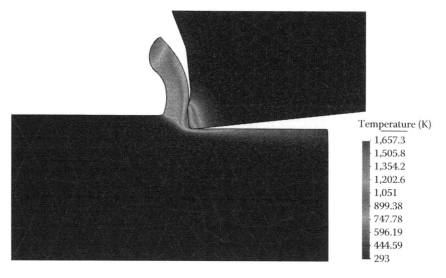

Figure 1.6 Temperature distribution.

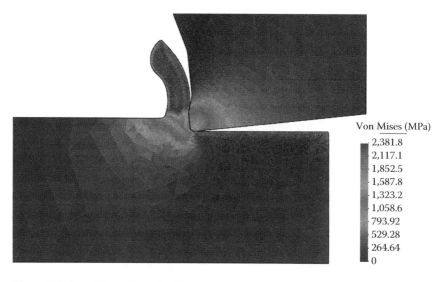

Figure 1.7 Von Mises stress field.

1.9   Conclusion

A Lagrangian formulation for analysis of metal cutting processes that involve thermally coupled interactions between deformable continua is presented. The governing equations for the generalized continuum are discretized using elements with equal linear interpolation for the displacement and the temperature. The merits of the formulation in terms of its general applicability have been demonstrated in the solution of three representative numerical simulations of orthogonal cutting using the PFEM.

Numerical results obtained using PFEM have been compared with experimental results. In addition, the numerical model developed within this work is in agreement with experimental results and can predict forces near the wanted precision.

In conclusion, PFEM is a suitable tool for machining processes simulation.

References

1. Ambati R, Pan X, Yuan H, Zhang X (2012) Application of material point methods for cutting process simulations. *Computational Materials Science* 57:102–110.
2. Limido J, Espinosa C, Salaün M, Lacome JL (2007) SPH method applied to high speed cutting modelling. *International Journal of Me Sciences* 49(7):898–908.
3. Illoul L, Lorong P (2011) On some aspects of the CNEM implementation in 3D in order to simulate high speed machining or shearing. *Computer and Structures* 89(11/12):940–958.
4. Eberhard P, Gaugele T (2012) Simulation of cutting processes using mesh-free Lagrangian particle methods. *Computational Mechanics*:1–18. doi:10.1007/s00466-012-0720-z.
5. Rodriguez JM, Carbonell JM, Cante JC, Oliver J (2015) The particle finite element method (PFEM) in thermo-mechanical problems. *International Journal for Numerical Methods in Engineering.* doi:10.1002/nme.5186.
6. Rodríguez JM, Jonsén P, Svoboda A (2016) Simulation of metal cutting using the particle finite-element method and a physically based plasticity model. *Computational Particle Mechanics*:1–17. doi:10.1007/s40571-016-0120-9.
7. Rodríguez JM, Cante JC, Oliver X (2015) *On the Numerical Modelling of Machining Processes via the Particle Finite Element Method (PFEM).* vol. 156. CIMNE, Barcelona.
8. Idelsohn SR, Oñate E, Pin FD (2004) The particle finite element method: A powerful tool to solve incompressible flows with free-surfaces and breaking waves. *International Journal for Numerical Methods in Engineering* 61(7):964–989.
9. Xu X, Harada K (2003) Automatic surface reconstruction with alpha-shape method. *The Visual Computer* 19(7/8):431–443.
10. Edelsbrunner H, M EP, #252, cke (1994) Three-dimensional alpha shapes. *ACM Trans Graph* 13(1):43–72. doi:10.1145/174462.156635.
11. Rodríguez JM (2014) *Numerical Modeling of Metal Cutting Processes Using the Particle Finite Element Method (PFEM).* Universitat Politècnica de Catalunya (UPC), Barcelona.
12. Oliver J, Cante JC, Weyler R, González C, Hernandez J (2007) Particle finite element methods in solid mechanics problems. *Computational Methods in Applied Sciences* 7:87–103.
13. Belytschko T, Liu WK, Moran B (2000) *Nonlinear Finite Element for Continua and Structures.* Wiley, Chichester, UK.

14. Fish J, Belytschko T (2007) *A First Course in Finite Elements*. Wiley, Chichester, UK.
15. Zienkiewicz OC, Taylor RL, Zhu JZ (2013) *The Finite Element Method: Its Basis and Fundamentals*. 7th ed. Elsevier, Oxford, UK.
16. Zienkiewicz OC, Taylor RL, Fox DD (2014) *The Finite Element Method for Solid and Structural Mechanics*. Elsevier, Oxford, UK.
17. Lal GK (2009) *Introduction to Machining Science*. 3rd ed. New Age, New Delhi, India.
18. Sekhon GS, Chenot JL (1993) Numerical simulation of continuous chip formation during non-steady orthogonal cutting simulation. *Engineering Computations* 10(1):31–48.
19. Simo JC, Hughes. TJR (1998) *Computational Inelasticity*. Springer-Verlag, New York.
20. Rodríguez J, Arrazola P, Cante J, Kortabarria A, Oliver J (2013) A sensibility analysis to geometric and cutting conditions using the particle finite element method (PFEM). *Procedia CIRP* 8:105–110. doi:10.1016/j.procir.2013.06.073.
21. Arrazola PJ, Ugarte D, Domínguez X (2008) A new approach for the friction identification during machining through the use of finite element modeling. *International Journal of Machine Tools & Manufacture* 48:173–183.
22. Arrazola PJ, Ozel T (2008) Numerical modelling of 3D hard turning using arbitrary Lagrangian Eulerian finite element method. *International Journal of Machining and Machinability of Materials* 4(1):14–25.
23. Svoboda A, Wedberg D, Lindgren L-E (2010) Simulation of metal cutting using a physically based plasticity model. *Modelling and Simulation in Materials Science and Engineering* 18(7):075005.
24. Arrazola PJ, Özel T (2010) Investigations on the effects of friction modeling in finite element simulation of machining. *International Journal of Mechanical Sciences* 52(1):31–42.
25. Özel T (2006) The influence of friction models on finite element simulations of machining. *International Journal of Machine Tools and Manufacture* 46(5):518–530.
26. Wedberg D, Svoboda A, Lindgren L-E (2012) Modelling high strain rate phenomena in metal cutting simulation. *Modelling and Simulation in Materials Science and Engineering* 20(8):085006.

chapter two

Smoothed particle hydrodynamics for modeling metal cutting

Mohamed N.A. Nasr

Contents

2.1 Overview... 25
2.2 Metal cutting: Background... 26
2.3 Finite element modeling 28
    2.3.1 Background... 28
    2.3.2 Finite element modeling of metal cutting.............................. 30
2.4 Smoothed particle hydrodynamics... 32
    2.4.1 Introduction.. 32
    2.4.2 Numerical discretization/particle approximation 33
    2.4.3 Solution procedure 35
    2.4.4 Smoothed particle hydrodynamics advantages and
            limitations 35
2.5 Smoothed particle hydrodynamics modeling of metal cutting....... 36
2.6 Summary and concluding remarks 45
References... 46

2.1 Overview

This chapter focuses on the use of smoothed particle hydrodynamics (SPH), as a mesh-free numerical technique, for simulating the cutting process, particularly metal cutting. The chapter is divided into six sections: Section 1.1 presents an overview of the chapter; Section 2.2 sheds some light on the basics of metal cutting; Section 2.3 covers finite element method (FEM) of metal cutting; Section 2.4 presents the numerical background of SPH; Section 2.5 covers the usage of SPH for modeling metal cutting, and Section 2.6 summarizes and concludes the chapter. In this chapter, the terms *machining*, *metal cutting*, and *cutting process* are used interchangeably.

2.2 Metal cutting: Background

Machining is currently the most widely used manufacturing process [1], and the machined products find their way in almost all industrial sectors. This includes, but not limited to, the aerospace, nuclear, automotive, and medical sectors. Accordingly, a massive body of literature has been dedicated to examine and understand different aspects of the machining process, with the utmost goal of understanding how different process parameters, as well as workpiece and tool material properties, would affect part performance. In addition to experimental investigations, different modeling techniques have been used to provide explanations for experimental findings, and for better understanding of the physical process [2]. The modeling efforts started by analytical modeling, which goes back to the work of Ernst and Merchant (1941) [3], followed by mechanistic modeling; later on, FEM has found its way as an effective tool for simulating the cutting process, which is capable of overcoming the limitations of analytical and mechanistic modeling [2]. In general, the majority of the modeling efforts has focused on orthogonal cutting, as a simple representation of the cutting process that does not alter the understanding of process mechanics.

Metal cutting is a clear example of severe plastic deformation (SPD) in which the workpiece material is subjected to plastic strains that may have magnitudes of up to 10 [1,4]. In addition, the material is subjected to very high strain rates, which can be in the order of 10^5–10^6 s^{-1}, and high temperatures, which can go up to 1000°C [1,2]. Experiencing such high values of plastic strain, strain rate, and temperature in a very confined region (chip generation region) makes the process one of the most complex processes to understand and model [2,4]. During cutting, the workpiece material experiences plastic deformation, and consequently heat generation, in three different zones: the primary deformation zone (PDZ), the secondary deformation zone (SDZ), and the tertiary deformation zone (TDZ), as schematically shown in Figure 2.1. As the tool progresses, the material is first deformed in the PDZ, where it is subjected to high shear stresses (along the shear plane, which makes an angle "φ" with the cutting direction) as it deforms forming the chip. In case of cutting with a sharp tool, the single-shear plane model applies in which plastic deformation is assumed to occur along one plane [5]. However, if a honed or chamfered tool is used (actual case), the single-shear plane model does not apply [6]. After the onset of chip formation, the material experiences further plastic deformation and heat generation as it moves along the tool rake face, due to friction, in the SDZ. Finally, as the machined surface is generated, further deformation and heat generation take place in the TDZ, which mainly depend on the geometry of the cutting edge [7]. Unless using a sharp-edged tool, ploughing takes place underneath the tool tip [6,7]. In addition

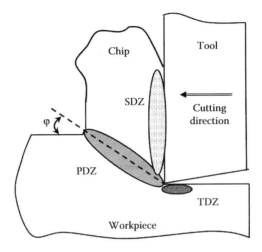

Figure 2.1 Deformation zones in metal cutting, with the shear plane angle (φ) shown.

to plastic deformation, it is important to note that friction between the tool and workpiece material plays an important role in heat generation during cutting [2,8]. The magnitudes of plastic strain, strain rate, and temperature in the three deformation zones depend on the used cutting conditions (feed rate, cutting speed, tool edge preparation, tool geometry, and cooling conditions) and on the workpiece and tool material properties [2,5,8–10].

The work of Ernst and Merchant [3] has paved the way to understand the mechanics of chip formation during metal cutting, and the Merchant's shear plane model (Merchant's circle) is still currently considered as a useful tool for understanding the cutting action, and for relating different cutting force components. After that, Oxley [5] developed an analytical model for dry orthogonal cutting using sharp tools, which is capable of predicting cutting forces, average workpiece temperatures, and strains in the PDZ and SDZ, based on the single-shear plane model. Oxley applied his model to low carbon steels and predicted cutting forces and chip thickness that are in good agreement with experimental measurements. More recently, Manjunathaiah and Endres [6] developed an analytical force model for predicting cutting forces when dry orthogonal cutting using honed tools. Moufki et al. [11] presented an oblique cutting model, which accounts for the viscoplastic and thermomechanical material properties as well as inertia effects in the PDZ. When the model was applied to oblique cutting of steels, the predicted chip flow angle and cutting forces were found to be in good agreement with the experimental data. In addition to the aforementioned models, several other analytical models have been developed, as can be found in the literature. However, as the focus of the current chapter is numerical modeling, particularly SPH, it is believed that the highlights presented earlier are enough.

Despite the analytical efforts made to model the cutting process, the assumptions and simplifications encountered in the developed models limited their capabilities from closely predicting the complex phenomena that take place during machining. For example, none of the developed analytical models is capable of predicting surface integrity parameters. Accordingly, FEM was sought to overcome the short comes of analytical modeling, as highlighted in Section 2.3 [1,2].

2.3 Finite element modeling

2.3.1 Background

In FEM, as shown in Figure 2.2, a finite number of elements (of predefined simple shapes) is used to discretize/mesh the part to be modeled, and these elements are interconnected using nodes. FEM was developed, as a numerical technique, in order to solve situations that encounter high degrees of complexity, which cannot be solved/addressed using analytical techniques. Such complexities could arise from geometrical, material, and/or boundary conditions. The advantage of discretization is that the governing equations (equilibrium, kinematic, and constitutive conditions) are solved for those simple elements, instead of the physical continuous part, taking into consideration the interconnection between elements. The nodes are used to apply known boundary conditions and to solve for unknown degrees of freedom (DOFs). On the other hand, the elements

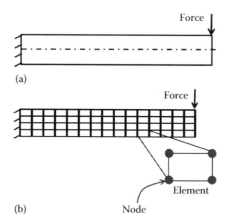

Figure 2.2 Concept of FEM. (a) Cantilever beam (physical case) and (b) finite element of a cantilever beam.

define how the nodes behave, where the derived quantities (stresses and strains) are calculated within each element at its integration points. Shape functions are used to define the shape of each element and its behavior; they are also used to interpolate different field variables within the element [12–14].

Based on how the elements and nodes (mesh) relate to the underlying material (to be modeled), different finite element (FE) formulation techniques exist. Figure 2.3 shows the two basic/classical modeling techniques: the Eulerian and Lagrangian techniques. This figure shows a material under shear loading during deformation, at different time (*t*). As shown, in case of an Eulerian mesh, the elements and nodes are totally fixed in space, whereas the underlying material is deformed under the applied load. In other words, the material point at a given integration point changes with time. On the other hand, a Lagrangian mesh is fully attached to the underlying material; therefore, element integration points remain coincident with material points. Accordingly, Lagrangian modeling is more suitable for history-dependent analyses as the material points, whose history variables are required, are coincident with the nodes, which are used for calculations. Furthermore, boundary conditions are easier to apply when a Lagrangian mesh is used, because boundary nodes remain on the boundaries of the material throughout the analysis [2,14].

The main disadvantage of Lagrangian formulation is mesh distortion, which takes place as the material deforms. Mesh distortion has a negative impact on the accuracy of results and becomes more evident as the analysis encounters nonlinearity [2,12]. It can even result in terminating the analysis in case if distortion exceeds the allowable limit during the analysis. Automatic remeshing can be used to overcome excessive mesh distortion, where a distorted mesh is replaced with a new one and the results are mapped between the two; however, the accuracy of results still

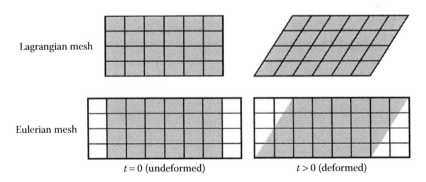

t = 0 (undeformed)   *t* > 0 (deformed)

Figure 2.3 Lagrangian versus Eulerian meshes—material under shear loading. (From Nasr, M.N.A., On modelling of machining-induced residual stresses, PhD thesis, McMaster University, Hamilton, Canada, 2008. With permission.)

deteriorates during mapping [2,7]. On the other hand, the main disadvantage of Eulerian formulation is the need for a large enough mesh that covers the undeformed and deformed shapes at the same time, in order to simulate transient effects, as shown in Figure 2.3. Otherwise, it can be only used to simulate steady-state conditions, and in such a case, the final deformed shaped needs to be known a priori [2,12]. Furthermore, material elasticity cannot be taken into consideration; accordingly, some modeling capabilities are lost. For example, residual stresses cannot be predicted, which is a significant drawback when simulating different manufacturing processes [2,7,15,16].

The arbitrary Lagrangian–Eulerian (ALE) technique, which is an arbitrary combination of the Lagrangian and Eulerian techniques, was developed in order to combine the advantages of the two classical techniques and to minimize their drawbacks [2,7]. In an ALE model, the mesh is neither attached to the underlying material nor it is fixed in space; that is, it is neither Lagrangian nor Eulerian and can be controlled independently. Accordingly, the word *arbitrary* refers to the fact that the user defines the combination of Lagrangian and Eulerian meshes by selecting the mesh motion in different parts of the model [2,14,16].

2.3.2   *Finite element modeling of metal cutting*

As mentioned earlier, metal cutting is one of the most challenging processes to model. This does not only apply to analytical modeling, but also applies to FEM. The main challenge with FEM is attributed to the high nonlinearity encountered in the process that arises from material, geometric, and status nonlinearities. Furthermore, with all these phenomena taking place simultaneously and in a very confined region (chip generation region), it becomes even more challenging [2,12].

FEM has played a significant role in simulating the cutting process and in understanding its different aspects. This applies to the classical approaches (Eulerian and Lagrangian), with a much wider use for the Lagrangian formulation, and ALE. In metal cutting, an Eulerian mesh is advantageous only around the tool tip, where severe material deformation takes place. This is because the Eulerian mesh can handle the material flow in that region without experiencing any mesh distortion. However, free surfaces and surface integrity cannot be predicted [2,7]. On the other hand, a Lagrangian mesh is suitable for predicting free surfaces (for example, chip generation) and surface integrity; however, it experiences severe mesh distortion. Automatic remeshing can be used to limit element distortion; however, due to the high nonlinearity encountered in the process, frequent remeshing is required, which would—along with the potential significant difference between two consecutive meshes—lead to accuracy degradation [2,7]. Furthermore, a Lagrangian mesh requires the use of a failure criterion in order to define chip generation and segmentation. In addition, cutting needs to be performed along a predefined path, typically referred to as *parting line*; accordingly, Lagrangian cutting models are more suitable for up-sharp tools [17,18]. As FEM is based on continuum mechanics, elements cannot be broken; therefore, the whole elements along the parting line need to be deleted, which leads to an inaccurate representation of the chip generation path [19]. Figure 2.4 schematically shows an

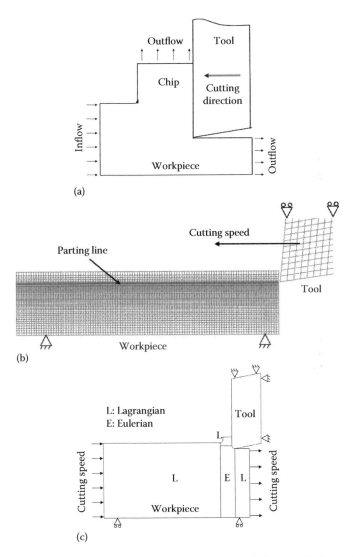

Figure 2.4 Orthogonal (2D) cutting models, using different FE formulations. (a) Eulerian model, (b) Lagrangian model, and (c) ALE model.

Eulerian and a Lagrangian cutting model for orthogonal cutting. It also shows how ALE can be used for modeling the cutting process. As shown in Figure 2.4c, an ALE workpiece is divided into different regions, where an Eulerian mesh is used around the tool tip, whereas a Lagrangian mesh is used elsewhere [7]. This is because, as mentioned earlier, an Eulerian mesh can handle the material flow around the tool tip, without experiencing any distortion and without the need for a failure criterion. At the same time, a Lagrangian mesh is suitable for modeling free surfaces and for predicting residual stresses [2,7].

FEM has been extensively used for modeling the cutting process, covering almost all its different aspects. This includes, for example, simulating the effects of tool edge geometry [7,15,17,18], tool wear [15,20,21], workpiece material properties [7,9,10,22,23] and tool–workpiece interaction [24,25] on the cutting process, and the generation of residual stresses. More recently, FEM has also been used to simulate relatively new machining techniques such as laser-assisted machining (LAM) [26–28] and cryogenic machining [18,29].

2.4   Smoothed particle hydrodynamics

2.4.1   Introduction

SPH is a mesh-free (or element-free) Lagrangian-based numerical method, which was originally developed by Lucy, Gingold, and Monaghan in 1977 for astrophysical problems. Since then, its use has been extended to simulate the dynamic response of solid materials and the dynamic fluid flows that experience large deformations [19,30]. Accordingly, in a general sense, the term *hydrodynamics* may be interpreted as *mechanics*. In SPH, each particle represents a specific material volume/mass, and accordingly SPH is a Lagrangian-based method [31]. In general, mesh-free methods were developed in order to numerically solve partial differential equations and/or integral equations, with all types of boundary conditions, using a set of arbitrarily distributed particles/nodes [30]. In other words, mesh-free methods do not use elements to discretize the problem domain, as in the case of FEM; rather, they use a set of nodes/particles that are scattered over the problem domain, as shown in Figure 2.5 [19,30]. As elements do not exist, mesh distortion is not an issue; therefore, mesh-free methods can handle cases with severe material deformations. As shown in Figure 2.5, the domain boundary is better represented using SPH particles, as compared to FEM. This is because, at any point between two boundary particles, one can interpolate using mesh-free shape functions, which are created using nodes/particles in a moving local domain. Accordingly, curved boundaries can be approximated very accurately even if linear polynomial shape functions are used [19]. On the other hand, with FEM,

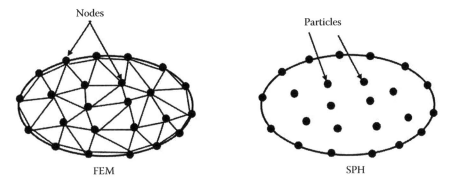

Figure 2.5 SPH versus FEM (linear elements)—geometrical representation.

the curved boundary is approximated as piecewise curves (straight lines) if linear elements are used, and higher order shape functions are required to accurately represent the boundary.

Furthermore, as the particles are not interconnected, no information on the relationship between them is required a priori, at least for field variable interpolation [19]. For the same reason, adaptive schemes can be easily developed and implemented in which nodes can be added or deleted at any location and at any time during the analysis. For example, in fracture mechanics problems, nodes can be simply added around the crack tip to capture stress concentration effects, and such refinement can move adaptively along with the crack as it propagates [19]. In addition to SPH, other examples of mesh-free methods include the element-free Galerkin (EFG) method, the meshless local Petrov–Galerkin (MLPG) method, and the point interpolation method (PIM) [19,30].

2.4.2 *Numerical discretization/particle approximation*

SPH is a continuum numerical method, which is based on the use of local interpolations from surrounding discrete particles to construct continuous field approximations. In SPH, field variables (temperatures, displacements, strains, and stresses) and their derivatives/integrals are approximated at a given particle location by interpolation of the respective values from the neighboring particles, using smoothing functions. The neighborhood of a particle includes those particles that influence its performance and is defined by the so-called influence, support or smoothing domain. The support domain is spherical or circular in shape in 3D and 2D simulations, respectively, and is defined by a smoothing length "*h*," as schematically shown in Figure 2.6.

Approximations are performed using interpolation/shape/smoothing functions that represent the shape of a Gaussian function in which higher

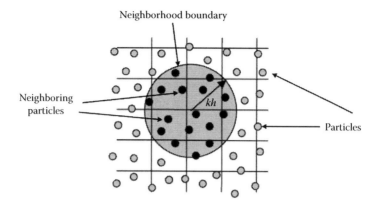

Figure 2.6 Smoothing/support domain.

weights are given to particles at the center of the support domain, and the weight diminishes as we move away from the particle of interest till it reaches zero at the boundary. In other words, weighted-average approximations are used that result in a smoothed approximation over the support domain. Such smoothed approximation is the reason behind the first term in the name *SPH*. The interpolation/smoothing process is performed for all particles, and as a result continuum distributions of field variables are obtained [30,32].

In SPH, the interpolated value of a field variable (function) "*f(x)*" at a particle "*i*," at location "x_i," is obtained as the summation of its values at particles "*j*" that fall within the support domain (of radius *kh*), where *k* is a positive constant, as given by Equation 2.1. In Equation 2.1, m_j and ρ_j are the mass and density of particle *j*, respectively, and *W* is the smoothing (or kernel) function and is given by Equation 2.2, where *D* represents the space dimension (for example, *D* = 2 for two-dimensional analyses). In order to define *W*, an auxiliary function "θ" is defined, as given by Equation 2.3 (as an example), where *C* is a constant of normalization that depends on the value of *D*. The spatial derivative of *f(x)* is defined by applying the variance operator on the smoothing length "*h*," as given by Equation 2.4. Using the aforementioned principle, the value of a continuous function, and its derivative, can be estimated at any location "x_i" based on known values at locations "x_j" that belong to the smoothing domain [1]. It is worth noting that the accuracy of the solution highly depends on the choice of the smoothing function "*W*" and the smoothing length "*h*" [30]. Furthermore, as the material deforms, *h* needs to be dynamically changed in order to avoid negatively affecting the results; it is increased as the material is stretched and is decreased as the material is compressed [31].

$$f(x_i) = \sum_j \frac{m_j}{\rho_j} f(x_j) W\left(x_i - x_j, h\right) \tag{2.1}$$

$$W\left(x_i - x_j, h\right) = \frac{1}{h^D} \theta\left(\frac{|x_i - x_j|}{h}\right) \tag{2.2}$$

$$\theta(y) = C. \begin{cases} 1 - 2/3y^2 + 0.75y^3, & y \le 1 \\ 0.25(2 - y)^3, & 1 < y \le 2 \\ 0, & y > 2 \end{cases} \tag{2.3}$$

$$\nabla \cdot f(x_i) = \sum_j \frac{m_j}{\rho_j} f(x_j) \nabla \cdot W\left(x_i - x_j, h\right) \tag{2.4}$$

2.4.3   Solution procedure

SPH is typically solved using explicit time integration methods, and the solution procedure is mainly similar to that of explicit FEM, except that the domain is represented using arbitrary distributed particles instead of elements. The solution procedure is as follows [30]:

1. Problem domain representation using particles (discretization).
2. Numerical discretization in which the derivatives or integrals in the governing equations (which are the same as those used in FEM) are represented using particle approximations.
3. Application of boundary conditions, and calculation of DOFs, strains, and then stresses at each material particle at time t.
4. Based on the calculated stresses, the acceleration at each particle is found.
5. The position of each particle is updated, based on the calculated accelerations, to find their new values after a time step Δt.
6. From the new positions, the new strains and stresses are calculated at time $t + \Delta t$.

2.4.4   Smoothed particle hydrodynamics advantages
and limitations

The main advantage of SPH over FEM is its adaptive nature, which arises from using local-based smoothing functions based on arbitrary distributed particles, and the fact that it is a mesh-free method, which is also attributed to its adaptive formulation. Such adaptivity is achieved at the very early stage of field variable approximation and can naturally handle problems with severe deformations [30]. However, special techniques are required to impose displacement boundary conditions, because the SPH shape functions do not satisfy the Kronecker delta conditions [19]. Furthermore, it is worth noting that standard SPH requires an equation of state (EOS) that governs the change in density based on pressure, which is essential only for solving compressible flows. This is because SPH was developed for fluid simulations in which particle motion is driven by the gradient of internal energy, which is function of pressure energy, density, and temperature. In general, an EOS is required in SPH simulations in order to accurately model the material hydrostatic behavior under high strain rates and pressures. The need for an EOS represents an issue for incompressible simulations, incompressible flows, and solid mechanics in which an EOS for pressure does not exist. Although it is possible to define a constant density in SPH formulations, as a constraint, the corresponding equations will be cumbersome to be solved. Instead, the artificial compressibility approach is used, which is based on the fact that any incompressible fluid is theoretically compressible to some extent. Accordingly, a quasi-incompressible EOS is used to model solids/incompressible fluid. The reason behind the need for an EOS, using artificial compressibility, is to generate a time derivative of pressure that is required in SPH simulations [19].

2.5   Smoothed particle hydrodynamics modeling of metal cutting

As SPH is highly capable of modeling severe material deformations, due to its adaptive and mesh-free nature, it was sought as a strong alternative to FEM for simulating the cutting process. In addition, the SPH contact control permits a *natural* and simple workpiece/chip separation, where the particles flow naturally around the tool tip. Another important advantage of SPH is that it avoids the need for remeshing [1]. The use of SPH in machining simulations goes back to 1997, when Heinstein and Segalman [33] examined the use of SPH for simulating orthogonal cutting. A close examination of the available literature shows that most of the work has been done using commercial FE software, where the majority of the work was performed using LS-DYNA, and very limited work could be found using Abaqus.

Umer et al. [34] used SPH to simulate chip morphology during high-speed cutting of AISI H13 steel and compared the results to Lagrangian FE, with special focus on the transition between continuous and serrated chips. The analysis was performed using the commercial software LS-DYNA. Only the workpiece was modeled using SPH, using a uniform particle spacing of 0.02 mm, whereas the tool was assumed to be rigid. The authors performed orthogonal high-speed cutting tests on AISI H13 tubes for model validation. Compared to traditional FEM, SPH was found to give a more realistic chip shape. This was attributed to the better capabilities of SPH in handling large deformations, and the natural separation of the chip without the need for a chip separation criterion (the case of FE Lagrangian simulation). Furthermore, the authors compared the default SPH, available in LS-DYNA, with the renormalized SPH formulation. Renormalized SPH was developed in order to allow for better distribution of particles around the contact boundaries, to result in better representation of domain boundaries, because the classical/default formulation struggles with particle distribution along the boundaries due to the lack of neighboring nodes. This is because the default SPH formulation treats all particles in the same way, without differentiating between internal and boundary particles. Normalized SPH was found to predict a more realistic chip shape and contact length. The default SPH predicted a significantly higher tool–chip contact length, as it is not capable of accounting for frictional effects that result in an unrealistic chip flow.

Madaj and Piska [35] modeled dry orthogonal machining of A2024-T351 aluminum alloy with the aid of SPH, using the commercial software LS-DYNA. The authors examined the effects of Johnson–Cook damage model parameters $(D_1–D_5)$ and SPH particle spacing on cutting forces, chip morphology, plastic strain, and strain rates. The developed model successfully predicted serrated chips, and the predicted results were in good agreement with experimental measurements. The density of SPH particles was found to affect chip segmentation, where higher density resulted in highly segmented chip. Moreover, the authors recommended the use of Johnson–Cook failure parameters to improve the predictability of chip morphology.

Limido et al. [1] developed an SPH cutting model, using the commercial software LS-DYNA, to simulate the process of dry orthogonal cutting of Al 6061-T6 aluminum alloy and AISI 4340 steel. In addition, the SPH results were compared to those obtained using AdvantEdge, a commercial FE code dedicated to machining. SPH was capable of predicting continuous chips as well as serrated chips, and cutting forces. It is worth noting that the frictional effects were not considered in the SPH model, and (by nature) no failure criterion was required for chip generation. The tool was assumed to be rigid, and a cutting speed that is 10 times higher than the actual speed was used in order to speed up the simulations. When cutting Al 6061-T6, continuous chips were generated, and the SPH model underestimated the chip thickness while the AdvantEdge model overestimated it, as reported in Table 2.1. With regard to the cutting force

Table 2.1 Al 6061-T6 chip thickness and cutting force component

	Experimental	AdvantEdge	SPH (LS-DYNA)
Chip thickness (μm)	400	490	295
Cutting force (N)	770	775	700

Source: Limido, J. et al., *Int. J. Mech. Sci.*, 49, 898–908, 2007.

Table 2.2 AISI 4340 serrated chip tooth thickness (μm)

	Experimental	AdvantEdge	SPH (LS-DYNA)
Feed rate = 220 μm/rev	140	170	140
Feed rate = 400 μm/rev	250	235	220

Source: Limido, J. et al., *Int. J. Mech. Sci.*, 49, 898–908, 2007.

component, AdvantEdge predicted better results as compared to SPH (Table 2.1). The authors attributed the underestimation of the cutting force to the frictional effects that were missing in the SPH model, whereas they were considered in the AdvantEdge model (using the simple Coulomb friction model). When cutting AISI 4340, serrated chips were generated, and a very good match was found between the SPH model and the experimental results, in terms of sawtooth thickness, as shown in Table 2.2. An important difference between the AdvantEdge and SPH models is that AdvantEdge adapts a fracture mechanics model, which allows crack initiation and propagation in order to simulate shear localization and serrated chips; however, the SPH model does not require the definition of a fracture model, as the material naturally flows around the tool tip generating the chip. Finally, the SPH model was capable of predicting the cutting force component for AISI 4340 within 15% of the experimental values.

Chieragatti et al. [36] examined the capabilities of SPH in modeling dry orthogonal cutting of the aerospace Ti-6Al-4V alloy. Their main focus was to predict cutting forces and chip morphology, using new and worn tools. Under the simulated cutting conditions, serrated chips were generated that agreed with what was found experimentally. In addition, the thickness of the shear band was found to increase with tool wear, accompanied with a decrease in segmentation frequency. SPH was also successful in capturing the dead metal zone around the tool tip, when worn tools were used. The feed force component was found to be significantly affected by tool wear as compared to the cutting component; this was predicted using the SPH model and confirmed experimentally. This phenomenon was mainly attributed to the formation of dead metal zone. Finally, friction between the tool and workpiece was not defined in the SPH model.

Zahedi et al. [37,38] presented a hybrid SPH–FEM model to simulate orthogonal micromachining of a copper single crystal, using the commercial FE software Abaqus/Explicit. The authors actually performed an indentation simulation rather than a cutting one. Their main focus was to evaluate the effects of crystallographic anisotropy on the machining response of FCC metals. The tool was considered as a rigid body, whereas the workpiece was modeled as a deformable body and was split into two regions: an SPH region around the tool tip and an FE region away from the cutting region. Even though the model was for orthogonal cutting, a 3D geometry was built, as 2D SPH modeling is not supported by Abaqus, and was meshed using linear brick elements (C3D8R) with reduced integration for the FE region and PC3D elements for the SPH region. The simple Coulomb friction model was used, and a coefficient of friction of 0.1 was assumed. The material subroutine VUMAT was used in order to implement crystal plasticity. The SPH particles (workpiece material) were found to rearrange themselves when the strain energy in the deformed lattice exceeded the binding energy.

Ghafarizadeh et al. [39] presented a hybrid SPH–FEM Lagrangian model for ball-end milling, using the commercial software LS-DYNA. They modeled the milling process of Al6061-T6 aluminum alloy under dry conditions, and the workpiece–tool friction was simulated using the simple Coulomb friction law. The effects of SPH particle spacing and friction coefficient on cutting forces were examined. Similar to the work of Zahedi et al. [37], SPH was only used around the cutting tool, whereas FEM was used away from the tool tip. The Mie–Grüneisen EOS, which defines the material pressure as a linear function of internal energy, density, and temperature, was used. The authors examined the effect of SPH particle spacing on cutting forces by varying it between 5 and 15 µm, and the average error was about 15% compared to experimental measurements. This applies only to the forces in the feed and normal directions; however, the axial force component experienced a significantly large error. On the other hand, particle spacing had a significant effect on the computational time. Accordingly, the authors selected an optimum particle spacing that provided acceptable results and reasonable computational time.

Islam et al. [32] used SPH to model nanomachining of copper in order to better understand the mechanisms involved in nanoscale deformation, and postmachined surface generation. An SPH nanomachining analysis was performed to simulate nanoindentation, using a conical tool, and the predictions were validated against experiments performed on a nanoindenter. The tool was first indented into the workpiece up to the given depth of cut, and then, cutting was performed. After cutting, the tool was withdrawn from the surface leaving a nanomachined surface. The authors reported that the feed force was found to be larger than the cutting component. The developed model captured the cutting and ploughing mechanisms, and it was consistent with the experimental observations. A larger negative rake angle was found to result in more ploughing and in higher residual stresses and strains. The ratio between the cutting and ploughing force components was found to be unaffected by the depth of cut. However, it was significantly affected by the rake angle.

Zhao et al. [40] developed an SPH model to study residual stresses after sequential cuts. The authors examined how chip formation, cutting forces, and residual stresses are altered by sequential cuts, when cutting OFHC copper. It was found that the first cut resulted in a work hardened subsurface that led to having thinner and more curled chips in the second cut. In addition, the minimum chip thickness was found to drop for the second cut that was attributed to the residual stresses induced by the first cut.

Cao et al. [41] investigated the process of material removal using ultrasonic-assisted grinding (UAG), in an effort to contribute to better understand the process. UAG is a promising machining technique, particularly for hard and brittle materials. In their study, the authors mainly focused on ultrasonic-assisted scratching (UAS) of SiC ceramics. They developed an SPH model to simulate the process and compared their findings to experimental measurements. The presented results demonstrated the ability of SPH to model UAS, and reported that the material deformation mechanism differs based on the scratching depth (depth of cut). Plastic deformation was found to prevail in case of low depths of cut, whereas brittle fracture was found to prevail at high depths of cut.

Akarca et al. [42,43] examined the large-strain deformation behavior of Al 1100 aluminum alloy, during orthogonal cutting, using experimental and numerical techniques. Based on careful examination of metallographic sections taken from the material ahead of the tool tip, the changes in flow lines orientation and shear angles were used to find out the local plastic strain distribution in the PDZ and SDZ. In addition, microhardness measurements were used to experimentally estimate local flow stresses. The authors also determined the parameters of Johnson–Cook plasticity model based on their experimental measurements. Two types of numerical models were used: an Eulerian FE model and an SPH model, and the numerical coefficient of friction was determined based on running a parametric study and comparing the predicted chip morphology to that found experimentally. For the simulated conditions, the numerical coefficient of friction was found to be 0.27 for the Eulerian model and 0.63 for the SPH model. In other words, the friction coefficient for SPH was found to be significantly higher than that for Eulerian FE model. The predicted stress and strain distributions were compared to those that were found out experimentally, and a good correlation was found for both the models. However, the Eulerian model was found to be more expensive, as it required a CPU time that is almost 2.75 times that of the SPH model.

Xi et al. [44] used SPH for studying thermally assisted machining of Ti-6Al-4V both in 2D and 3D configurations. The authors focused on chip formation and cutting forces, and thermally assisted cutting tests were performed for model validation. A very good agreement was found between the predicted chip morphology and experimental results, where segmented chips were observed in all cases. The simulations showed that, during cutting, cracks initiate and propagate inside the PZD, which was considered to be the main cause of chip segmentation; however, no clear correlation (neither numerically or experimentally) was found between the segmentation pattern and the workpiece initial temperature. At the same time, the predicted cutting forces were in good agreement with the measured ones, where an increase in workpiece initial temperature resulted in reducing cutting forces. It was also shown that the cyclic frequency of cutting forces was in direct correlation with the segmentation frequency. Higher forces were recorded when a sawtooth was fully generated, and the shear stress was localized in the PDZ, while the force started to drop at the incidence of crack propagation.

Geng et al. [45] used SPH to simulate dry orthogonal cutting of stainless steel AISI 316L, and investigated the effects of sequential cuts on residual stresses. The built model was able to capture shear banding and the correct chip morphology, similar to what was found experimentally. A slight drop was noticed in the cutting force component with sequential cuts; however, the thrust force component was increased. Surface residual stresses were found to increase with sequential cuts; however, no explanation was provided. The same authors [46] employed SPH to simulate dry orthogonal cutting of OFHC copper, and examined the effects of friction coefficient along the tool–workpiece contact length on the predicted cutting forces and chip morphology. Surprisingly, the cutting force was found to drop with the increase of friction coefficient (from 0.1 to 0.3). The predicted forces were compared to experimental results and were found to be in good agreement, where the maximum error was found to be about 15%. Furthermore, the authors compared their SPH predictions to those obtained using the ALE FE technique, and a good agreement was found. In addition, the predicted average strain value in the chip was found to match well with the analytical value that was estimated using the classical theory of orthogonal cutting. Finally, the cutting force was found to increase in the cases with less chip curl.

Spreng and Eberhard [47] investigated the capabilities of SPH to simulate the machining process, with special focus on chip morphology, cutting forces, workpiece stresses, and temperatures. They built 2D and 3D models and compared the predicted results to experimental measurements when cutting steel C45E. The authors presented some improvements over the standard SPH formulation in order to better simulate the cutting process. They used the Johnson–Cook plasticity model, which is widely used in metal cutting simulations, and implemented the Johnson–Cook damage fracture model in order to better simulate the cutting process. Furthermore, they developed and implemented a boundary force model to improve simulating the tool–workpiece interaction. Their model also considered heat generation due to plastic deformation and friction, which are the two main sources of heat generation in metal cutting. Finally, a local adaptive resolution strategy was introduced in order to improve the accuracy of spatial discretization, and to reduce the required computational time. Orthogonal and oblique cutting simulations were performed using the improved SPH formulation, and the predicted results showed a good agreement with experimental measurements, in terms of cutting forces and workpiece temperatures.

Calamaz et al. [48] studied the wear of tungsten carbide tools under dry conditions, when machining Ti-6Al-4V, using experimental testing and numerical simulations. Dry orthogonal cutting tests were performed, and cutting forces were recorded along with chip morphology and tool wear. SPH simulations were performed using new and worn tools. The predicted chip morphology and cutting forces, for both tools, showed a good agreement with the experimental trends. Cutting forces were found to increase with tool wear, especially the thrust component. This was explained in terms of the formation of a dead metal zone ahead of the tool tip, which was evident in the SPH simulations.

Heisel et al. [49] used SPH to model the process of dry orthogonal cutting of AISI 1045 steel, using the commercial FE software LS-DYNA. The authors examined the effects of different SPH parameters, including the initial smoothing length and particle density, on cutting forces, chip compression ratio, and computational time. The optimum parameters were then selected and recommended as a starting point for future simulations.

Xi et al. [50] used SPH to simulate the process of LAM of Ti-6Cr-5Mo-5V-4Al titanium alloy. First, the laser heating effects were modeled, and after that the developed temperatures were used as initial conditions for the SPH cutting model. The authors used two different material constitutive equations: the Johnson–Cook and Zerilli–Armstrong models and obtained the parameters of both models, using experimental data that was obtained using the split-Hopkinson pressure bar (SHPB). Both conventional machining (CM) and LAM were simulated, and cutting forces were predicted using the two different material models and compared to experimental measurements. Furthermore, workpiece temperature predictions were compared to experimental results. The laser model successfully predicted the workpiece temperatures, which were found to decrease with laser speed. Based on the obtained results, the Johnson–Cook material model was found to predict cutting forces more accurately than the Zerilli–Armstrong material model. Furthermore, the Zerilli–Armstrong model did not succeed in predicting the effect of laser assistance on cutting forces, which dropped in case of LAM compared to CM. Finally, the cutting speed was found to have insignificant effect on cutting forces within the tested range.

Umer et al. [51] used the renormalized SPH formulation, available in Abaqus/Explicit, to predict chip morphology, which is a key factor in the assessment of any machining operation, during hard turning of steels. The renormalized formulation was used because, as reported by the authors, it can account for frictional effects along the tool–chip interface; accordingly, it provides a better representation of the chip shape and material flow. The developed model successfully captured the phenomenon of shear localization. In addition, the SPH results were compared to those predicted using the traditional FE Lagrangian formulation, and SPH was found to provide better predictions without the need for a chip separation criteria. However, both techniques (SPH and FEM) successfully predicted the transition from continuous to serrated chips, as the cutting speed was increased. Cutting forces were found to be almost unaffected by the cutting speed; however, they were significantly affected by the feed rate. Finally, the SPH model predicted lower cutting forces and more chip curling compared to FEM, for the same material properties and friction parameters.

Shchurov et al. [52] presented the first attempt to model the machining process of unidirectional fiber-reinforced composites using SPH. The issue with fibrous composites is the significant difference in strength and stiffness between fibers and matrix, which mainly results in debonding during machining. Two different methodologies were proposed that may reduce debonding during machining, and their applicability was examined using SPH–FE modeling. Both methodologies depend on using a supporter, one in the form of a wedge and another in the form of a roller, to the workpiece ahead of the tool in order to suppress debonding. The two approaches were compared to free cutting in order to evaluate their effectiveness. The use of an advanced roller was found to be a more promising technique in limiting debonding during machining, particularly under oblique cutting conditions as compared to orthogonal cutting.

Mir et al. [53] built an SPH model in order to numerically investigate tool wear during single-point diamond turning (SPDT) of silicon. The main focus was to contribute to a better understanding of the ductile-to-brittle transition (DBT) of the machining mode that results from tool wear. A set of experimental tests was performed, which included a series of facing and plunging cuts, and the profile of the machined surface was evaluated along with the progression of tool wear. The transition from ductile to brittle machining mode was identified by analyzing the surface profiles using a scanning electron microscope (SEM), a 2D contact profilometer, and a white light interferometer. The SPH model was used to provide a better understanding of the stress distribution along the cutting edge due to tool wear, and how it affects the DBT.

Zhao et al. [54] used SPH as an effective tool to simulate the effects of sequential cuts and tool edge radius during microcutting of OFHC copper on residual stresses, chip formation, and cutting forces. The presented results showed that the second cut experienced a significant increase in chip curling, compared to the first cut, accompanied by a drop in the minimum chip thickness. Such results were attributed to the residual stresses remaining from the first cut. The cutting force component was also found to drop in the second cut; on the other hand, the thrust force component was almost unaffected. Furthermore, subsurface tensile residual stresses that were generated after the first cut were found to change to compressive stresses after the second cut.

Nam et al. [55] focused on the behavior of brittle materials during cutting, with the aid of SPH. SPH was used to investigate the mechanics involved during cutting brittle materials, as compared to ductile materials. The Johnson–Holmquist material model was implemented in the developed SPH model. The presented model was able to capture crack initiation and propagation during cutting that resulted in discontinuous chips, a main characteristic of brittle materials during machining. Furthermore, the model was used to investigate surface roughness, and its dependence on different cutting parameters (cutting speed, cutting depth, and rake angle) was examined. Optimal cutting conditions were considered to be those that resulted in the best surface finish. Such conditions were found to be high cutting speed, low cutting depth, and a zero rake angle.

Balbaa and Nasr [56] examined how LAM would affect the residual stresses induced in the machined surface, after dry orthogonal cutting of Inconel 718. The authors built an SPH model, using the commercial FE software Abaqus/Explicit. Inconel 718 was selected as the workpiece material, as a representative for hard-to-cut materials. First, the laser preheating effects were modeled using a transient thermal analysis, and a Lagrangian thermal FE model. After that, cutting was simulated using SPH; finally, an implicit Lagrangian model was used to predict residual stresses. The predicted cutting forces and residual stresses in the cutting direction were compared to the experimental results of Shi et al. [57], whose experimental cutting conditions were used for model validation. The predicted results were found to be in good agreement with the experimental ones, where LAM was found to induce surface compressive residual stresses, as compared to CM (resulted in surface tensile residual stresses). This was explained in terms of the thermal softening effects of the laser beam ahead of the tool tip, which resulted in higher tensile plastic strains, and accordingly more compressive (or less tensile) residual stresses.

Olleak et al. [58] developed an SPH model, using the commercial software LS-DYNA, to examine the effects of the Johnson–Cook plasticity parameters on the cutting process. Cutting forces and residual stresses were predicted during dry orthogonal cutting of stainless steel AISI 316L at different cutting conditions. The predicted results were validated by comparing them to the experimental results of Outerio et al. [59]. SPH was found to be capable of predicting cutting forces and residual stresses, and the presented results confirmed the significant role that the material model plays in metal cutting simulations. In addition, the authors highlighted the need for a detailed investigation on how frictional effects are modeled in SPH cutting simulations.

Parle et al. [60] employed SPH to simulate orthogonal microcutting of steel AISI 1045, using the commercial software LS-DYNA. Orthogonal microcutting experimental tests were performed in which cutting forces were measured for model validation. The authors examined the stress and strain distributions as well as cutting forces and specific cutting energy. The predicted results were in good agreement with the experimental measurements, which demonstrated the capabilities of SPH to model micromachining. In addition, the results demonstrated the fundamental behavior of ductile materials during cutting. The cutting force increased with feed rate, whereas it decreased with an increase in rake angle and cutting speed. The specific cutting energy was found to increase as the feed rate and cutting speed decreased. Finally, the specific cutting energy was found to drop with an increase in rake angle, which was explained in terms of the decrease in cutting forces due to reduced ploughing effects at higher rake angles.

2.6   Summary and concluding remarks

This chapter has focused on the use of SPH for modeling metal cutting. It can be concluded that SPH is a promising numerical technique for simulating the cutting process, as it can simply handle high degrees of nonlinearity, which is typical in case of metal cutting. However, SPH still requires significant efforts in order to improve its boundary condition capabilities (including friction modeling). Furthermore, currently, only 3D SPH analyses are supported by the commercial FE software, ANSYS, and Abaqus/Explicit; accordingly, it is highly recommended to adapt SPH for 2D analyses, as this will have a significant effect on cutting down the computational cost associated with the use of SPH. Finally, it has been noticed that a minimal focus has been given for the use of SPH for predicting residual stresses, which is another important point that is worth of investigation.

References

1. Limido, J., Espinosa, C., Salaun, M., Lacome, J.L., 2007, SPH method applied to high speed cutting modelling, *International Journal of Mechanical Sciences* 49: 898–908.
2. Nasr, M.N.A., 2008, On modelling of machining-induced residual stresses, PhD thesis, McMaster University, Hamilton, Canada.
3. Ernst, H., Merchant, M.E., 1941, Chip formation, friction and high quality machined surfaces, *Transactions of the American Society for Metals* 29: 299–378.
4. Nasr, M.N.A., 2017, On the role of different strain components, material plasticity, and edge effects when predicting machining-induced residual stresses using finite element modelling, *Transactions of the ASME Journal of Manufacturing Science & Engineering* 139: 071014.
5. Oxley, P.L.B., 1989, *The Mechanics of Machining: An Analytical Approach to Assessing Machinability*, E. Horwood, New York.
6. Manjunathaiah, J., Endres, W.J., 2000, A new model and analysis of orthogonal machining with an edge-radiused tool, *Journal of Manufacturing Science & Engineering* 122(3): 384–390.
7. Nasr, M., Ng, E.-G., Elbestawi, M., 2007, Modelling the effects of tool-edge radius on residual stresses when orthogonal cutting AISI-316L, *International Journal of Machine Tools and Manufacture* 47(2): 401–411.
8. Nasr, M.N.A., Ghandehariun, A., Kishawy, H.A., 2016, A physics-based model for metal matrix composites deformation during machining: A modified constitutive equation, *Transactions of the ASME, Journal of Engineering Materials and Technology* 139(1): 011003.
9. Nasr, M., Ng, E.-G., Elbestawi, M., 2007, Effects of strain hardening & initial yield strength on machining-induced residual stresses, *Transactions of the ASME, Journal of Engineering Materials & Technology* 129(4): 567–579.
10. Nasr, M., Ng, E.-G., Elbestawi, M., 2007, Effects of workpiece thermal properties on machining-induced residual stresses—Thermal softening & conductivity, *Journal of Engineering Manufacture, Proceedings of the IMechE, Part B* 221(9): 1387–1400.
11. Moufki, A., Dudzinski, D., Molinari, A., Rausch, M.A., 2000, Thermoviscoplastic modelling of oblique cutting. *International Journal of Mechanical Sciences* 42: 1205–1232.
12. Nasr, M.N.A., Ng, E.-G., Elbestawi, M.A., 2008, A modified time-efficient FE approach for predicting machining-induced residual stresses, *Finite Elements in Analysis and Design* 44(4): 149–161.
13. Cook, R., Malkus, D., Plesha, M., 1989, *Concepts & Applications of Finite Element Analysis*, John Wiley & Sons, New York.
14. Belytschko, T., Liu, W.K., Moran, B., 2001, *Nonlinear Finite Elements for Continua and Structures*, John Wiley & Sons, New York.
15. Chen, L., El-Wardany, T.I., Nasr, M., Elbestawi, M.A., 2006, Effects of edge preparation and feed when hard turning a hot work die steel with polycrystalline cubic boron nitride tools, *CIRP Annals—Manufacturing Technology* 55(1): 89–92.
16. Movahhedy, M.R., Gadala, M.S., Altintas, Y., 2000, Simulation of the orthogonal cutting process using an Arbitrary Lagrangian–Eulerian finite-element method, *Journal of Materials Processing Technology* 103: 267–275.
17. Nasr, M.N.A., Outeiro, J.C., 2015, Sensitivity analysis of cryogenic cooling on machining of magnesium alloy AZ31B-O, *Procedia CIRP* 31: 264–269.

18. Nasr, M.N.A., 2015, Effects of sequential cuts on residual stresses when orthogonal cutting steel AISI 1045, *Procedia CIRP* 31: 118–123.
19. Liu, G.R., 2003, *Mesh Free Methods, Moving beyond the Finite Element Method*, CRC Press.
20. Yen, Y.-C., Sohner, J., Lilly, B., Altan, T., 2004, Estimation of tool wear in orthogonal cutting using finite element analysis, *Journal of Materials Processing Technology* 146(1): 82–91.
21. Kishawy, H.A., Elbestawi, M.A., 2004, Tool wear and surface integrity during high-speed turning of hardened steel with polycrystalline boron nitride (PCBN) tools, *Proceedings of the Institution of Mechanical Engineers, Part B* 125: 755–767.
22. Nasr, M.N.A., 2014, Predicting the effects of grain size on machining-induced residual stresses in steels, *Advanced Materials Research* 996: 634–639.
23. Pu, Z., Umbrello, D., Dillon, Jr. O.W., Lu, T., Puleo, D.A., Jawahir, I.S., 2014, Finite element modelling of microstructural changes in dry and cryogenic Machining of AZ31B magnesium alloy, *Journal of Manufacturing Processes* 16: 335–343.
24. Liu, C.R., Guo, Y.B., 2000, Finite element analysis of the effect of sequential cuts and tool-chip friction on residual stresses in a machined layer, *International Journal of Mechanical Sciences* 42(6): 1069–1086.
25. Ozel, T., 2006, The influence of friction models on finite element simulations of machining, *International Journal of Machine Tools and Manufacture* 46(5): 518–530.
26. Nasr, M., Balbaa, M., 2014, Effect of laser power on residual Stresses when laser-assisted turning of AISI 4340 steel, *Canadian Society for Mechanical Engineering (CSME) International Congress*, Toronto, Canada, June 1–4.
27. Nasr, M.N.A., Balbaa, M., Elgamal, H., 2014, Modelling machining-induced residual stresses after laser-assisted turning of steels, *Advanced Materials Research* 996: 622–627.
28. Attia, H., Tavakoli, S., Vargas, R., Thomson, V., 2010, Laser-assisted high-speed finish turning of superalloy Inconel 718" *CIRP Annals—Manufacturing Technology* 59(1): 83–88.
29. Pu, Z., Umbrello, D., Dillon, Jr. O.W., Jawahir, I.S., 2014, Finite element simulation of residual stresses in cryogenic machining of AZ31B Mg alloy, *Procedia CIRP* 13: 282–287.
30. Liu, G.R., Liu, M.B., 2003, *Smoothed Particle Hydrodynamics: A Mesh-free Particle Method*, World Scientific.
31. Su, C., Zhang, Y., Hou, J., Wang, W., 2008, Numerical simulation and analysis for metal cutting processes based on FEM and SPH, *2008 Asia Simulation Conference—7th International Conference on Systems Simulation & Scientific Computing*, pp. 1325–3128.
32. Islam, S., Ibrahim, R., Das, R., Fagan, T., 2012, Novel approach for modelling of nanomachining using a mesh-less method, *Applied Mathematical Modelling* 36, 5589–5602.
33. Heinstein, M., Segalman, D., 1997, *Simulation of Orthogonal Cutting with Smooth Particle Hydrodynamics*, Sandia National Laboratories, Livermore, CA.
34. Umer, U., Mohammed, M.K., Abu Qudeiri, J., Al-Ahmari, A., 2016, Assessment of finite element and smoothed particles hydrodynamics methods for modelling serrated chip formation in hardened steel, *Advances in Mechanical Engineering* 8(6), 1–11.

35. Madaj, M., Piska, M., 2013, On the SPH orthogonal cutting simulation of A2034-T351 alloy, *Procedia CIRP* 8: 152–157.

36. Chieragatti, R., Espinosa, C., Lacome, J.L., Limido, J., Mabru, C., Salaun, M., 2008, Modelling high speed machining with the SPH method, *10th International LS-DYNA Users Conference*, pp. 1–13.

37. Zahedi, A., Li, S., Roy, A., Babitsky, V., Silberschmidt, V.V., 2012, Application of smooth-particle hydrodynamics in metal machining, *Journal of Physics* 382, 012017: 1–5.

38. Zahedi, A., Demiral, M., Roy, A., Silberschmidt, V.V., 2013, FE/SPH modelling of orthogonal micro-machining of F.C.C. single crystal, *Computational Materials Science* 78: 104–109.

39. Ghafarizadeh, S., Tahvilian, A.M., Chatelain, J.-F., Liu, Z., Champliaud, H., Lebrun, G., 2017, Numerical simulation of ball-end milling with SPH method, *International Journal of Advanced Manufacturing Technology* 88: 401–408.

40. Zhao, H., Zhang, P., Liu, H., Liu, C., Tong, D., Zhang, L., Ren, L., Dong, X., Liang, S., 2014, Influences of residual stress induced by cutting on subsequent scratch using smooth particle hydrodynamic (SPH), *Materials Transactions* 55(9): 1440–1444.

41. Cao, J., Wu, Y., Li, J., Zhang, Q., 2016, Study on the material removal process in ultrasonic-assisted grinding of sic ceramics using smooth particle hydrodynamic grinding of sic ceramics using smooth particle hydrodynamic (SPH) method, *International Journal of Advanced Manufacturing Technology* 83: 985–994.

42. Akarca, S.S., Song, X., Altenhof, W.J., Alpas, A.T., 2008, Deformation behaviour of aluminium during machining: Modelling by eulerian and smoothed-particle hydrodynamics methods, *Proceedings of the IMechE Part L: Journal Of Materials: Design and Applications* 222: 209–221.

43. Akarca, S.S., Altenhof, W.J., Alpas, A.T., 2008, A smoothed-particle hydrodynamics (SPH) model for machining of 1100 aluminium, *10th International LS-DYNA Users Conference*, 1–8.

44. Xi, Y., Bermingham, M., Wang, G., Dargusch, M., 2014, SPH/FE modelling of cutting force and chip formation during thermally assisted machining of Ti6Al4V alloy, *Computational Materials Science* 84: 188–197.

45. Geng, X., Dou, W., Deng, J., Yue, Z., 2016, Simulation of the cutting sequence of AISI 316L steel based on the smoothed particle hydrodynamics method, *International Journal of Advanced Manufacturing Technology*, 1–8.

46. Geng, X., Dou, W., Deng, J., Ji, F., Yue, Z., 2017, Simulation of the orthogonal cutting of OFHC copper based on the smoothed particle hydrodynamics method, *International Journal of Advanced Manufacturing Technology* 91: 265–272.

47. Spreng, F., Eberhard, P., 2015, Machining process simulations with smoothed particle hydrodynamics, *Procedia CIRP* 31: 94–99.

48. Calamaz, M., Limido, J., Nouari, M., Espinosa, C., Coupard, D., Salaün, M., Girot, F., Chieragatti, R., 2009, Toward a better understanding of tool wear effect through a comparison between experiments and SPH numerical modelling of machining hard materials, *International Journal of Refractory Metals & Hard Materials* 27: 595–604.

49. Heisel, U., Zaloga, W., Krivoruchko, D., Storchak, M., Goloborodko, L., 2013, Modelling of orthogonal cutting processes with the method of smoothed particle hydrodynamics, *Production Engineering Research and Development* 7: 639–645.

50. Xi, Y., Zhan, H., Rashid, R.A., Wang, G., Sun, S., Dargusch, M., 2014, Numerical modelling of laser assisted machining of a beta titanium alloy, *Computational Materials Science* 92: 149–156.

51. Umer, U., Abu Qureiri, J., Ashfaq, M., Al-Ahmari, A., 2016, Chip morphology predictions while machining hardened tool steel using finite element and smoothed particles hydrodynamics methods, *Applied Physics and Engineering* 17(11): 873–885.

52. Shchurov, I.A., Nikonov, A.V., Boldyrev, I.S., 2016, SPH- simulation of the fibre-reinforced composite workpiece cutting for the surface quality improvement, *Procedia Engineering* 150: 860–865.

53. Mir, A., Luo, X., Sun, J., 2016, The investigation of influence of tool wear on ductile to brittle transition in single point diamond turning of silicon, *Wear* 364–365, 233–243.

54. Zhao, H., Liu, C., Cui, T., Tian, Y., Shi, C., Li, J., Huang, H., 2013, Influences of sequential cuts on micro-cutting process studied by smooth particle hydro-dynamic (SPH), *Applied Surface Science* 284: 366–371.

55. Nam, J., Kim, T., Cho, S.W., 2016, A numerical cutting model for brittle materials using smooth particle hydrodynamics, *International Journal of Advanced Manufacturing Technology* 82:133–141.

56. Balbaa, M., Nasr, M.N.A., 2015, Prediction of residual stresses after laser-assisted machining of Inconel 718 using SPH, *Procedia CIRP* 31: 19–23.

57. Shi, B., Attia, H., Vargas, R., Tavakoli, S. 2008, Numerical and experimental investigation of laser-assisted machining of Inconel 718, *Machining Science and Technology* 12(4): 498–513.

58. Olleak, A., Nasr, M.N.A., El-Hofy, H., 2015, The influence of Johnson-Cook parameters on SPH modeling of orthogonal cutting of AISI 316L, *10th European LS-DYNA Conference*, pp. 1–8.

59. Outeiro, J.C., Umbrello, D.M. Saoubi, R., 2006, Experimental and numerical modelling of the residual stresses induced in orthogonal cutting of AISI 316L steel. *International Journal of Machine Tools & Manufacture*: 1786–1794.

60. Parle, D., Singh, R., Joshi, S., 2014, Modelling of specific cutting energy in micro-cutting using SPH simulation, *9th International Workshop on Microfactories*, October 5–8, Honolulu, HI, pp. 1–6.

chapter three

Failure analysis of carbon fiber reinforced polymer multilayer composites during machining process

Sofiane Zenia and Mohammed Nouari

Contents

3.1   Introduction 52
3.2   Numerical modeling 54
    3.2.1   Machining parameters and boundary conditions 54
    3.2.2   Combined elastoplastic damage behavior law and
        interface delamination 57
        3.2.2.1   Progressive damage analysis 57
        3.2.2.2   Plastic model 59
    3.2.3   Interface delamination modeling 61
3.3   Numerical results: Simulation of the orthogonal cutting 62
    3.3.1   Chip formation process 62
        3.3.1.1   Orientation case of $\theta = 45°$ 62
        3.3.1.2   Orientation case of $\theta = 90°$ 63
        3.3.1.3   Orientation case of $\theta = -45°$ 63
    3.3.2   Prediction of cutting forces 64
    3.3.3   Prediction of the induced subsurface damage 66
    3.3.4   Effect of tool rake angle 67
    3.3.5   Effect of clearance angle 68
    3.3.6   Effect of tool edge radius 70
    3.3.7   Effect of the depth of cut a_p 71
    3.3.8   Effect of the cutting speed on the machining forces 73
    3.3.9   Effect of fiber orientations on the interlaminar
        delamination 74
3.4   Simulation of the drilling operation 76
3.5   Conclusion 78
References 79

The machining process of the carbon fiber reinforced polymer (CFRP) multilayer composite structures leads to four failure modes: matrix cracking, fiber matrix debonding, fiber rupture, and interlaminar delamination. The latter occurs at the interface between two adjacent layers and can generate the total failure of the composite structure. In the current work, cohesive-zone elements (CZE) are used to analyze the interlaminar delamination and simulate the machining of multilayer composites. The other failure modes mentioned earlier, and that appear within the composite layers, are analyzed through three-dimensional numerical simulations. A VUMAT subroutine, providing the capability for implementing combined elastoplastic-damage models, has been performed under Abaqus/Explicit code. Damage variables have been calculated for each type of damage that appears in the workpiece: fiber rupture, matrix cracking, and fiber–matrix debonding. The proposed approach is primarily focused on the understanding of interactions between the fiber orientation, machining parameters, and physical phenomena governing the behavior of CFRP composites materials under high mechanical loading in machining.

3.1   Introduction

Generally, damage mechanisms induced by machining of CFRP composites include four types of failure modes: transverse matrix cracking, fiber–matrix interface debonding, fiber rupture, and interply delamination. Compared with metals, relatively little research has been carried out on the failure analysis of composites. The current state of knowledge in this area of research is mainly limited to experimental studies, and only few theoretical models have been developed from the last few years.

The experimental observations conducted by several authors such as Koplev [1] and Wang et al. [2] showed that the CFRP composite chips are formed during machining through a series of brittle fractures under high mechanical loading. These authors then considered that the brittle behavior of CFRP workpieces dominates during machining. The main conclusion of the work of Koplev [1] and Wang et al. [2] is that the fiber orientation plays a key role in the chip formation process. Other machining tests on edge trimming and orthogonal cutting of graphite/epoxy composites were conducted by Arola et al. [3], who observed the existence of a primary fracture and a secondary fracture. In all test cases, these authors showed that the orientation of fracture strongly depends on the fiber orientation. The secondary fracture occurs along the matrix–fiber interface and follows the fiber orientation. Consequently, it has been concluded from the work of Arola et al. [3] that the chip formation, the cutting forces, and the surface morphology were highly dependent on the fiber orientation. However, the optimization of these parameters only by experimental approaches often requires long and very expensive trials. So, numerical simulation and modeling can be very helpful to characterize and validate optimal domains of cutting parameters.

Modeling of machining composites was first developed by Arola and Ramulu [4]. They presented a finite element (FE) model with a predefined fracture plane to simulate the chip formation in orthogonal cutting configuration. They explained the mechanism of the chip formation, which is composed into primary and secondary ruptures. Other works have focused on the mechanisms of chip formation, cutting forces calculation, induced subsurface damage, and surface roughness [5–13]. These different research works show that numerical simulations and theoretical modeling can be interesting tools for the analysis of the physics that governs the cutting composites and for studying the most influential parameters. In addition, these approaches help us understand the physical mechanisms of failure and have a clear idea about the state of the induced subsurface damage in the machined part.

Different models have been proposed to analyze the failure of CFRP composites during machining. Micromechanical modelings of Gopala Rao et al. [7] proposed, for example, a quasi-static approach based on the Abaqus/Explicit software. Lasri et al. [8] and Soldani et al. [9] opted for a macroscopic model, where the workpiece is considered a homogeneous equivalent material (HEM). Iliescu et al. [10] proposed another work based on discrete element method (DEM) to simulate the mechanisms of chip formation and calculate machining forces in orthogonal cutting of unidirectional (UD)-CFRP composites.

In the current investigation, a complete model with different physical aspects of machining composites has been development. The proposed approach is based on a three-dimensional (3D) mesomechanic model, with a combination of the stiffness degradation effect in the response material behavior, plasticity using the effective stress concept, and evolution laws, to predict damage initiation and progression during the chip formation process. Besides, the delamination, which can occur at the interply interface, was taken into account, using the CZE procedure available in the Abaqus package [11]. The model proposes a dynamic approach based on Abaqus/Explicit software, and a damaged-mechanical behavior was implemented in 3D numerical models, using a VUMAT subroutine. In this work, the workpiece is modeled as an HEM. The model allows a better understanding of the physical phenomena observed during the cutting operation and gives an accurate numerical tool to simulate the real chip formation, cutting forces, and induced subsurface damage. The obtained numerical results were compared with the results of the experiments performed by Iliescu et al. [10]. The comparison shows a good agreement.

3.2   Numerical modeling

The machining FE model developed in this work consists of an HEM composite for the workpiece with a damaged-elastoplastic behavior law and a rigid body law for the cutting tool and the twist drill. The numerical simulations were conducted using a CFRP composite with different fiber orientations (0°, 45°, 90°, and −45°) for the orthogonal cutting study and a multilayers composite (0_4°, 90_8°, 0_4°) for the drilling study.

3.2.1   Machining parameters and boundary conditions

The geometry of the part and boundary conditions are shown in Figure 3.1. As regards the orthogonal cutting operation (Figure 3.1a) nodes on the vertical surfaces, right and left sides are constrained to move along the horizontal direction (X). Nodes on the horizontal bottom surface are restrained to move along the horizontal and vertical directions, (X), (Y),

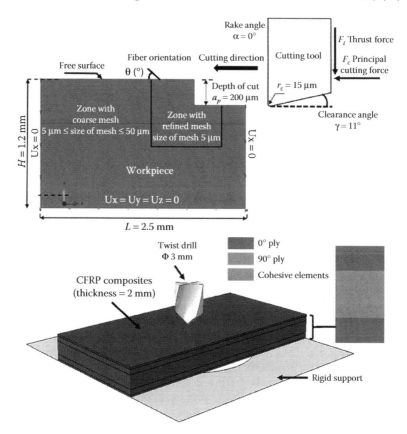

Figure 3.1 Boundary condition and geometry of the tool–workpiece couple.

and (Z), respectively. The values of cutting parameters and tool dimensions are the same to those defined and used in [10], in order to compare between predicted numerical results and experiments. The rake angle α is stated equal to 0°, the clearance angle γ is fixed at 11°, the tool edge radius r_ε is equal to 15 μm, and the depth of cut $a_p = 0.2$ mm. The cutting speed V_c is about 60 m/min.

For the drilling operation, the tool geometry and the boundary conditions are shown in Figure 3.1b. The workpiece is laid on a rigid support. The values of the machining parameters are selected from the work of Phadnis et al. [12], in order to validate the results obtained by simulation with the experimental work. The tool is a twist drill with a 3-mm diameter; the point angle is taken equal to 120°; and the clearance angle $\gamma = 30°$. The feed rate is equal to 150, 300, and 500 m/min, and the spindle speed is equal to 2500 rpm.

The cutting tool for the orthogonal operation and the twist drill for the drilling operation are modeled as a rigid body and controlled by a reference point, where the cutting speeds (cutting speed for the orthogonal cutting operation and the feed rate and the spindle speed for the drilling operation) are applied and the machining forces are measured as reaction forces in the output. The properties of a CFRP ply of the T300/914 composite are taken from the work of Iliescu et al. [10] and are listed in Table 3.1. The workpiece is considered as a HEM with a longitudinal modulus in the fiber direction more than 10 times higher than the transverse modulus.

Numerical simulations are conducted using Abaqus/Explicit code [11]. A 3D modeling was performed using eight-node linear brick elements with reduced integration, C3D8R, available within Abaqus.

The near zone of the tool tip where the chip would be formed was finely meshed. In a previous work, Zenia et al. [13] proved that when the element size is less than or equal to 7 μm, the differences in numerical results are negligible. For the orthogonal cutting operation, the size of elements in this zone is taken about 5 μm, whereas the remaining part is

Table 3.1 Mechanical properties of the aeronautical CFRP composite T300/914

Mechanical properties	
E_1^0 (MPa)	136,600
E_2^0 (MPa)	9,600
G_{12}^0 (MPa)	5,200
v_{12}^0	0.29
ρ (Kg/m^3)	1,578

meshed coarsely with an element size in the range of 5 μm in the vicinity of the finely meshed area and 50 μm on the edges of the workpiece. In the case of drilling operation, the element size of the near zone is taken about 150 μm and 1 mm on the edges.

A VUMAT subroutine, providing a very general capability for implementing elastoplastic damage models, was used in Abaqus/ Explicit. In addition, the element deletion approach is applied to represent the process of chip formation, based on initiation and damage evolution in the workpiece. The set of the plastic-damage model parameters reported by Feld in [14] have been adopted for the simulations in this work (Table 3.2).

The interaction between the node set of the workpiece surface and the tool surface is modeled using surface-to-nodes contact algorithm coupled to kinematic predictor/corrector contact algorithm with finite sliding formulation, both of which are available in the Abaqus/Explicit package. The latter allows to have the chip formation.

The contact between the tool and the workpiece is done at two contact zones. The first is located between the cutting face and the produced chip. The second is located between the flank face and the machined surface. The interaction between the surfaces (tool/workpiece) is controlled by the Coulomb's friction law, and the friction coefficient, μ, is assumed to be constant during the cutting operation, as in various numerical studies,

Table 3.2 Plastic and damage parameters of UD-CFRP T300/914

Damage parameters	
Y_{12}^C (MPa)	8
Y_{12}^0 (MPa)	0.03
b	0.5
b'	0.8
Y_{11}^t (MPa)	15
Y_{11}^c (MPa)	12
a	1
τ_c (μs)	6
Plastic parameters	
α	0.54
β (MPa)	1000
c	0.7
R_0 (MPa)	64

Table 3.3 Material parameters used to model interface cohesive elements

K_n (N/mm³)	$K_s = K_t$ (N/mm³)	G_n (N/mm)	$G_s = G_t$ (N/mm)	t_n^0 (MPa)	$t_t^0 = t_s^0$ (MPa)	η
4×10^6	4×10^6	0.2	1	60	90	1.8

Nayak et al. [6], Gopala et al. [7], and Lasri et al. [8]. In the present study, a coefficient of friction equal to 0.4 was used.

The interply interface was modeled with cohesive elements of type COH3D8, with a thickness of 5 μm. According to the literature, different values were used. To simulate the interface degradation, Phadnis et al. [12] and Shin et al. [15] used a thickness of 10 μm and 5 μm, respectively. In our work, the thickness was chosen equal to that used by Shin et al. [15]. However, the use of cohesive elements with a thickness of 5 μm or 10 μm has no effect on the behavior of the interface.

These cohesive elements are controlled by damage criteria discussed later. The removal of the element is performed once the degradation parameters reach the limit value of 0.99, and the failed cohesive elements are removed from the FE model. Mechanical properties of cohesive zone are reported by Phadnis et al. [12] and Shin et al. [15] (see Table 3.3).

3.2.2 Combined elastoplastic damage behavior law and interface delamination

3.2.2.1 Progressive damage analysis

In the proposed model, different degradation modes were considered: fiber breakage in traction, and in compression, matrix cracking and fiber-matrix debonding. The strain energy density of the damaged ply is defined as follow [16–19]:

$$
\begin{aligned}
E_D = \frac{1}{2}\Bigg(& \frac{1}{(1-D_{11})}\Bigg[\frac{(\sigma_{11})^2}{E_{11}^0} - \left(\frac{v_{12}^0}{E_{11}^0} + \frac{v_{21}^0}{E_{11}^0}\right)\sigma_{11}\sigma_{22} - \left(\frac{v_{13}^0}{E_{11}^0} + \frac{v_{31}^0}{E_{33}^0}\right)\sigma_{11}\sigma_{33} \\
& - \left(\frac{v_{23}^0}{E_{22}^0} + \frac{v_{32}^0}{E_{33}^0}\right)\sigma_{22}\sigma_{33}\Bigg] + \frac{\langle\sigma_{22}\rangle_-^2}{E_{22}^0} + \frac{\langle\sigma_{33}\rangle_-^2}{E_{33}^0} + \frac{1}{(1-D_{12})} \\
& \times \Bigg[\frac{(\sigma_{12})^2}{G_{12}^0} + \frac{(\sigma_{23})^2}{G_{23}^0} + \frac{(\sigma_{13})^2}{G_{13}^0}\Bigg] + \frac{1}{(1-D_{22})}\Bigg[\frac{\langle\sigma_{22}\rangle_+^2}{E_{22}^0} + \frac{\langle\sigma_{33}\rangle_+^2}{E_{33}^0}\Bigg]\Bigg)
\end{aligned}
\tag{3.1}
$$

where $E_{22}^0 = E_{33}^0, G_{12}^0 = G_{13}^0, v_{12}^0 = v_{13}^0, G_{23}^0 = E_{33}^0 / (2(1 + v_{23}))$. For a 3D stress state, the symbols $\langle \bullet \rangle_-$ and $\langle \bullet \rangle_+$ in Equation 3.1 mean the negative and positive part of $\bullet$, respectively, introduced to model the unilateral effect for the effective transverse stress.

The formula shows which terms of the stiffness are influenced by the damage. From this formula, we derive the thermodynamic force vector **Y** conjugated to damage, in order to describe the initiation and progression of degradation mechanisms:

$$\mathbf{Y} = \frac{\partial \langle\langle E_D(\sigma, \mathbf{D}) \rangle\rangle}{\partial \mathbf{D}} \tag{3.2}$$

The symbol $\langle\langle \bullet \rangle\rangle$ in Equation 3.2 means the average value of the quantity $\bullet$ within the thickness. In the present study, the strain energy density is computed locally at each integration point across the ply thickness.

The activation of damage and its evolution are governed by the square root of a linear combination of the two thermodynamic forces Y_{22} and Y_{12}:

$$Y = \sup_{\tau \leq t} \left(\sqrt{Y_{12} + bY_{22}} \right) \tag{3.3}$$

where b is a coupling term between the transverse and shear damages

The variables Y_{22} and Y_{12} are defined according to relation (3.4):

$$Y_{22} = \frac{\partial \langle\langle e_d \rangle\rangle}{\partial D_{22}} = \frac{\langle \sigma_{22} \rangle_+^2}{2(1 - D_{22})^2 E_{22}^0} \left(\frac{\langle \sigma_{22} \rangle_+^2}{E_{22}^0} + \frac{\langle \sigma_{33} \rangle_+^2}{E_{33}^0} \right)$$

$$Y_{12} = \frac{\partial \langle\langle e_d \rangle\rangle}{\partial D_{12}} = \frac{1}{2(1 - D_{12})^2} \left(\frac{(\sigma_{12})^2}{G_{12}^0} + \frac{(\sigma_{23})^2}{G_{23}^0} + \frac{(\sigma_{13})^2}{G_{13}^0} \right) \tag{3.4}$$

The transverse and shear damage variables D_{22} and D_{12} are defined as:

$$\Rightarrow D_{12} = \begin{cases} \dfrac{\left\langle \sqrt{Y} - \sqrt{Y_{12}^0} \right\rangle_+}{\sqrt{Y_{12}^c} - \sqrt{Y_{12}^0}} & \text{si } D_{12} < 1, \\[4mm] D_{12} = 1 & \text{sinon.} \end{cases} \tag{3.5}$$

$$D_{22} = \begin{cases} b'D_{12} & \text{si } D_{22} < 1 \text{ et } D_{22} < 1 \\[2mm] D_{12} = 1 & \text{sinon.} \end{cases}$$

where b' is a coupling term between the transverse and shear damages. Y_{22}^c and Y_{12}^0 are the limit strength for damage and the threshold strength for the initiation of damage, respectively. These material parameters are identified experimentally.

In addition to the previous equations, the model is completed by a brittle failure criterion that takes into account failure of the fiber in tension and compression. This is governed by two critical damage thresholds Y_{11}^t and Y_{11}^c for the variable Y_{11}:

$$Y_{11} = \frac{\partial \langle\langle e_d \rangle\rangle}{\partial D_{11}} = \frac{1}{2(1-D_{11})^2}\left(\frac{(\sigma_{11})^2}{E_{11}^0} - \sum_{i=1}^{2}\sum_{j>i}^{3}\left(\frac{v_{ij}^0}{E_i^0} + \frac{v_{ji}^0}{E_j^0} \right)\sigma_{ii}\sigma_{jj}\right) \quad (3.6)$$

The damage fiber is introduced in the model, by considering the Young's modulus E_{11} as a nonlinear, and it does depend on stresses σ_{11}:

$$\begin{cases} \text{si } \sigma_{11} > 0 \;\rightarrow\; \begin{cases} \text{si } Y_{11} > Y_{11}^t \;\; D_{11} = 1 \\ \qquad\quad D_{11} = 0 \quad \text{sinon} \end{cases} \\[3mm] \text{si } \sigma_{11} < 0 \;\rightarrow\; \begin{cases} \text{si } Y_{11} > Y_{11}^c \;\; D_{11} = 1 \\ \qquad\quad D_{11} = 0 \quad \text{sinon} \end{cases} \end{cases} \quad (3.7)$$

To limit the maximum damage rate and avoid numerical localization of damage, regularization parameters are introduced [18,19], and the damage variables are corrected as follows:

$$\dot{D}_{ij}^{k+1} = \frac{1}{\tau_c}\left(1 - e^{-a\left(D_{ij}^{k+1} - D_{ij}^k \right)} \right) \quad (3.8)$$

The same material constants, τ_c and a, are taken for the three damage evolution laws. For this model with delay effects, the variation of the forces Y_i does not lead to instantaneous variations of the damage variables D_i. There is a certain delay, defined by the characteristic time τ_c.

3.2.2.2   Plastic model

The elastoplastic-damage model is based on the effective stresses concept, as shown by Lemaitre and Chaboche [20]. In the current work, the yield function is written considering an isotropic hardening, and it is assumed that there is no plastic flow in the fiber direction. The layer is assumed to be in three dimensions. The elasticity domain is defined according to the following plastic activation function:

$$F(\tilde{\sigma}, \sigma_y) = f^p(\tilde{\sigma}) - \sigma_y(p) \quad (3.9)$$

where f^p is the plastic potential, and σ_y is the current yield stress, which represents the isotropic hardening law and is defined in function of the cumulated plastic strain p:

$$\sigma_y(p) = R_0 + R(p) = R_0 + \beta p^\alpha \tag{3.10}$$

where R_0 is the initial yield stress, and the quantities β and α are the hardening parameters.

The plastic potential function is defined considering a plane stress condition and does not depend on stresses σ_{11} in the fiber direction, because the fiber behavior is assumed to be elastic brittle under tension or compression:

$$f^p(\tilde{\sigma}) = \sqrt{\tilde{\sigma}_{12}^2 + \tilde{\sigma}_{23}^2 + \tilde{\sigma}_{13}^2 + c^2 \left[\tilde{\sigma}_{22}^2 + \tilde{\sigma}_{33}^2 \right]} \tag{3.11}$$

where c is a coupling parameter, and the effective stresses are defined as follows:

$$\tilde{\sigma}_{12} = \frac{\sigma_{12}}{1 - D_{12}} ; \tilde{\sigma}_{23} = \frac{\sigma_{23}}{1 - D_{12}} ; \tilde{\sigma}_{13} = \frac{\sigma_{13}}{1 - D_{12}}$$

$$\tilde{\sigma}_{22} = \frac{\langle \sigma_{22} \rangle_+}{1 - D_{22}} + \langle \sigma_{22} \rangle_- ; \tilde{\sigma}_{33} = \frac{\langle \sigma_{33} \rangle_+}{1 - D_{22}} + \langle \sigma_{33} \rangle_-$$

where D_{12} and D_{22} denote the damage developed in the transverse direction and under shear stress condition, respectively. The transverse behavior in compression is indefinitely elastoplastic, due to the introduced unilateral effect. So, transverse damage affects only the tensile behavior.

The effective inelastic part of the deformation is defined by the flow rule (or normality rule) as:

$$\mathbf{d\tilde{\varepsilon}^P} = d\lambda \frac{\partial F}{\partial \tilde{\sigma}} \text{ et } dp = -d\lambda \frac{\partial F}{\partial \sigma_y} = d\lambda \tag{3.12}$$

where $d\lambda$ is a nonnegative plastic consistency parameter (plastic multiplier).

The plastic strain increment is obtained from the equivalence principle of the plastic work increment dW^p, presented as follows:

$$dW^P = \tilde{\sigma} : \mathbf{d\tilde{\varepsilon}^P} = \sigma : \mathbf{d\varepsilon^P} \tag{3.13}$$

In addition, the consistency condition ($dF = 0$) should be satisfied, which leads to compute the cumulated plastic increment:

$$dp = \frac{\left(\partial F/\partial \tilde{\sigma}\right)\mathbf{C}_{(D)}}{\left(\partial F/\partial \tilde{\sigma}\right)\mathbf{C}_{(D)}\left[\left(\partial F/\partial \tilde{\sigma}\right)+\left(\partial \sigma_y/dp\right)\right]}\, d\varepsilon = \mathbf{a}\, d\varepsilon \qquad (3.14)$$

An algorithm based on a radial returns predictor [21] is implemented in order to return the stresses to the yield surface. In fact, for an increment strain, an initial elastic prediction step is carried out. If the yield function is greater than zero, an iterative correction procedure uses the normal of the last yield surface, until the yield function vanishes.

3.2.3 Interface delamination modeling

In this section, a focus is put on the interply (or interlaminar) delamination, which can be generated during machining operations and causing the complete failure of the workpiece. Delamination mechanisms are often characterized by the separation between plies in the thickness of the composite. They are characterized by the formation of interlaminar cracks in the material. The delamination damage is particularly exhibited when the two adjacent plies are not oriented in the same direction.

To consider this damage mode, interply delamination was modeled using CZE available in the Abaqus/Explicit package. The damage is assumed initiated when a quadratic interaction of a function involving interaction nominal stress ratio reaches a value of one. This criterion can be represented as follows [11]:

$$\left\{\frac{\langle t_n\rangle}{t_n^0}\right\}^2 + \left\{\frac{t_s}{t_s^0}\right\}^2 + \left\{\frac{t_t}{t_t^0}\right\}^2 = 1 \qquad (3.15)$$

where:
   t is a nominal traction stress vector.
   The subscripts n, s, and t represent the normal, first shear, and second shear direction, respectively.
   The superscript 0 represents the peak value of nominal stress.
   The symbol $\langle\ \rangle$ is a Macaulay bracket and denotes the positive part.

The coupling between the stresses and damage is done as follows:

$$t_n = \left(1-d\right)\langle \bar{t}_n\rangle$$

$$t_s = \left(1-d\right)\bar{t}_s \qquad (3.16)$$

$$t_t = \left(1-d\right)\bar{t}_t$$

where d represents the damage variable and $\bar{t}$ the stress components without damage. These stress components are predicted by the elastic traction-separation behavior, as shown in [22]. Here, the damage evolution is defined by Benzeggagh–Kenane criterion [23], which is based on the dissipated energy by the damage process.

$$G_n^C + \left(G_s^C - G_n^C \right) \left(\frac{G_s + G_t}{G_n + G_s + G_t} \right)^\eta = G^C \tag{3.17}$$

where:
   G is the fracture energy
   η is a material parameter [23]
   C represents the critical fracture energy

3.3 Numerical results: Simulation of the orthogonal cutting

3.3.1 Chip formation process

As said earlier, the mechanical properties of the element degrade when one of the three damage variables increases, because the elastic modulus E_i and shearing modulus G_{ij} are directly coupled to the damage variables. Consequently, the element finds itself into loss of total rigidity when one of the three damage variables reaches the maximum value, D_{max}, which causes loss of total rigidity of the module to which it has been coupled. The element with very low stiffness (in the vicinity of zero) can then be removed. This procedure allows following the progression of the primary and secondary cracks up to the chip formation.

3.3.1.1 Orientation case of $\theta = 45°$

Figure 3.2 shows the state of damage caused by the cutting tool edge in the workpiece oriented at 45° during the machining operation. The chip is produced by a succession of two failures. The first failure is called *primary failure*, (Figure 3.2a), and this is caused by compression-induced shear perpendicular to the fiber axis (Figure 3.2b).

The second failure is called secondary fracture. This is produced along the fiber–matrix interface, which is caused by the fiber–matrix debonding (Figure 3.2b), until it reaches the free surface of the workpiece, then forming the complete chip. These results are in good agreement with the experimental results of Iliescu [10] (Figure 3.2c), Wang et al. [2], and Arola and Ramulu [3].

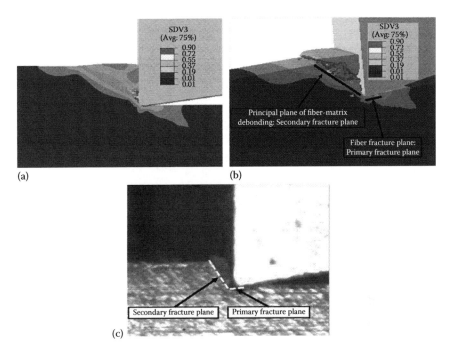

Figure 3.2 Progressive failure analysis of the chip formation with 3D model for 45° fiber orientation. (a) Primary rupture. (b) Secondary rupture and complete chip formation. (c) Experimental result of Iliescu et al. [10].

3.3.1.2   Orientation case of θ = 90°

For the 90° fiber orientation in Figure 3.3, the chip formation is also produced by a succession of two failures. Primary failure is produced by tearing of fibers under the tool advancement (Figure 3.3a), whereas the secondary failure (Figure 3.3b) that propagates perpendicularly to the cutting direction is caused by the fiber–matrix debonding under the effect of shear stress. The latter propagates toward the free surface of the workpiece, giving rise to the total chip formation (Figure 3.3b). These results are in good agreement with the experimental results of Teti [24] (Figure 3.3c) and Iliescu et al. [10].

3.3.1.3   Orientation case of θ = −45°

For a fiber orientation at −45°, the chip formation is produced by a primary rupture along the fiber–matrix interface toward the interior of the workpiece (Figure 3.4a). The fibers, being negatively oriented, bend under the effect of the advancement of the cutting tool.

Therefore, a secondary rupture appears and takes the direction of the free surface (Figure 3.4b).

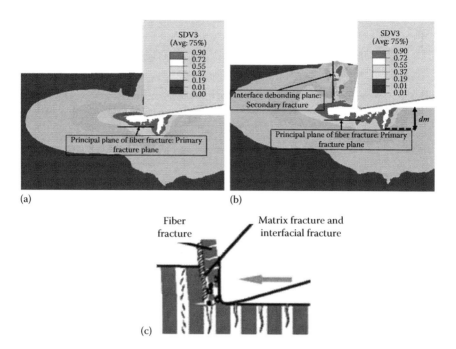

Figure 3.3 Progressive failure analysis of chip formation with 3D model for 90° fiber orientation. (a) Primary rupture. (b) Secondary rupture and complete chip formation. (c) Schematization of the experimental chip formation process by Teti [24].

The total chip formation occurs when the secondary rupture reaches the free surface of the part, as shown in Figure 3.4b. These results are in good agreement with the experimental results of Arola et al. [4] (Figure 3.4c).

3.3.2 Prediction of cutting forces

Cutting forces are calculated at each increment of time during the displacement of the cutting tool following the cutting direction. The cutting force is measured in the cutting direction (Figure 3.1). The effect of the fiber orientation on cutting forces is shown in Figure 3.5. The conclusion that can be drawn from this graph is that the fiber orientation affects the cutting forces very significantly. The evolution of the cutting forces according to the fiber orientation found in this study correlates with the findings made in different studies such as Koplev et al. [1], Wang et al. [2], Arola et al. [3], and Iliescu et al. [10].

The values of the cutting force F_c obtained by simulations using 3D model for different fiber orientations are in good agreement with those obtained experimentally by Iliescu [10] (Figure 3.5).

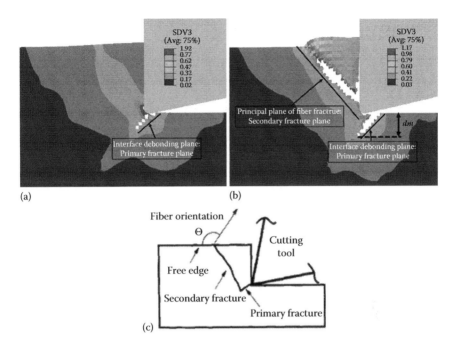

(a)   (b)

(c)

Figure 3.4 Progressive failure analysis of chip formation with 3D model for −45° fiber orientation. (a) Primary rupture. (b) Secondary rupture and complete chip formation. (c) Schematization of the experimental chip formation process [4].

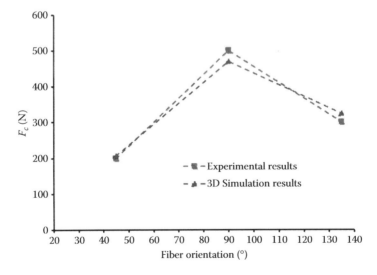

Figure 3.5 Cutting force F_c obtained during FE simulation for different fiber orientations with unidirectional composite compared with experimental results [10] ($V_c = 60$ m/min, $a_p = 0.2$ mm, $\alpha = 0°$).

We note that the cutting forces F_c are important for orientations at 90° and −45°. This is explained by the fact that the fibers tend to bend before they are cut by the tool; therefore, orientation at 90° requires cutting forces greater than those required for the orientation at 45°, where the chip formation is mainly due to the fiber–matrix debonding phenomenon.

3.3.3   Prediction of the induced subsurface damage

The main objectives of the study are to predict the subsurface damage induced by the machining operation and to analyze its interaction with the fiber orientation. Experimental studies previously conducted by Wang and Zhang [25] showed that the 90° fiber orientation is a critical orientation that exhibits severe and deep subsurface damage. Other authors also showed that the fiber orientation plays an important role in the damage induced by the machining of FRP workpieces.

For all studied orientations (45°, 90°, and −45°), the damage is initiated at the contact zone between the cutting tool tip and the machined part. After the initiation stage, the damage propagates following perpendicular and parallel directions to the fibers' orientation.

The damage tends to increase with the advancement of the tool in the machined material. Figure 3.6 shows the evolution of the damage depth *dm* inside the workpiece versus the fiber orientation. A deeper damage (large value of damage) can be observed in the part with fibers

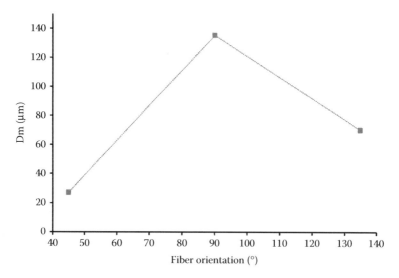

Figure 3.6 Depth of damage dm obtained during FE simulation for different fiber orientations with unidirectional composite ($V_c = 60$ m/min, $a_p = 0.2$ mm, $\alpha = 0°$).

oriented at 90°. This orientation also generates the highest cutting force, as shown in the evolution of cutting forces in Figure 3.5.

3.3.4   Effect of tool rake angle

This part is devoted to the effect of tool rake angle on the machining forces and chip formation process. Simulations were carried out by varying the cutting angle. Studied tool rake angle are −5°, 0°, 10°, and 20°. This was motivated according to the chip formation mechanisms observed during the cutting operation [5,6,8,26]. Indeed, for a positive rake angle, the preponderant mechanism is shearing, whereas with a negative rake angle, the dominant mechanism is buckling.

The obtained cutting forces are reported in Figure 3.7. From these results, it can be concluded that the cutting force has a trend to decrease in a moderate way in passing of a negative rake angle to a positive rake angle. This tendency is also observed in the experimental works of Arola et al. [3] and numerical studies of Lasri et al. [8] and Zenia et al. [27].

Furthermore, tool rake angle affects the chip formation process and its shape; this has been shown by several authors [2,6,28]. The latter highlights the existence of two cutting mechanisms, which are shear and buckling fibers, respectively. The shear of fibers is observed for positive tool rake angles (Figure 3.8a), whereas the buckling mechanism is present more in the case of negative tool rake angles (Figure 3.8b). Indeed, in the

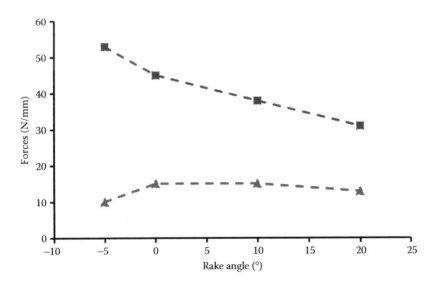

Figure 3.7 Effect of tool rake angle on machining forces, $V = 60$ m/min, $a_p = 200$ μm, $R = 15$ μm, $\gamma = 11°$.

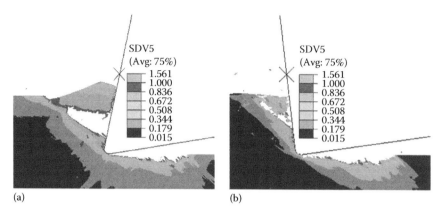

Figure 3.8 Effect of tool rake angle on the chip formation process during cutting of CFRP composites and for fiber orientation at 45°: (a) by shear $\alpha = 10°$, and (b) by buckling $\alpha = -5°$.

latter case, the fracture is done in *zigzag* along the shear plane in the perpendicular direction to fibers' axes.

Figure 3.8 shows the results obtained with the numerical simulation model. It shows the mechanisms of chip formation by shear (Figure 3.8a) and by buckling (Figure 3.8b). Although the effect of tool rake angle on the chip-forming mode is minor, its effect on the topography of the surface and the quality of machining, in general, is clear [28]. The spread of the matrix on the machined surface decreases with an increase in tool rake angle [26]. An increase in tool rake angle also allowed to improve the overall quality of the machined surface, and this is because the fibers cutting and chip release are occurring easily.

The conclusion that can be drawn for this work part is that the tools with positive tool rake angles facilitate the chip formation and its release, unlike tools with negative rake angles. Furthermore, tools with positive rake angles generate cutting forces and tool wear [28] lower than those obtained with negative tools rake angles.

Finally, cutting operation perpendicular to fibers' direction, made with tools having positive toll rake angle, forms the chip by cutting the fibers. Whereas, in the case where the tools have zero or negative rake angles, the chip is formed by macrocracking caused by the buckling of the fibers.

3.3.5   Effect of clearance angle

The clearance angle is generally solicited by the elastic return of fibers (bouncing-back phenomenon reported by Wang et al. [2] [Figure 3.9]), which bounce back after the passage of the cutting edge, as described by Wang et al. [2] and Jamel [28].

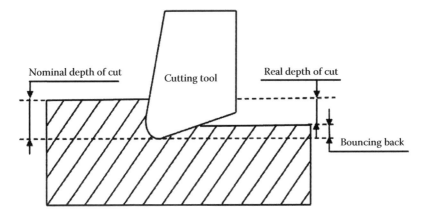

Figure 3.9 Illustration of the bouncing-back phenomenon.

Figure 3.10 shows the evolution of the machining forces according to the clearance angle. The latter almost does not affect the cutting forces, and this is in good agreement with what was reported by Arola et al. [3] and Wang et al. [2]. According to Jamal [28], the clearance angle affects the thrust forces because it controls the fibers' bounce back on the clearance tool surface. Small clearance angles allow the brushing of the fibers on the clearance surface during cutting operation, and therefore, this increases the thrust forces.

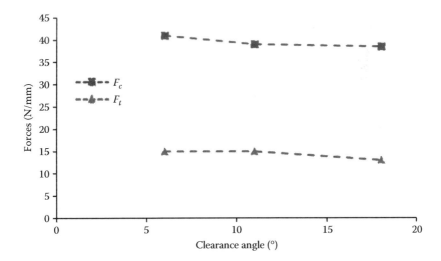

Figure 3.10 The effect of clearance angle on machining forces, $V = 60$ m/min, $a_p = 200$ μm, $\alpha = 10°$, $r_\varepsilon = 15$ μm.

On the other hand, the clearance angle does not appear to have any significant effect on the chip formation mode or on the topography of the machined surface, with the exception of a slight improvement in edge quality when a wide clearance angle is used [3]. This only confirms what has been reported by Jamal [28].

3.3.6   Effect of tool edge radius

This point focuses on the role that can be played by the tool edge radius on the machining forces and the damage generated during the orthogonal cutting operation. Indeed, the choice of materials to use in tools edge manufacture is very important. The CFRPs have the carbon fibers as reinforcement, which are natural abrasive and thus cause the tool edge wear during machining; this increases the machining forces and the damage induced in the workpiece. However, this study focuses only on the influence of the radius of the tool on the machining forces, because the tool is modeled as a rigid body and it is not damaged during machining operation.

The investigation was carried out on the edge radius at 5, 15, 30, and 50 μm. The results obtained are shown in Figure 3.11; these show that the cutting forces increase with increasing the tool edge radius. These results are in good agreement with the results obtained experimentally by Nayak et al. [6].

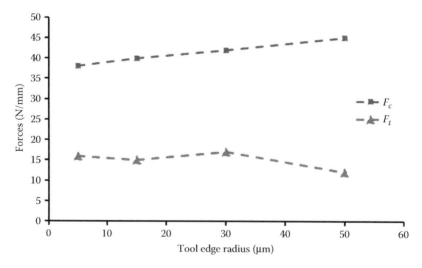

Figure 3.11 The effect of tool edge radius on machining forces, $V = 60$ m/min, $a_p = 200$ μm, $\alpha = 10°$, $\gamma = 11°$.

3.3.7 Effect of the depth of cut a_p

A study was conducted to show the effect of depth of cut a_p on the machining forces. This study was also interested in the influence that the depth of cut can have on the chip size and the subsurface damage caused by the machining operation.

Indeed, series of simulation were made in orthogonal section under the same conditions as the study presented previously; that is, the boundary conditions and the tool-piece geometry were the same as those reported in Figure 3.1. Only the depth of cut a_p varied. A fiber orientation was chosen at 45° for all simulations. The cutting depths studied were 50, 100, 150, 200, 250, and 300 μm. The results obtained for the cutting F_c and thrust F_h forces have been reported in Figure 3.12. The main observation was that the cutting forces increase with increasing depth of cut in a constant manner.

These results are in good agreement with those obtained experimentally by Wang et al. [2] and Nayak et al. [6] and numerically by Lasri [8] and Zenia et al. [27]. Indeed, one of the parameters that have an important effect on the value of the machining forces is the depth of cut, regardless of the machined material (organic or metal). The machining forces increase with increasing depth of cut a_p.

As regards the chip size, the numerical model allowed to measure the size of the latter for each test. Figure 3.13 shows the evolution of chip size versus the depth of cut a_p. The conclusion that can be drawn from this graph is that the depth of cut affects the chip size very significantly. Those results

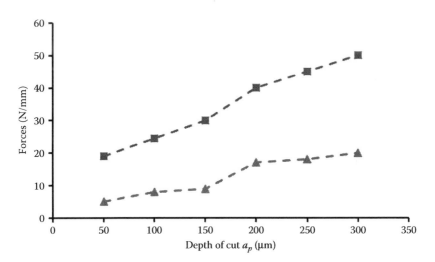

Figure 3.12 Cutting depth effect on machining forces, $V = 60$ m/min, $r_\varepsilon = 15$ μm, $\alpha = 10°$, $\gamma = 11°$.

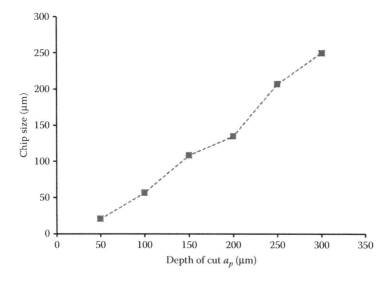

Figure 3.13 Cutting depth effect on chip size, $V = 60$ m/min, $r_\varepsilon = 15$ μm, $\alpha = 10°$, $\gamma = 11°$.

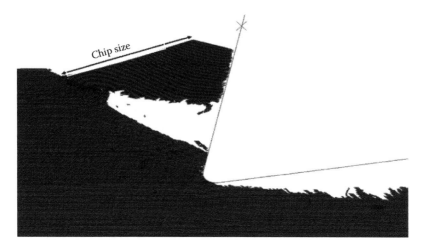

Figure 3.14 Size chip measurement: fiber orientation 45°, $V = 60$ m/min, $r_\varepsilon = 15$ μm, $\alpha = 10°$, $\gamma = 11°$.

found in this study correlates with the findings made in the literature [2]. The manner how the chip is measured is shown in Figure 3.14.

During the results analysis, it was found that the subsurface damage increases significantly with the increase in depth of cut, as shown in Figure 3.15. These results are consistent with those obtained by Nayak et al. [6] and Lasri [8].

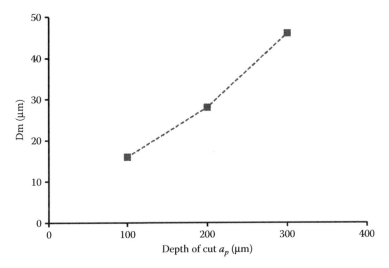

Figure 3.15 Cutting depth effect on the damage depth, $V = 60$ m/min, $r_\varepsilon = 15$ µm, $\alpha = 10°$, $\gamma = 11°$.

3.3.8   Effect of the cutting speed on the machining forces

Figure 3.16 shows the influence of the cutting speed V_c on the cutting forces F_c. Three cutting speeds, 6, 30, 60 m/min, were examined, and the results obtained showed that there is no significant influence of the

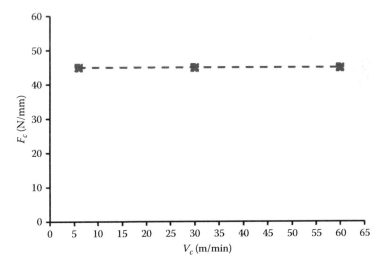

Figure 3.16 Velocity effect on cutting forces for fiber orientation at 45°: $a_p = 200$ µm, $\alpha = 10°$.

cutting speed V_c on the cutting forces F_c during orthogonal cutting of composite materials. Iliescu [10] reached to the same conclusion during his experimental work. This is explained by the fact that the cutting speeds taken during the study are considered to be of the same speed range. Moreover, in the numerical model, the tool is considered as a rigid body that does not wear out. Therefore, it remains healthy throughout the simulation.

3.3.9 *Effect of fiber orientations on the interlaminar delamination*

To investigate the effect of fiber orientation on interlaminar delamination, four simulations were carried out, with two adjacent layers oriented at 45°/0°, 45°/−45°, 90°/−45°, and 45°/45°, respectively. The aim of these orientation pairs is to see what are the directions of adjacent layers (Figure 3.17) that generate the largest delamination. The interface between two adjacent layers has a thickness of 5 μm. This was chosen according to various works found in the literature that treat the interface delamination. Phadnis et al. [12] used an interface with the thickness of 10 μm and Feito et al. [29] an interface of 5 μm. The latter is modeled using cohesive elements COH3D8 available in Abaqus. The two adjacent plies have the same geometric dimensions as shown in Figure 3.1a, with a thickness of 125 μm and the same mesh. In addition, the boundary conditions are the same as given in Section 3.2.1.

Figure 3.18 shows the delamination scope for four pairs of pleats. It can be seen that the delamination does not spread to other counterparties when the two adjacent layers have the same fiber orientation as in the case of the 45°/45° pair. However, when one pair of adjacent layers has different

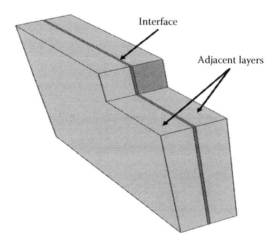

Figure 3.17 Two adjacent layers with interlaminar interface.

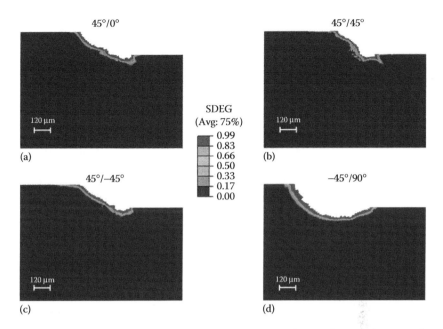

Figure 3.18 Damage of the interface between two adjacent layers, showing the delamination process for four configurations: (a) 45°/0°, (b) 45°/45°, (c) 45°/−45°, and (d) −45°/90°.

fiber orientations, as in the case of the 90°/−45° pair, this generates a very broad delamination (Figure 3.18d). These results are in good agreement with what has been reported in the literature. Ladevèze [18] reported an absence of interlaminar delamination between two plies having the same fiber orientation. Therefore, they can be considered as forming one and the same fold. Moreover, he found that delamination increased with an increase in the difference between fiber orientation angles of two adjacent plies forming the pair, leading to a greater delamination, and this was observed in the case of the configuration −45°/90°. This is explained by an increase in the shear stresses at the interlaminar interface exceeding the critical threshold of interface failure.

The study of the machining parameters' effect can prove very costly in terms of time and money. Indeed, the large number of parameters makes these studies complex. To optimize and also to highlight the effect of the interactions between the parameters on the machining product, Zenia et al. decided to carry out a numerical study that would investigate the effect of the machining parameters on the cutting forces and the induced damage. This, using experimental plans, more precisely the orthogonal design of experiments (DoE) $L_{27}(3^{13})$ of Taguchi, has been applied to investigate the effect of the fiber orientation, the tool rake angle, the depth of cut, and the tool edge radius. Conclusions drawn from this work show that the major factors that control the cutting force and the induced damage are (1) the fiber orientation, (2) the depth of cut, and (3) the tool rake angle. These results are in good agreement with the results obtained in this work.

3.4   Simulation of the drilling operation

This section shows the numerical results obtained during a conventional drilling operation of CFRP composite laminates. The plates are produced from the stack of unidirectional layers $[0_4, 90_8, 0_4]$, which give a total plate thickness of 2 mm. This work focuses on thrust forces Ft and the interlaminar delamination that occurs between two adjacent layers. The parameters of the used tool are previously mentioned in Section 3.2.1.

Figure 3.19 shows the different stages of a drilling operation with a conventional tool. It also shows the chip morphology obtained in this

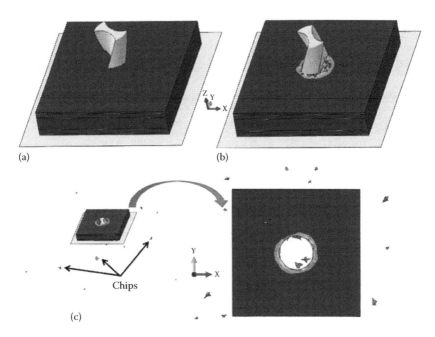

(a)   (b)

(c)   Chips

Figure 3.19 Steps of hole drilling (a) contact between the tool and the workpiece, (b) material removal, and (c) hole completely drilled.

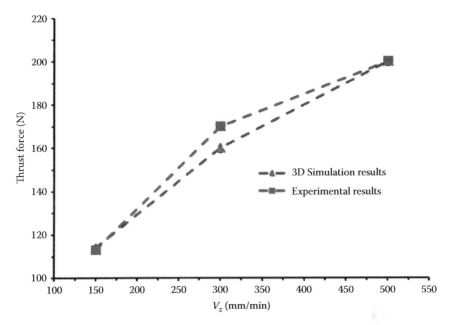

Figure 3.20 Comparison between experimental [12] and 3D simulation thrust forces.

cutting process. The latter is in the form of powder, and this is due to brittle behavior of this type of material.

Figure 3.20 shows the values of the thrust force Ft obtained using 3D model for different drill feed rate. It can be noticed that the cutting forces increase when increasing the drill feed rate. The numerical results are in good agreement with those obtained experimentally by Phadnis et al. [12] (Figure 3.20).

Figure 3.21 shows the delamination predicted numerically with the elastoplastic model and experimentally [12] in the input and output of the drilled hole. The conclusion that can be made is that the delamination is more important at the last interface, which is located between the two latest layers. These results are in good agreement with those obtained by Phadnis et al. [12], as shown in Figure 3.21a and b. The appearance of this type of delamination is due to the vacuum that is there under the drilled workpiece and the so-called drilling operation, *in the air.*

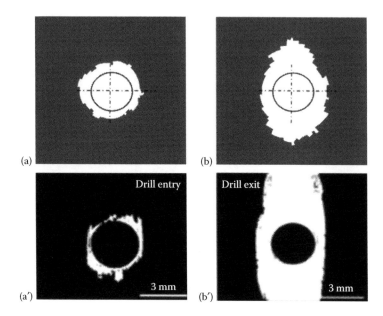

Figure 3.21 Drill entry delamination: (a) simulation result and (a′) experimental result [12]. Drill exit delamination: (b) simulation result and (b′) experimental result [12].

3.5   Conclusion

The main contribution of the current work concerns the development of a complete mechanical approach that integrates coupling between the damage and elastic–plastic behaviors to accurately simulate the cutting process of FRP composites. The original point of this work is the consideration of the interply interface using CZE and the prediction of interply damage. The comparison of the obtained results with experiments shows an accurate and realistic prediction of the chip formation process, cutting forces, and induced cutting damage. The chip formation process can be clearly described and analyzed by the simulation of the physical mechanisms such as the primary and secondary ruptures. Moreover, the proposed model allows to predict the accurate cutting forces, as shown by the validation with experimental results taken from the literature.

Finally, the model allows studying the effect of the drilling parameters on the multilayers CFRP composites and defines the delamination that can occur at the interply interface. Furthermore, we intend to include the temperature in the FE model, in order to investigate the thermal effect on the damage initiation and growth. That will be done in future work.

References

1. Koplev, A., Lystrup, A., Vorm, T. The cutting process, chips and cutting forces in machining CFRP. *Composites* 1983; 14:371–376.
2. Wang, D.H., Ramulu, M., Arola, D. Orthogonal cutting mechanisms of graphite/epoxy composite. Part I: Unidirectional Laminate. *Int J Mach Tool Manuf* 1995; 35:1623–1638.
3. Arola, D., Ramulu, M., Wang, D.H. Chip formation in orthogonal trimming of graphite/epoxy. *Compos Part A* 1996; 27:121–133.
4. Arola, D., Ramulu, M. Orthogonal cutting of fiber-reinforced composites: a finite element analysis. *Int J Mech Sci* 1997; 39:597–613.
5. Zitoune, R., Collombet, F., Lachaud, F., Piquet, R., Pasquet, P. Experimental calculation of the cutting conditions representative of the long fiber composite drilling Phase. *Compos Sci Techno* 2005; 65:455–466.
6. Nayak, D., Bhatnagar, N., Mahajan, P. Machining studies of UD-FRP composites part 2: finite element analysis. *Machining Sci Technol* 2005; 9:503–528.
7. Venu Gopala Rao, G., Mahajan, P., Bhatnagar, N. Micro-mechanical modelling of machining of FRP composites-cutting force analysis. *Compos Sci Techno* 2007; 67:579–593.
8. Lasri, L., Nouari, M., El-Mansori, M. Modelling of chip separation in machining unidirectional FRP composites by stiffness degradation concept. *Compos Sci Technol* 2009; 69: 684–692.
9. Santiuste, C., Soldani, X., Miguélez, H.M. Machining FEM model of long fiber composites for aeronautical components. *Compos Struct* 2010; 92: 691–698.
10. Iliescu, D., Gehin, D., Iordanoff, I., Girot, F., Gutiérrez, M.E. A discrete element method for the simulation of CFRP cutting. *Compos Sci Technol* 2010; 70:73–80.
11. ABAQUS Documentation for version 6.11-2 Dassault systems Simulia, 2011.
12. Phadnis, V.A., Makhdum, F., Roy, A., Silberschmidt, V.V. Drilling in carbon/epoxy composites: Experimental investigations and finite element implementation. *Compos Part A* 2013; 47: 41–51.
13. Zenia, S., Ben Ayed, L., Nouari, M., Delamézière, A. Numerical prediction of the chip formation process and induced damage during the machining of carbon/epoxy composites. *Int J Mech Sci* 2015; 90: 89–101
14. Feld, N. Vers un pont micro-méso de la rupture en compression des composites stratifiés. PhD thesis 2011; 115.
15. Shin, D.K., Kim, H.C., Lee, J.J. Numerical analysis of the damage behavior of an aluminum/CFRP hybrid beam under three point bending. *Compos: Part B* 2014; 56:397–407.
16. Ladeveze, P., LeDantec, E. Damage modelling of the elementary ply for laminated composites. *Compos Sci Technol* 1992; 43: 257–267.
17. Lubineau, G., Ladevèze, P. Construction of a micromechanics-based intralaminar mesomodel, and illustrations in ABAQUS/Standard. *Comput Mater Sci* 2008; 43:137–145.
18. Ladevèze, P., Allix, O., Deü, J.F., Lévêque, D. A mesomodel for localisation and damage computation in laminates. *Comput Meth Appl Mech Eng* 2000; 183:105–122.
19. Allix, O., Feissel, P., Thévenet, P. A delay damage mesomodel of laminates under dynamic loading basic aspects and identification issues. *Comput Struct* 2003; 81:1177–1191.

20. Lemaitre, J., Chaboche, J.L. *Mechanics of Solid Materials*. Cambridge University Press, Cambridge, UK, 1990.

21. Crisfield, M.A. *Non-Linear Finite Element Analysis of Solids and Structures*. Vol 1: Essentials 1991.

22. Turon, A., Dávila, C.G., Camanho, P.P., Costa, J. An engineering solution for mesh size effects in the simulation of delamination using cohesive zone models. *Eng Fract Mech* 2007; 74(10):1665–1682.

23. Benzeggagh, M., Kenane, M. Measurement of mixed-mode delamination fracture toughness of unidirectional glass/epoxy composites with mixed-mode bending apparatus. *Compos Sci Technol* 1996; 56: 439–449.

24. Teti, R. Machining of composite materials. *CIRP Annals—Manuf Techno* 2002; 51(2): 611–634.

25. Wang, X.M., Zhang, L.C. An experimental investigation into the orthogonal cutting of unidirectional fibre reinforced plastics. *Int J of Meach Tools Manuf* 2003; 43: 1015–1022.

26. Kaneeda, T., CFRP cutting mechanism. Transaction of North American Manufacturing Research Institute of SME 1991; 19: 216–221.

27. Zenia, S., Ayed, L.B., Nouari, M., Delamézière, A. Numerical analysis of the interaction between the cutting forces, induced cutting damage, and machining parameters of CFRP composites. *Int J Adv Manuf Technol* 2015: 78(1–4):465–480

28. Jamal, Y.C.A. *Machining of Polymer Composites*. Springer 2009.

29. Feito, N., López-Puente, J., Santiuste, C., Miguélez, M.H. Numerical prediction of delamination in CFRP drilling. *Compos Struct* 2014:108: 677–683.

chapter four

Numerical modeling of sinker electrodischarge machining processes

Carlos Mascaraque-Ramírez and Patricio Franco

Contents

4.1 Introduction 82
4.2 Objectives of electrodischarge machining numerical modeling..... 83
4.3 Basic formulation for electrodischarge machining numerical
    modeling 86
    4.3.1 Heat transfer produced by electrical discharges.................... 87
    4.3.2 Heat transfer in the workpiece... 89
        4.3.2.1 Conduction heat transfer 90
        4.3.2.2 Convection heat transfer 91
4.4 General structure of electrodischarge machining numerical
    model 92
    4.4.1 Definition of simulation mesh 93
    4.4.2 Temperature transfer equation and equivalent
        temperature concept.. 94
    4.4.3 Boundary conditions.. 96
        4.4.3.1 End points of the mesh... 97
        4.4.3.2 Number of simultaneous sparks............................... 98
        4.4.3.3 Definition of discharge cycles and cooling cycles...... 99
        4.4.3.4 Maximum discharge gap 100
    4.4.4 Process parameters 100
        4.4.4.1 Constant parameters.. 100
        4.4.4.2 Random parameters 101
        4.4.4.3 Output parameters.. 102
4.5 Main difficulties for electrodischarge machining numerical
    modeling 102
    4.5.1 2D and 3D modeling 102
    4.5.2 Modeling of large parts (utilization of progressive mesh) 103

4.5.3   Limits in the precision of meshing.. 104
4.6   Conclusions.. 105
References... 105

4.1   Introduction

The electrodischarge machining (EDM) is widely used for manufacturing molds, dies, and other different products and tools. The detailed analysis of these nonconventional machining processes provides to the manufacturing engineers the knowledge required for decision making about the optimum technologies for each industrial application. This knowledge is also the basis to enhance these manufacturing processes, in accordance with the technical requirements related to design and fabricability of mechanical parts to be produced.

To deduce the optimum process conditions for these machining technologies, numerous experimental and computational studies have been carried out by different researchers. The advantage of works focused on the theoretical modeling of these material-removal processes consists of the reduced costs of computational techniques in comparison with experimental analysis. This chapter is dedicated to explain the fundamentals of numerical modeling and simulation of EDM processes and, more specifically, sinker EDM processes.

The EDM consists of a cutting technology based on the application of controlled electric discharges that occur between the cutting tool (electrode) and the workpiece to be machined. These electric discharges provoke some sparks that generate a strong temperature increase on the workpiece within the domain of the cutting zone, causing the fusion on a specific proportion of the workpiece material and thus the desired material removal.

To control the electrical discharges and guarantee the formation of stable arcs, the entire process is carried out by immersing the workpiece and electrode in a dielectric fluid; this process is responsible for removing the molten material and keeping the workpiece surface clean.

Some references present a review about the advances produced in EDM during the last years [1,2]. It is crucial to focus the analysis of these machining processes on heat transfer, including the temperature increase caused by the electric discharges, as well as the cooling facilitated by the contact with the dielectric fluid and the conduction heat transfer to the rest of the workpiece.

Different studies were focused on exhaustive experimental tests, with the purpose of determining the influence of the main factors of this process on the final properties of machined part. For example, there are experimental works dedicated to study the surface finish that can be achieved in the EDM processes [3] and the energy distribution in elements of the system [4].

The development of numerical models that allow the prediction of the expected results of this process began in the late 1980s, with studies such as those of DiBitonto et al. [5] and Patel et al. [6]. Since then, the theoretical modeling of these machining processes has evolved parallel to the power of computers and the reduction of computational times.

Among the numerous numerical works that were carried out in the last years, we can highlight, for example, some studies that propose combined models for material removal and surface finish [7], the analysis of the transient temperature distribution, state material transformation and residual stresses [8], and theoretical approaches based on a simplified matrix calculation perspective [9].

In this chapter, the main aspects of numerical models that can serve to predict the expected results of sinker EDM are described. These approaches will be based on the finite difference method; however, the principles and instructions provided in this chapter can also be easily extended to models based on finite element method and other possible methodologies.

In the next sections, the main objectives of the numerical modeling of EDM processes will be defined. The fundamental equations that must be implemented in these approaches will also be reviewed. After that, the most common difficulties in the simulation of these machining processes and possible solutions about these questions will be explained, followed by some conclusions about EDM modeling and simulation.

4.2   Objectives of electrodischarge machining numerical modeling

The numerical modeling of EDM processes pursues the objective of predicting the phenomena that occur in these machining processes and the expected properties of machined workpiece. More specifically, the aim of this numerical modeling is to determine the process parameters that allow optimization of the quality of final parts, manufacturing costs, and productivity level.

These three objectives are closely linked, since the increase in the part quality usually affects the production cost and/or the time required for these processes. The minimization of manufacturing costs can be carried out by reducing the total time for which the EDM machine is utilized or by preventing an excessive wear rate in the electrodes. The cost per unit will depend on the resources needed to produce the parts, which comprise the number of electrodes to be used.

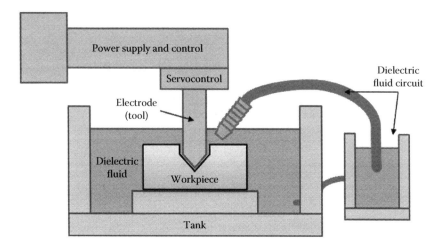

Figure 4.1 Schematic representation of the sinker EDM process.

A schematic representation of the structure of EDM machines can be observed in Figure 4.1. To execute these nonconventional machining processes, EDM machines are composed of the following main elements:

- *Power supply*: With the aim of rectifying the input current to the machine and generating the desired voltage between the part and electrode, according to the voltage and intensity selected for each machining process.
- *Servo control*: Responsible for the movement of the electrode. It must provide the constant feed of the electrode, maintaining a gap between the workpiece and the cutting tool to ensure that the spark could be produced correctly. If the gap is too large, the electric discharge will not occur, but if the part and electrodes come in contact, a short circuit will befall, which would not produce machining in the part and would be detrimental for the surface finish of the workpiece.
- *Tank*: Whose function is to house the dielectric fluid, where the part and cutting tool will be submerged. The workpiece will be fixed on the working table located inside the tank.
- *Dielectric fluid circuit*: Will serve to recirculate the dielectric fluid, as well as to filter and remove the debris generated during this machining process.

As can be seen in Figure 4.2, when the spark is originated inside the dielectric fluid, a plasma channel is produced, which transmits the energy provided by the electric discharges. After that, there will be an increase

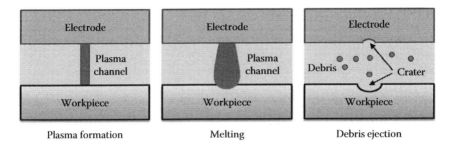

Figure 4.2 Different states of plasma channel during the EDM process.

in the temperature along the area of influence of the plasma channel, provoking the partial melting of both the part and the cutting tool. This process is independent of the mechanical properties of workpiece material, since it is only influenced by its thermal properties.

Figure 4.3 shows an example of the texture found in the surface of the part and electrode from certain process conditions. The typical crater shape of the part surfaces subjected to EDM process is depicted in Figure 4.3a, whereas Figure 4.3b illustrates the progressive wear of the electrode and the protuberances that are originated on the electrode surface due to melting of some material portions and their later solidification during the process.

The EDM is a machining process based on the material removal from the initial geometry of workpiece by means of electric discharges that provoke the melting and even the evaporation of this material inside the cutting zone. It is indeed a thermal process, since the material removal is a consequence of the heat input provided by these electric discharges.

For this reason, the numerical modeling of EDM process is focused on the analysis of heat transfer from the plasma channel to the workpiece and the prediction of temperature distribution on the workpiece surface

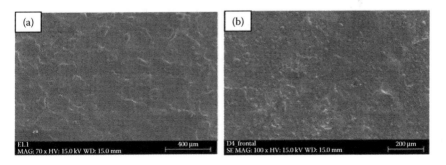

Figure 4.3 Examples of scanning electron microscope (SEM) images for workpiece and electrode in the EDM process: (a) Stainless steel workpiece and (b) copper electrode.

as a function of cutting time, which is the cause of the material removal and the instantaneous geometry of workpiece surface.

Other of the main objectives of the simulation of these nonconventional machining processes should be the prediction of the final geometry and the surface finish of the manufactured part, not only at the end of this process but also at intermediate phases during the material removal. It is also interesting to include the prediction of progressive wear and material adhesion on the electrode surface.

The numerical models should be properly validated by experimental results to guarantee the accuracy of their predictions. In addition, it is convenient that the proposed models comprise agile routines, in order to minimize the computational costs associated with the simulation of this material-removal process.

4.3   Basic formulation for electrodischarge machining numerical modeling

The first step to develop a mathematical model for simulation of the EDM processes should be the appropriate definition of basic formulation to be applied. These equations must be implemented in the proposed model and will serve as the basic equations for process simulation.

In this sense, it is important to identify the physical and thermodynamic principles that govern the EDM process. For this purpose, in this section, the different phases of the process will be described, along with the thermodynamic phenomena related to them. The impact of each phase of the process and the mathematical formulation required will also be explained.

The thermodynamic aspects involved in these machining processes can be divided into two major concepts:

- Electrical discharges originated between the electrode and the part
- Heat transfer in the workpiece material

The material removal during the EDM processes will start from an initial state of thermal stability, in which the workpiece and dielectric fluid are found at the same temperature. This initial temperature will correspond to the environmental temperature at the beginning of the process, for which a reference value of about 24°C is usually adopted.

From the initial geometry of workpiece, the material processing will be executed, thanks to the electric discharges generated between the electrode and the workpiece. The expressions to be applied to describe the discharges provoked on the workpiece will be explained in a later section.

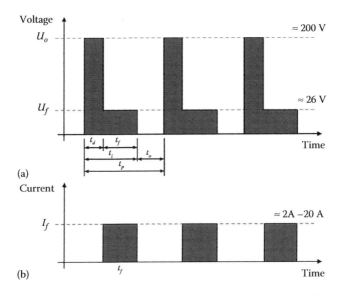

Figure 4.4 Different phases of the EDM processes: (a) Voltage diagram and (b) current intensity diagram.

Figure 4.4 illustrates the relative levels of voltage and current intensity for each one of the phases contained in the EDM process. Indeed, as can be seen in this figure, the machining process is divided into two major phases, the pulse phase and the pause phase.

The parameter U_o of this figure consists of the voltage during the ionization phase, whereas U_f and I_f correspond to the voltage and current intensity, respectively, during the discharge phase. In addition, t_d is the ionization time, t_f the discharge time, and t_0 the pause time. The sum of the ionization time and discharge time is known as the pulse time, which when added to the pause time represents the cycle time t_p.

4.3.1 Heat transfer produced by electrical discharges

The pulse phase is characterized by the application of a certain electrical voltage between the electrode and the workpiece. It is divided between the ionization phase and the discharge phase, the first of which is responsible for generating the plasma channel, whereas the second one represents the stabilization of the plasma channel and therefore the transfer of thermal energy to the workpiece.

The discharge phase can be considered the main phase of the EDM process, as it is responsible for material removal. Conversely, during the pause phase, there is no transfer of thermal energy between the electrode and the workpiece, and so, this phase is characterized by the heat transfer within the workpiece itself and also between the workpiece and dielectric fluid. The mathematical expressions to be applied with regard to each one of these phases will be described as follows.

Once the plasma channel becomes stable at the end of ionization phase, the material removal begins, thanks to the heat input provided to the workpiece material and the temperature increase caused by this thermal energy. The supply of thermal energy to the part to be machined can be described by the following expression:

$$q(r,d) = a \cdots q_0 \cdots \exp\left[b \cdots \left(r^2 + d^2\right)/\left(R_p^2\right)\right] \tag{4.1}$$

where:

> a and b are the heat input constants
> q_0 is the total thermal energy provoked by each spark
> r is the radial distance to the plasma center for each point of the workpiece material
> d is the distance between the spark center and the two-dimensional (2D) simulation area in traverse direction (which will depend on the random position that could correspond to the successive sparks to be considered for each calculation step during the modeling of EDM)
> R_p is the radius of plasma channel

The heat input constants a and b depend basically on the thermal properties of the workpiece material, and the values found in literature can be considered for this generalized expression.

Figure 4.5 illustrates the heat flux distribution on the workpiece surface according to this mathematical expression. This figure is

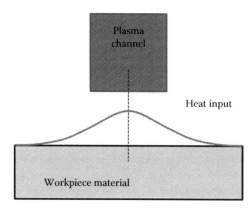

Figure 4.5 Heat input distribution on the workpiece surface during the EDM process.

located on the 2D simulation area, and so, only the effect of the distance to central position of plasma channel is represented. As can be seen in this figure, the greatest amount of thermal energy will be received at the nodes closer to the plasma axis, whereas an exponential decay will be produced as the radial distance to the plasma axis is increased.

For computation of the thermal energy supply to the machining area, according to Equation 4.1, the overall heat input generated by each spark could be estimated using the following expression as a function of electrical parameters assumed for the EDM process:

$$q_0 = F_w \cdots U \cdots I \cdots t_i \, / \, \pi \cdots R_p^2 \qquad (4.2)$$

where:

F_w is the proportion of the heat input transmitted to the workpiece material

U and I are the burning voltage and current intensity, respectively, to be programmed in the numerical control (CNC) of the EDM machine

t_i is the pulse time during which the electrical circuit will be active for each burning cycle

4.3.2   *Heat transfer in the workpiece*

One of the phases mentioned in the previous section is the pause phase. During this phase, the heat transfer within the workpiece and between the part and other elements of the EDM machine is performed. The heat transfer that takes place during the pause phase can be divided into following three concepts:

- Heat transfer along the workpiece itself
- Heat transfer from the workpiece to the working table of the EDM machine
- Heat transfer from the workpiece to the dielectric fluid

The first two correspond to heat transfer by conduction, inside the own workpiece or through the contact surface between the part and the working table of the machine. On the other hand, the transmission of thermal energy between the part and dielectric fluid corresponds to heat transfer by convection.

This section is not dedicated to an exhaustive thermodynamic study about heat transfer by conduction and convection but to explain the mathematical formulation that should be applied for this purpose inside the EDM numerical model.

4.3.2.1   *Conduction heat transfer*

The heat transfer by conduction could be defined as a thermal energy transmission process based on direct contact between different bodies and without exchange of matter, whereby the heat flows from a node at a higher temperature to another node at a lower temperature that is in contact with the previous one. The physical property of materials that determines their ability to conduct heat is the thermal conductivity.

Figure 4.6 shows a basic diagram of conduction heat transfer, where the node on the left is at higher temperatures than the rest of the nodes, and therefore, the heat transfer takes place to nodes on the right. The temperature level of the nodes of this simplified system will change as a function of time according to the second principle of thermodynamics, which determines that the thermal energy can only flow from a hot body to a cold one.

The adjacent nodes will transmit heat by conduction to the rest of the nodes in this chain, until the total extension of the workpiece is covered. Owing to the closed contact between the part and the working table of the EDM machine, the thermal energy will also be transmitted from the end nodes of the workpiece to the working table and finally to the rest of elements of the whole machine, which suppose the dissipation of a certain proportion of the total heat input throughout the system.

By transferring the principles of conduction heat transfer to the formulation necessary for numerical modeling of the EDM, the following equation of temperature increase between two adjacent nodes can be applied:

$$(\Delta T)_{i,k,s} = \Delta t \cdots k \cdots \Delta A \cdots \left(T_{p,q,s-1} - T_{i,k,s-1} \right) / \Delta m \cdots c_p \cdots \Delta s \qquad (4.3)$$

where:

$(\Delta T)_{i,k,s}$ is the temperature increase by conduction at a node with coordinates (i, k) in 2D for a simulation instant s

Δt is the time interval between instants s and $s-1$

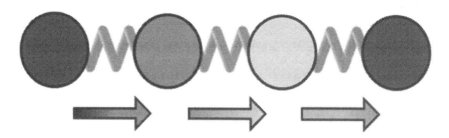

Figure 4.6 Basic diagram of conduction heat transfer.

k is the thermal conductivity of the workpiece material expressed in
W/m K (e.g., in the case of stainless steel AISI 316, a value of
17 W/m K can be adopted)

Δm is the effective mass for the active node

c_p is the specific heat capacity that corresponds to the phase transition
of the workpiece material

Δs is the distance between the two adjacent nodes (which will be
affected by the position step adopted for the calculation mesh)

ΔA is the heat transfer area between both nodes

$T_{p,q,s-1}$ is the temperature at the adjacent node with coordinates (p, q)
for the simulation instant before the current instant in which the
calculation of the heat transfer is carried out

$T_{i,k,s-1}$ is the temperature at the node of study at the previous instant

4.3.2.2 Convection heat transfer

The heat transfer by convection is produced by the movement of a fluid
that transports the thermal energy between zones with different temperatures. In the case of EDM, the heat transfer by convection occurs
between the dielectric fluid and the workpiece surface. The fluid is in
motion due to the feed pump of the EDM machine, which generates a turbulent regime over the machining area, as can be observed in Figure 4.7.

The flow of dielectric fluid impacting the workpiece surface will provoke a turbulent regime that will facilitate the heat transfer by convection.
Figure 4.8 illustrates the convection heat transfer around the workpiece
surface, thanks to the dielectric fluid, in addition to heat transfer produced by conduction along the different zones of the part to be machined.

Finally, the following mathematical expression could be assumed to
estimate the temperature increment to be produced as a consequence of

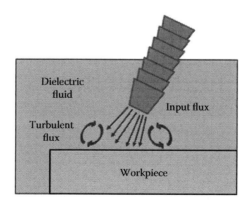

Figure 4.7 Dielectric fluid turbulence around the workpiece surface.

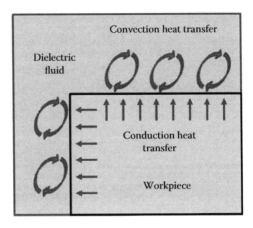

Figure 4.8 Convection heat transfer throughout the dielectric fluid in EDM processes.

the heat transfer by convection between the machined surface and dielectric fluid:

$$(\Delta T)_{i,k,s} = \Delta t \cdots h \cdots \Delta A \cdots \left(T_d - T_{i,k,s-1}\right) / \Delta m \cdots c_p \qquad (4.4)$$

where:

h is the convection coefficient

ΔA is the contact area for thermal energy convection between the computed nodes that correspond to the workpiece surface and the surrounding fluid dielectric

T_d is the temperature achieved by the dielectric fluid during the EDM

The other parameters were described previously in the section dedicated to conduction heat transfer. This equation will be implemented in the numerical model, together with the equations that have also been described for conduction heat transfer and heat input provided by the electric discharge.

4.4 General structure of electrodischarge machining numerical model

The computational modeling that will be presented in this section assumed a matrix calculation based on the finite different method. The objective of this model must be to predict the surface finish of the machined surface, the progressive wear of the electrode, and the production time required for this machining process.

In the following sections, a method for calculating the temperature matrix and the boundary conditions to be established will be explained, and after that, the procedure to define the simulation mesh will be also described.

4.4.1   Definition of simulation mesh

Although some of the properties of the mesh and solutions to reduce the processing times without affecting the mesh properties will be discussed later, before beginning with the definition of the temperature matrix, it is convenient to define some characteristics of the simulation mesh to be considered for an appropriate simulation of material removal on the workpiece.

Figure 4.9 shows the simulation mesh to be considered, in this case represented over a machined surface. The lines in dark gray color represent the outer edge of the mesh, whereas the white lines are utilized for the zone that corresponds to the cutting area and the light gray lines serve to identify the remaining material that is located outside the area to be machined.

The nodes in white color will be those that undergo the material removal and therefore are expected to disappear during the process. These nodes are the most important ones for simulation of the EDM, since they will define the resultant surface of the machined part. On the other hand, the light gray nodes are intrinsically necessary for computation of heat transfer by conduction with the rest of the workpiece and material removal on the zone composed of the cutting area. Finally, the dark gray nodes will serve to delimit the simulation zone and establish the boundary conditions to be assumed by the numerical model, as will be described in another section.

The simulation mesh must be sufficiently fine to ensure the required accuracy in the numerical predictions. Nevertheless, if an excessively

Figure 4.9 Example of simulation mesh for an EDM process.

fine meshing is considered, an unacceptable computational cost can be encountered, since the calculation times grow in a quadratic form. For this purpose, some solutions will be proposed in Section 4.5.3.

Once the definition of the simulation mesh is explained, the following section will be dedicated to describe the temperature matrix; each point of the matrix will correspond to a node of the mesh.

4.4.2 *Temperature transfer equation and equivalent temperature concept*

The formulation presented in this section is oriented to obtain the temperature at each node of the mesh for each time step during the process. The phase change at each node should be analyzed in order to know the nodes that are melted or evaporated for each cutting time from the heat transfer matrix and therefore will disappear from the workpiece surface. Nevertheless, this computation is certainly complex and requires great processing times. To avoid this problem, the concept of equivalent temperature is usually assumed.

The model works with the concept of equivalent temperature, in order to discard the heat input needed for the phase change from solid to liquid, and therefore, the temperature at which the material is in its liquid phase is extrapolated. In this way, it is not necessary to consider the latent heat of fusion or evaporation but simply determine the cutting temperatures from the specific heat in the solid phase. From this assumption, a certain value of equivalent temperature is adopted when the material is instantaneously melted. According to the literature, this value oscillates between 1400°C and 1800°C for different types of stainless steels.

The concept of equivalent temperature is illustrated in Figure 4.10. The gray lines represent the relationship between the heat input and temperature for a stainless-steel workpiece, whereas the dotted gray line shows the concept of equivalent temperature, and the gray circle corresponds to the temperature to which the material will be melted instantly.

The concept of equivalent temperature allows to operate using a temperature matrix instead of a heat transfer matrix. Thus, a temperature transfer equation will be assumed, which will serve to evaluate the temperature of each node (i, k) of the workpiece for a process instant s as follows:

$$T_{i,k,s} = T_{i,k,s-1} + (\Delta T_L)_{i,k,s} + (\Delta T_R)_{i,k,s} + (\Delta T_T)_{i,k,s} + (\Delta T_B)_{i,k,s} \qquad (4.5)$$

where $T_{i,k,s}$ and $T_{i,k,s-1}$ are the equivalent temperatures for node (i, k) for the current instants s and previous instant $s-1$, respectively. The terms $(\Delta T_L)_{i,k,s}$, $(\Delta T_R)_{i,k,s}$, $(\Delta T_T)_{i,k,s}$, and $(\Delta T_B)_{i,k,s}$ represent the variation of cutting

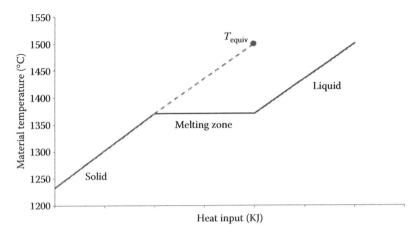

Figure 4.10 Diagram with the concept of equivalent temperature.

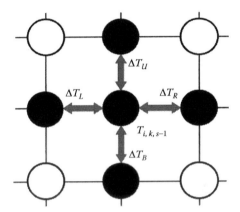

Figure 4.11 Variation of equivalent temperature at the node of study from heat transfer with adjacent nodes.

temperature at the active node as a consequence of heat transfer, with the adjacent nodes situated on the left, right, top, and bottom, respectively.

The relationship between the node of study and these adjacent nodes is illustrated in Figure 4.11.

Depending on the situation of the node of study within the workpiece, different types of circumstances can be found, which are described as follows:

- *Type I*: Nodes located inside the workpiece material and so are uniquely influenced by conduction heat transfer in the workpiece itself.

- *Type II*: Nodes located at the workpiece surface, with heat transfer by convection with the dielectric fluid and by conduction with the adjacent nodes of the workpiece.
- *Type III*: Nodes on the workpiece surface that are affected by the plasma channel during the electrical discharge, which can also experience conduction heat transfer with other nodes of the workpiece and convection with the dielectric fluid.

Figure 4.12 shows the different options that can be present in each node of the simulation mesh. Convention heat transfer will always be considered from the workpiece to the dielectric fluid, whereas the plasma channel will be assumed as a heat input resource to the nodes located inside the cutting zone. In the case of nodes of type I, it will be necessary to evaluate if a positive or negative variation will be produced according to each one of the adjacent nodes.

The theoretical model will proceed to the resolution of Equation 4.5 for each cutting time, using the equations explained in the previous section. An iterative process will be needed until the equivalent temperature for each node of the mesh for the current instant is achieved. At the end of each calculation cycle, the equivalent temperature matrix $T_{i,\,k,s-1}$ will be initialized from the data of the matrix $T_{i,k,s}$, and so, the next cycle can be computed.

4.4.3   Boundary conditions

The definition of boundary conditions for a system to be simulated is always complex, and in fact, it will be a key factor for the results provided

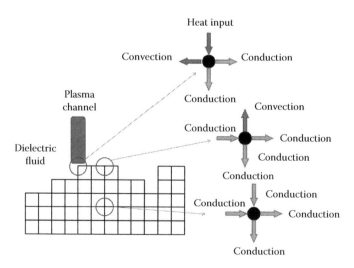

Figure 4.12 Examples of heat transfer at different nodes of the simulation mesh.

by the model. Thus, the boundary conditions to be adopted must be analyzed carefully, and they should be established from experimental information. The instructions provided as follows will serve to determine
preliminary values to be assumed for EDM simulation.

4.4.3.1 *End points of the mesh*

According to the equation for calculation of equivalent temperature, the
variation of cutting temperature will depend on the temperature of adjacent nodes. When the node of study receives a temperature increase due
to convection or electric discharge, there is no problem, since all the information necessary to solve the formulation is available.

Meanwhile, in the case of nodes subjected to heat transfer by conduction, it is necessary to know the temperature of the adjacent nodes
beforehand. The problem arises when there are no adjacent nodes in the
meshing, because the active node is located at the extremity of the calculation zone. If adjacent nodes do not exist in the calculation meshing,
the boundary condition must be applied. These end points are illustrated
using a thick line in pale white color in Figure 4.13.

To apply Equation 4.3 related to conduction heat transfer, it is also
necessary to know the equivalent temperature at the adjacent node for
the previous instant ($T_{p,q,s-1}$). If this information is not available, a boundary condition could be assumed. For conduction heat transfer calculations, the equivalent temperature for nodes external to the meshing at
the previous instant shall be equal to the temperature of the two previous
cycles at the nearest node in the mesh, which can be expressed by the following equation:

$$T_{p,q,s-1} = T_{k,i,s-2} \tag{4.6}$$

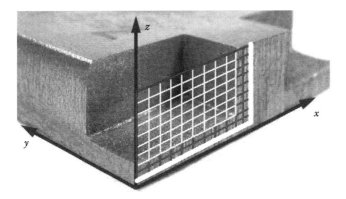

Figure 4.13 Example of end points in the workpiece meshing.

Therefore, for the end points of the mesh, the temperature variation according to thermal conduction can be expressed as:

$$(\Delta T)_{i,k,s} = \Delta t \cdots k \cdots \Delta A \cdots \left(T_{i,k,s-2} - T_{i,k,s-1}\right) / \Delta m \cdots c_p \cdots \Delta s \qquad (4.7)$$

In the numerical model, it will be convenient to include a matrix containing the information of the equivalent temperatures at each node for a process instant $s-2$, previously to start the computation of a new study cycle. Consequently, a total of three temperature matrices, T_s, T_{s-1}, and T_{s-2}, will be needed.

4.4.3.2   Number of simultaneous sparks

It is practically impossible to predict the number of sparks that can be present in each pulse. According to the literature, it is known that the surface geometry of the workpiece and the electrode wear have repercussions in this value, but they cannot be formulated.

Therefore, in this section, this criterion is proposed to be treated as a boundary condition, at the beginning of the simulation process.

The energy transmitted to the workpiece will be divided between the different sparks produced simultaneously. As a criterion, this heat input could be divided equally between all the sparks originated, and so, Equation 4.2 would be modified by dividing F_w between the number of simultaneous discharges n_w:

$$q_0 = \left(F_w / n_w\right) \cdots U \cdots I \cdots t_i / \pi \cdots R_p^2 \qquad (4.8)$$

With regard to the number of discharges to evaluate, it is possible to choose between more simplified or complicated criteria for this purpose. Of course, if a complex criterion is assumed, the calculation time will be increased. The options for this boundary condition can be the following ones:

- Define an exact number of sparks for each discharge cycle, which can be usually between 1 and 5. The main advantage of choosing this approach is that it greatly reduces the difficulty of programming, and besides, the results are very satisfactory.
- Generate a random variable with a uniform distribution between 1 and 5 for each discharge cycle. This option complicates the calculation of each cycle but is a slightly better option in terms of physical meaning.
- Generate a random variable with a nonuniform distribution between 1 and 5, giving a greater probability of a unique discharge and a

lower probability of five simultaneous discharges. Definitely, this is the smartest option, but the calculation becomes more complex, and therefore, the process time is increased.

4.4.3.3   Definition of discharge cycles and cooling cycles

In practice, the durations of the discharge cycles (t_i) and the cooling cycles (t_0) are neither constant nor equal, as illustrated in Figure 4.14. For this reason, during the numerical simulation of EDM process, a time step t_n must be defined for each calculation cycle, and it can be constant throughout the process simulation.

It is therefore necessary to include in the model a condition that states whether the current cycle corresponds to an electric discharge phase or a cooling phase.

In the case of a discharge phase, it is needed to compute the node (or nodes) in which the sparks will act and the heat transfer provoked by these sparks, according to the expression (4.2). Nevertheless, in the cooling phase, no sparks will occur, and therefore, the entire surface of the workpiece will be affected only by convection heat transfer with the dielectric fluid, according to the expression (4.4).

To establish the phase that corresponds to each simulated cycle, one of the following criteria can be selected:

- Define a series of one discharge cycle and one cooling cycle.
- Define a series of one discharge cycle and two cooling cycles.
- Define a series of a discharge cycle, a random discharge or a cooling cycle, and a cooling cycle.

The third option is the most realistic but also the most complicated in terms of programming. In general, it would be convenient to conduct experimental tests and thus to select the criterion that could be considered more adequate.

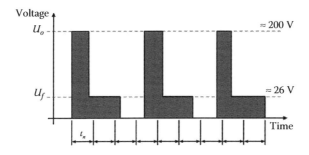

Figure 4.14 Variable duration of the discharge and cooling cycles.

4.4.3.4   Maximum discharge gap

It is well known that electric discharges will most likely occur in the zones with a minimum distance between the workpiece and the electrode, but it cannot be assured if two or more consecutive sparks will be produced at the same node (or nodes), even though it represents the nearest zone between the cutting tool and the part to be machined.

For this reason, it is necessary to establish a criterion about the maximum gap between the electrode and the workpiece for discharge to occur. This will prevent erroneous sparks in the central areas of the craters and excessively sharp peaks in the workpiece, two phenomena that are not produced in experimental tests and consequently must be avoided in the numerical model.

For this purpose, it is enough to include a criterion about the maximum gap that allows the electrical discharge. As a guideline, values between 50 μm and 200 μm could be adopted for this maximum gap.

4.4.4   Process parameters

For numerical simulation of the EDM process, one needs to know the main parameters that govern the material removal. In this sense, one should distinguish among the parameters to be introduced as a constant value, the parameters of a random character, and the parameters to be provided by the model as output results of the process simulation.

4.4.4.1   Constant parameters

These values could be divided into two large groups, the parameters that are specific for the material used in the workpiece and the electrode and the parameters that are intrinsic for the EDM process.

With regard to the material, there is certain information that must be known, which appears in the expressions described in the previous sections. Some of the most relevant physical and thermal properties of the material for EDM processes are listed in Table 4.1.

On the other hand, there are also some process parameters that must be properly identified to make the numerical simulation of the EDM

Table 4.1 Material properties for EDM numerical modeling

Parameter	Units
Equivalent temperature	°C
Convection heat transfer coefficient	$W/m^2\,K$
Density	kg/m^3
Specific heat	$J/K\,kg$
Thermal conductivity	$W/m\,K$

Table 4.2 Process parameters for EDM numerical modeling

Parameter	Units
Workpiece geometry	Mm
Voltage during the ionization	V
Voltage during the discharge	V
Intensity during the discharge	A
Ionization time	μs
Discharge time	μs
Pulse time	μs
Pause time	μs
Room temperature	°C
Gap	μm
Radius of the plasma channel	μm
Dielectric fluid temperature	°C
Percent of pulse energy transferred to workpiece	%

process possible. Many of these parameters will have a constant value during the entire simulation, since they correspond to parameters to be selected in the EDM machine. Some of the most relevant process parameters are listed in Table 4.2.

Other variables to be defined in the model are the parameters related to process simulation. In general, they can be maintained constant during the numerical simulation of these machining processes. The most relevant parameters of this type are listed in Table 4.3.

4.4.4.2 Random parameters

There are some parameters of random character, such as the incidence point for each spark. The number of simultaneous discharges per cycle is also a random variable, and it was previously defined in the section of boundary conditions.

About the incidence point of the electric discharge, first, all the nodes that satisfy the boundary condition about the maximum discharge gap

Table 4.3 Parameters for process simulation

Parameter	Units
Mesh dimensions	mm
Mesh size	μm
Time step for process simulation	μs
Maximum discharge gap	μm
Number of cycles for output of results	

should be identified by the theoretical model, and then, the specific position in which the discharge is produced can be estimated by means of a uniform probability distribution, without discriminating any of these nodes.

4.4.4.3 Output parameters

The computational model must contain some parameters that will serve to express the results obtained for each simulation instant, including not only the final instant of process simulation but also the intermediate cutting times of special interest.

The main variable to be provided by the model for each simulation cycle is the equivalent temperature at each node.

From the temperature distribution for each cutting time, the points that define the instantaneous surface of the workpiece are also considered as output parameters. These points are deduced by removing the nodes that exceed the reference limit previously established for the equivalent temperature.

Other output parameters of EDM simulation are the points that constitute the electrode surface, caused by the electrode degradation during this nonconventional machining process. The vertical position of the electrode is changing continuously during the process, as the workpiece is machined. A constant vertical speed can be assumed for the cutting tool; however, it can also go back if a short circuit is produced. If the gap between the part and the cutting tool is greater than the position of highest node at the workpiece, then the electrode can continue its move, penetrating in the workpiece, but in the opposite case, it would go back in the next simulation cycle.

4.5 Main difficulties for electrodischarge machining numerical modeling

In the previous sections, the basis for numerical modeling of EDM has been introduced. The mathematical formulations to be implemented in the model were explained, and the main parameters that affect the results of these machining processes were described. Nevertheless, the development of a numerical model to simulate the behavior of EDM process implies multiple difficulties, the main of which will be described in the following sections.

4.5.1 2D and 3D modeling

Before proceeding to modeling and simulation of these machining processes, one must decide if the model should execute a 2D or 3D simulation.

Of course, 3D simulations are more complicated to perform and suppose higher computational costs; however, they provide valuable additional information.

Three-dimensional simulations will require to create a temperature matrix $T_{i,k}$ for each plane, and the size of these matrix will depend on the mesh size and the extension of the cutting zone to be analyzed. Meanwhile, if 2D simulation is selected, the matrix of equivalent temperatures will be calculated for a unique plane, and so, the processing times will be considerably reduced.

Another possible solution is to perform a 3D simulation for electric shocks and 2D simulation for meshing of the cutting zone; this combined solution provides a good accuracy in the numerical calculations. In this case, the model will use a central plane in the workpiece, with a mesh adequate for 2D simulation, and all the temperature calculations would be performed in that plane.

To choose the option of 3D simulation, it is necessary to assume two random variables in order to express the incidence point for the electric discharges. These two variables can be located in planes x and y, and then, the nodes in plane x can also be established for computation in any other possible plane y. Figure 4.15 illustrates the plane for 2D simulation in black color, whereas the rest of planes are depicted in white.

4.5.2   Modeling of large parts (utilization of progressive mesh)

When attempting to model pieces of a relevant size, the size of the calculation arrays may become too extensive and slow down the simulation. To maintain adequate execution times, there are some alternatives that can be adopted.

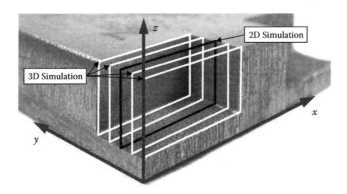

Figure 4.15 Planes defined for a 2D/3D simulation.

One alternative to this problem can be the utilization of an adaptive mesh, and so, a higher mesh size can be selected for the initial meshing, while the mesh can be refined later. To facilitate the transition between the thick mesh and the fine mesh, it can be necessary to define transformation conditions with the purpose to extract the calculation of temperatures at the new nodes more easily.

Another alternative would be the application of a progressive mesh. In this case, the mesh could be moved together with the electrode, and, from the analysis of the results of temperature matrix at intermediate calculation steps, it will be possible to determine the depth to which the temperature is practically constant, and so, these nodes can be deleted from the matrix in order to save computational costs.

Figure 4.16 illustrates an example of temperature matrix, at an inter-mediate time during the machining process. As shown in this figure, from a certain depth, the workpiece is found at room temperature, and therefore, the computation of these nodes does not provide interesting information. Conversely, it would be interesting to delete these nodes to save the computing time.

4.5.3   Limits in the precision of meshing

To improve the accuracy of the numerical model, it could be convenient to decrease the time step Δt and mesh size Δx in the simulation mesh. Nevertheless, this is not exactly the solution, because a very small mesh size Δx can provoke mistakes in the calculations, based on the basic for-mulations. The general equation of conduction (4.3) is oriented to relevant distances between the adjacent nodes, where the increment in the tem-perature is a function of the thermal gradient, material properties, time step, and distance between the adjacent nodes.

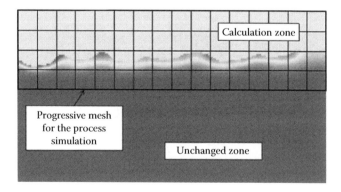

Figure 4.16 Example of equivalent temperature matrix to define the progressive mesh.

It is recommended to use a mesh size (or distance between adjacent nodes) suitable to the part and conditions to be considered.

Likewise, one should avoid to reduce the mesh size without reducing the time between calculation cycles, since it would provide a worse prediction accuracy and computation time. Despite that, a certain balance between the mesh size and time step should be maintained, for each numerical simulation of the EDM.

4.6 Conclusions

The numerical modeling of EDM processes is a complex subject, due to the existence of multiple factors that must be properly evaluated. A good knowledge about the different parameters that are involved in this machining process is essential to define correctly the numerical model. The explanations contained in this chapter can help develop an EDM model with a favorable balance between the accuracy of results and the computational costs associated with numerical simulations. In this chapter, the mathematical fundamentals for numerical modeling of the EDM have been stated and the difficulties that can be faces during the development and execution of the model have been described. One must always verify the numerical model with experimental results and so check the validity of boundary conditions, mathematical equations, and process meshing to be applied.

References

1. Abbas, N.M., Solomon, D.G. and Bahari, M.F. 2007. A review on current research trends in electrical discharge machining (EDM). *International Journal of Machine Tools and Manufacture* 47 (7–8):1214–1228. doi:10.1016/j.ijmachtools.2006.08.026.
2. Ho, K.H. and Newman, S.T. 2003. State of the art electrical discharge machining (EDM). *International Journal of Machine Tools and Manufacture* 43(13):1287–1300. doi:10.1016/S0890-6955(03)00162-7.
3. Mascaraque-Ramirez, C., and Franco, P. 2015. Experimental study of surface finish during electro-discharge machining of stainless steel. *Procedia Engineering* 132:679–685. doi:10.1016/j.proeng.2015.12.547.
4. Singh, H. 2012. Experimental study of distribution of energy during EDM process for utilization in thermal models. *International Journal of Heat and Mass Transfer* 55(19–20):5053–5064. doi:10.1016/j.ijheatmasstransfer.2012.05.004.
5. DiBitonto, D.D., Eubank, P.T., Patel, M.R., and Barrufet, M.A. 1989. Theoretical models of the electrical discharge machining process. I. A simple cathode erosion model. *Journal of Applied Physics* 66(9):4095–4103.10.1063/1.343994.
6. Patel, M.R., Barrufet, M.A., Eubank, P.T., and DiBitonto, D.D. 1989. Theoretical models of the electrical discharge machining process. II. The anode erosion model. *Journal of Applied Physics* 66(9):4104–4111. doi:10.1063/1.343995.

7. Aich, U., and Banerjee, S. 2014. Modeling of EDM responses by support vector machine regression with parameters selected by particle swarm optimization. *Applied Mathematical Modelling* 38(11/12):2800–2818. doi:10.1016/j.apm.2013.10.073.
8. Das, S., Klotz, M., and Klocke, F. 2003. EDM simulation: finite element-based calculation of deformation, microstructure and residual stresses. *Journal of Materials Processing Technology* 142(2):434–451. doi:10.1016/S0924-0136(03)00624-1.
9. Mascaraque-Ramírez, C., and Franco, P. 2015. Numerical modelling of surface quality in EDM processes. *Procedia Engineering* 132:671–678. doi:10.1016/j.proeng.2015.12.546.

chapter five

Modeling of interaction between precision machining process and machine tools

Wanqun Chen and Dehong Huo

Contents

5.1 Introduction 108
5.2 Integrated method for machine tool dynamic performance
    analysis 109
    5.2.1 Establishment of the state space model 110
    5.2.2 Theoretical basis 112
5.3 Case study—interaction between the machining process and
    the machine tool structure of fly-cutting machining 115
    5.3.1 Modeling process of the integrated method 116
        5.3.1.1 Modeling of potassium dihydrogen phosphate
                crystal 116
        5.3.1.2 Modeling of cutting force 119
        5.3.1.3 Cutting path generation 120
    5.3.2 Establishment of state space model based on finite
          element model 123
        5.3.2.1 Configuration of the fly-cutting machine tool 123
        5.3.2.2 Finite element modeling and state space
                establishment 124
        5.3.2.3 The stiffness equivalence principle based on
                the pressure distribution 127
        5.3.2.4 Finite element modeling of the air spindle 130
5.4 Finite element modeling of the machine tool 131
5.5 Simulation of interaction between the machining process
    and the machine tool structures 134
5.6 Concluding remarks 137
References 138

5.1   Introduction

In order to meet stringent tolerance and surface finish, a great deal of attention has been paid to the machining process and the machine tool (MPMT) itself [1–3]. Most of the improvement was made on the MPMT performance separately, while the interaction between them has been always overlooked. It has been recognized that the interaction between the machining process and the machine tool structures (IMPMTS) plays an essential role in the machining performance, which directly affects the material removal rate, and workpiece surface quality, as well as dimensional and form accuracy [4–7]. A better understanding of the IMPMTS becomes increasingly important, not only for engaging in ultraprecision manufacturing, but also for the precision machine tool design and machining parameters optimization. Therefore, it is essential to study the IMPMTS systematically.

The machine tool dynamic performance analysis is the key step to study IMPMTS. There are two main methods for the machine tool dynamic performance analysis; one method is the numerical method, based on lumped mass models. It is suitable for single and simple component analysis and has been applied to the design of spindles [8–10] and slides [11]. Eric et al. [12] simplified the machine tool to a single degree of freedom (DOF) system to study the waviness generation. Chen et al. [13] studied the multimode frequency vibration in fly cutting by the lumped mass models and pointed that the multi-mode vibration of the machine tool has an important influence on surface generation. Although this method is less time-consuming, the simulation accuracy is limited due to the finite dimension approximation. Another method is the finite element (FE) method, which is widely employed as an effective approach to studying statics, dynamics, and thermal aspects of single components or whole complex machine tools [14–17] due to its high computational accuracy, convenience of result interpretation, and capture of time/spatial details. Piendl et al. [18] simulated the chip formation by using the FE method. In their model, the structural–mechanical behavior of a machine and a workpiece was considered, but this approach is complex, time-consuming, and costly, which made it only suitable for the small-sized simulation. However, an FE model of the machine tool usually has more than 10,000 DOFs. This causes the transient response calculation of the machine tool under the cutting force to be very time-consuming and hence affects the computational efficiency and the design cycle of the machine tool development. For studying the IMPMT, a more flexible, simple, but accurate simulation method is urgently needed, which can realize the IMPMT quickly and accurately.

In this chapter, a novel method for machine tool dynamic response analysis under the cutting force is proposed by integrating the state space model with the FE method. The proposed integrated method has the advantages of both the lumped mass method and the FE method. In this method, a dynamic model of the machine tool in state space is established, based on the results of complex FE models representing the machine tool system, which reduces the amount of computational significantly but still provides correct responses for the forcing function input and desired output points. The dynamic response of the machine tool is used to simulate the contour profile of the machined surface, and in order to realize the precise simulation in the whole size of the workpiece, the control system is also considered to generate the cutting trajectory in the whole machining process; the cutting force generated in the manufacturing process introduces the IMPMT. Thus, the influence of IMPMT on the machined results is achieved rapidly and precisely.

5.2 Integrated method for machine tool dynamic performance analysis

By considering the interaction between process forces and machine structural dynamics, an integrated simulation method that integrates the advantages of the cutting force simulation, the machine tool dynamic performance simulation, and the motion control system is proposed in this chapter. Then, the waviness and the profile simulation of workpiece can be realized.

The application flowchart of this method is illustrated in Figure 5.1. To begin with, the cutting simulation is carried out. The cutting force, cutting heat, and residual stresses are obtained. Next, the FE model of the machine tool is established, and a simple state space model of the machine tool is obtained from the FE model. After that, the cutting force is inputted to the state space model. Then, the response of the tool tip is outputted to couple with the machine tool control system, for generating the machined surface profile.

In this method, both the material property and the cutter shape are considered, which makes the cutting force forecasting more precise. Besides, the state space model of the machine tool is obtained from the FE model as a multidegree freedom system rather than a single-degree freedom system, which can get more accurate transfer function. The cutting force extracted from the FE simulation is used as the input force in the state space model. In addition, the cutting path generated by the control system is considered to achieve the surface simulation.

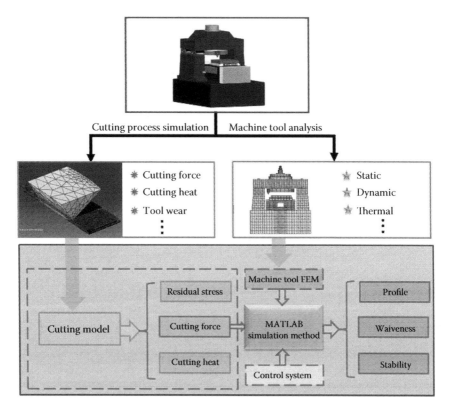

Figure 5.1 Flowchart of the integrated method.

5.2.1 Establishment of the state space model

The establishment of the state space model is the critical step in this method. As shown in Figure 5.2, it starts with the modal analysis of the FE model to obtain eigenvalues and eigenvectors (resonant frequencies and mode shapes). The block Lanczos method is used as eigenvalue extraction technique, which can calculate all the eigenvalues and eigenvectors in a specific frequency range. The sizes of the mass matrix M, the stiffness matrix K, and the damping matrix c are $n \times n$, respectively, where n denotes the DOFs of the model. There are as many eigenvalues and eigenvectors as DOFs for the model. Although they provide considerable insight into the system dynamics, the extraction of all these eigenvalues and eigenvectors would account for extremely high computational resource in the simulation, and it is typically too large to be used in a mathematical model. To reduce computational resource and to obtain the solutions in an efficient manner, only the DOFs of the nodes

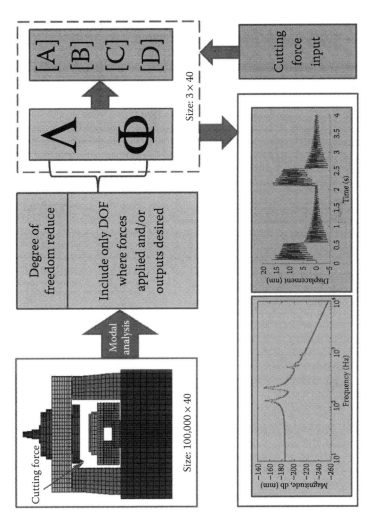

Figure 5.2 Establishment of state space model based on the FE model.

where forces are applied and outputs of interests are extracted. Then, these eigenvalue results obtained from the modal analysis are used to build the state space.

5.2.2 Theoretical basis

In the second step, the objective is to provide an efficient, *small* model representing the mechanical and servo system of the machining system. It needs to reduce the size of the model and keep the accuracy of calculation, as well as maintain the desired input–output relationship. In order to reduce the number of DOFs of the model, only those DOFs are reserved in the new model where forces are applied and responses are desired.

The theoretical basis of establishing a *small* state space model by using limited eigenvalues and eigenvector information is explained as follows (eigenvector entries for all modes only for input and output DOFs). For a given structure, an FE model can be established with mass matrix $\mathbf{M}$, damping matrix $\mathbf{C}$, and stiffness matrix $\mathbf{K}$; the size of each matrix is $n \times n$. Therefore, the fundamental equation describing the dynamic behavior of a structure discredited by FE can be written as:

$$\mathbf{M}\ddot{x} + \mathbf{C}\dot{x} + \mathbf{K}x = \mathbf{F} \tag{5.1}$$

where:

F denotes an n-dimensional vector, designating the force applied on each DOF

x denotes the displacement vector, caused by the force

In order to uncouple Equation 5.1, the damping matrix can be replaced by using the Lord Rayleigh's hypothesis:

$$\mathbf{C} = \alpha\mathbf{M} + \beta\mathbf{K} \tag{5.2}$$

where:

$\boldsymbol{\alpha}$ is the constant mass matrix multiplier for alpha damping
$\boldsymbol{\beta}$ is the constant stiffness matrix multiplier for beta damping

Thus, Equation 5.1 can be rewritten as:

$$\mathbf{M}\ddot{x} + (\alpha\mathbf{M} + \beta\mathbf{K})\dot{x} + \mathbf{K}x = \mathbf{F} \tag{5.3}$$

After the coordinate decoupling transformation, Equation 5.3 is changed as:

$$\ddot{x}_p + (\alpha I + \beta \Omega)\dot{x}_p + \Omega x_p = F_p \tag{5.4}$$

which can also be written as:

$$\ddot{x}_{pi} + 2\varepsilon_i \dot{x}_{pi} + \omega_i^2 x_{pi} = F_{pi}, \quad i = 1, 2, \ldots, n \tag{5.5}$$

where:

ω_i is the value of the ith natural frequency

Ω is the diagonal matrix, $\Omega = \mathrm{diag}(\omega_i^2)$

ε_i is the percentage of critical damping for the ith mode, $\varepsilon_i = (\alpha + \beta \omega_i^2)/2\omega_i$

The state space method can be convenient to tackle the problem of frequency domain response and time domain response. So, the dynamic equation described by state space is:

$$\begin{cases} \dot{x} = Ax + Bu \\ y = Cx + Du \end{cases} \tag{5.6}$$

Defining states:

$$\begin{cases} x_1 = x_{p1}, \\ x_2 = \dot{x}_{p1}, \\ x_3 = x_{p2}, \\ x_4 = \dot{x}_{p2}, \\ \vdots \\ x_{2n-1} = x_{pn}, \\ x_{2n} = \dot{x}_{pn}. \end{cases} \tag{5.7}$$

Differentiating Equation 5.7 gives:

$$\begin{cases} \dot{x}_1 = x_2, \\ \dot{x}_2 = -\omega_1^2 x_1 - 2\varepsilon_1 \omega_1 x_2 + F_{p1}, \\ \dot{x}_3 = x_4, \\ \dot{x}_4 = -\omega_2^2 x_3 - 2\varepsilon_2 \omega_2 x_4 + F_{p2}, \\ \vdots \\ \dot{x}_{2n-1} = x_{2n}, \\ \dot{x}_{2n} = -\omega_n^2 x_{2n-1} - 2\varepsilon_n \omega_n x_{2n} + F_{pn}. \end{cases} \tag{5.8}$$

Equation 5.8 can be written in the matrix form:

$$\begin{Bmatrix} \dot{x}_1 \\ \dot{x}_2 \\ \dot{x}_3 \\ \dot{x}_4 \\ \vdots \\ \dot{x}_{2n-1} \\ \dot{x}_{2n} \end{Bmatrix} = \begin{bmatrix} 0 & 1 & 0 & 0 & \cdots & 0 & 0 \\ -\omega_1^2 & -2\varepsilon_1\omega_1 & 0 & 0 & \cdots & 0 & 0 \\ 0 & 0 & 0 & 1 & \cdots & 0 & 0 \\ 0 & 0 & -\omega_2^2 & -2\varepsilon_2\omega_2 & \cdots & 0 & 0 \\ \vdots & \vdots & \vdots & \vdots & \vdots & \vdots & \vdots \\ 0 & 0 & 0 & 0 & \cdots & 0 & 1 \\ 0 & 0 & 0 & 0 & \cdots & -\omega_n^2 & -2\varepsilon_n\omega_n \end{bmatrix}$$

$$\times \begin{Bmatrix} x_1 \\ x_2 \\ x_3 \\ x_4 \\ \vdots \\ x_{2n-1} \\ x_{2n} \end{Bmatrix} + \begin{Bmatrix} 0 \\ F_{p1} \\ 0 \\ F_{p2} \\ \vdots \\ 0 \\ F_{pn} \end{Bmatrix} \tag{5.9}$$

and its short form is:

$$\dot{x} = Ax + Bu \tag{5.10}$$

where system matrix A is constituted by the natural frequency and the damping ratio; the ω_i and x_n are obtained from modal analysis. The input matrix B is formed by F_p.

$$\begin{Bmatrix} F_{p,1} \\ F_{p,2} \\ \vdots \\ F_{p,2n} \end{Bmatrix} = \begin{bmatrix} x_{1,1} & x_{2,1} & \cdots & x_{2n,1} \\ x_{1,2} & x_{2,2} & \cdots & x_{2n,2} \\ \vdots & \vdots & \vdots & \vdots \\ x_{1,2n} & x_{2,2n} & \cdots & x_{n,2n} \end{bmatrix} \begin{Bmatrix} F_1 \\ 0 \\ \vdots \\ 0 \end{Bmatrix} = \begin{Bmatrix} x_{1,1}F_1 \\ x_{1,2}F_2 \\ \vdots \\ x_{1,2n}F_1 \end{Bmatrix} \tag{5.11}$$

For the space state Equation 5.9, by applying F_1 on the DOFs of the nodes where forces applied, the other DOFs are set to be zero. So, when calculating F_p, only the first column is needed, as shown in Equation 5.11.

The x solved in Equation 5.9 are in principal coordinates; they should be transferred to physical coordinates, $y = x_n x$. Rewriting this in the matrix form:

$$
\begin{Bmatrix} y_1 \\ y_2 \\ y_3 \\ y_4 \\ \vdots \\ y_{2n-1} \\ y_{2n} \end{Bmatrix} = \begin{bmatrix} x_{n11} & 0 & x_{n12} & 0 & \cdots & x_{n1n} & 0 \\ 0 & x_{n11} & 0 & x_{n12} & \cdots & 0 & x_{n1n} \\ x_{n21} & 0 & x_{n22} & 0 & \cdots & x_{n2n} & 0 \\ 0 & x_{n21} & 0 & x_{n22} & \cdots & 0 & x_{n2n} \\ \vdots & \vdots & \vdots & \vdots & \vdots & \vdots & \vdots \\ x_{nn1} & 0 & x_{nn2} & 0 & \cdots & x_{nnn} & 0 \\ 0 & x_{nn1} & 0 & x_{nn2} & \cdots & 0 & x_{nnn} \end{bmatrix} \begin{Bmatrix} x_1 \\ x_2 \\ x_3 \\ x_4 \\ \vdots \\ x_{2n-1} \\ x_{2n} \end{Bmatrix}
\tag{5.12}
$$

which is:

$$
y = Cx + D \tag{5.13}
$$

In the response analysis, the DOFs where forces are applied and outputs of interests are assumed as the first m DOFs. For modal analysis, the modal vector of these m DOFs is then used to form the modal matrix $X_{m\times n}$.

For the state space Equation 5.12, only the displacement response of the first m DOF is needed.

First m DOF:

$$
\begin{Bmatrix} y_1 \\ y_2 \\ \vdots \\ y_m \end{Bmatrix} = \begin{bmatrix} x_{n11} & 0 & x_{n13} & 0 & \cdots & x_{n1n} & 0 \\ x_{n21} & 0 & x_{n22} & 0 & \cdots & x_{n2n} & 0 \\ \vdots & \vdots & \vdots & \vdots & \vdots & \vdots & \vdots \\ x_{nm1} & 0 & x_{nm2} & 0 & \cdots & x_{nmn} & 0 \end{bmatrix} \begin{Bmatrix} x_1 \\ x_2 \\ x_3 \\ x_4 \\ \vdots \\ x_{2n-1} \\ x_{2n} \end{Bmatrix}
\tag{5.14}
$$

Therefore, the size of the matrix is reduced significantly.

After this, the state space model is used for frequency and time domain analyses. This step realizes the transformation of the FE model to the state space model, which can reduce the time and cost in the following frequency and time domain analyses.

5.3   Case study—interaction between the machining process and the machine tool structure of fly-cutting machining

The proposed IMPMT model has been implemented in fly-cutting machining of potassium dihydrogen phosphate (KDP for short) crystals.

Ultraprecision fly cutting, as a single-point cutting process, is the main manufacturing method to generate optical quality surface finishes

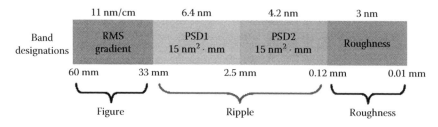

Figure 5.3 The topography requirements of the KDP crystal.

on flat workpieces [19]. An excellent example of ultraprecision fly cutting is the production of KDP crystals for the inertial confinement fusion (ICF) program [20–22]. The KDP crystals are widely used in the ICF program for frequency multiplication and Pockels cells [23–25]. The KDP crystal has extremely harsh requirements of the topography in the ICF program, not only for the root mean square (RMS) but also for the power spectral density (PSD). The specific indexes for KDP crystal are shown in Figure 5.3. It requires roughness values less than 3 nm in 0.01 ~ 0.12 mm, RMS of no more than 4.2 nm, and PSD2 better than 15 nm²mm in 0.12 ~ 2.5 mm, the RMS less than 6.4 nm and PSD1 better than 15 nm²mm in 2.5 ~ 33 mm, and gradient RMS gradient better than 11 nm/cm in the range over 33 mm. The size of this crystal is up to 410 mm; therefore, the fly-cutting method with a large fly-cutting head is adopted for machining this crystal both in the United States and in China [26,27]. However, unreasonable defects are found after testing the machined surface morphology of the KDP crystal processed by the fly-cutting method. These defects will cause the distortion and initiation of the changes of the incident angle and refractive index of the laser, inducing the phase mismatch and ultimately reducing the frequency conversion efficiency [28–30]. Therefore, studying the influence of the IMPMT on the defects of the KDP crystal optical and finding the main sources that lead to them can not only provide the theoretical basis for increasing the optical performance but also guide the KDP processing and machine structure improvement.

5.3.1   Modeling process of the integrated method

5.3.1.1   Modeling of potassium dihydrogen phosphate crystal

In order to model the KDP crystal in the cutting simulation method, its material property must be obtained. In this study, the nano-indentation experiment is used to obtain the material property of the KDP crystal. The KDP surface is machined by the fly-cutting machine tool to generate a smooth surface with roughness RMS less than 5 nm; then, the nano-indentation experiments are carried out on the machined surface by using Nano-Indenter XP. In this experiment, a Berkovich diamond tip is adopted, with the displacement resolution and force loading resolution of 0.01 nm and 50 nN, respectively. The maximal indentation depth and the impressing velocity in the experiment are set as 1000 nm and 10 nm/s, respectively. As a result, the indenting loads-displacement curve of KDP crystal is obtained as shown in Figure 5.4. It can be found that the nano-indentation is divided into two processes of loading and unloading, and there is a degree of residual indentation depth after unloading. In order

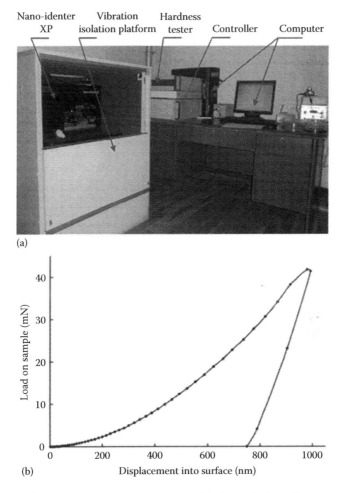

Figure 5.4 Nano-indentation experiment. (a) Nano-indentation experiment system and (b) the curve of load-displacement sampled on the KDP crystal surface. (Reproduced with permission from Chen, W. et al., *Int. J. Adv. Manuf. Technol.*, 1–8, 2016.)

Table 5.1 Measured Young's modulus E and micro-hardness H of type II KDP crystal

No	E (GPa)	H (GPa)
1	49.352	2.262
2	56.056	2.293
3	42.959	2.417
4	44.414	2.226
5	52.73	2.277
Average	49.1022	2.295

to guarantee the reliability of the experimental results, five points on the machined surface are selected to carry out the nano-indentation process, and the average value is used in the following FE model [31].

The measured results of Young's modulus and microhardness of the KDP crystal are listed in Table 5.1; the tested Young's modulus is 49.1022 GPa, and the microhardness is 2.295 GPa.

Considering that the Berkovich diamond tip is employed in the indentation experiments, a given secondary function and an Oliver–Pharr power function are used to fit the loading and unloading curves [32]:

$$\begin{cases} P = Ah^2 \\ P_u = B\left(h - h_f\right)^m \end{cases} \tag{5.15}$$

where:

    P is the applied load
    h is the indentation depth
    A is the fitting coefficient independent of the loading force
    P_u is the unloading force
    h_f is the residual indentation depth
    B and m are the fitting coefficients

Substituting for all the sampled data plotted in Figure 5.2 into Equation 5.15, the loading and unloading curves are fitted as:

$$\begin{cases} P = 4.6072 \times 10^{-5} h^2 \\ P_u = 0.02588\left(h - 754.38592\right)^{1.34882} \end{cases} \tag{5.16}$$

Following, the load–displacement curve is converted to stress–strain curve based on dimensional analysis [33,34]. The elastic–plastic behavior is assumed to be:

$$\sigma = \begin{cases} E\varepsilon & (\sigma \le \sigma_y) \\ R\varepsilon^n & (\sigma \ge \sigma_y) \end{cases} \tag{5.17}$$

$$\sigma_y = E\varepsilon_y = R\varepsilon_y^n \tag{5.18}$$

where:

σ is the engineering stress in megapascal
E is Young's modulus, and $E = 49.1022$ Gpa
ε is the engineering strain
R is the strength coefficient
n is the strain hardening exponent
σ_y is the initial yield stress

Finally, fitted Equation 5.16 is used to calculate the unknown parameters R and n in Equations 5.17 and 5.18 through the dimensional analysis [35]. The results are listed as follows:

$$\sigma = \begin{cases} 49.1022 \times 10^3 \varepsilon & (\sigma \le 198.563\,\text{MPa}) \\ 6.7371\varepsilon^{0.63955} & (\sigma \ge 198.563\,\text{MPa}) \end{cases} \tag{5.19}$$

$$\varepsilon_y = 4.04387 \times 10^{-3} \tag{5.20}$$

The relationship between the true flow stress and plastic strain should be defined in the FE simulation accurately. According to Equations 5.19 and 5.20:

$$\begin{cases} \sigma = 6.7371\varepsilon^{0.63955} & (\sigma \ge 181.724\,\text{MPa}) \\ \sigma_{\text{ture}} = \sigma(1+\varepsilon) \\ \varepsilon_p = \ln(1+\varepsilon) - \left(\dfrac{\sigma_{\text{ture}}}{E}\right) \end{cases} \tag{5.21}$$

where:

σ_{ture} is the true flow stress
ε_p is the plastic strain

5.3.1.2   Modeling of cutting force

After obtaining the material parameters of the KDP crystal, the FE simulation method is used to predict the cutting force of KDP crystal. The cutting model is established, as shown in Figure 5.5. In this work, the commercially available FE software, that is, Abaqus, is employed to simulate the diamond fly-cutting procedure. Both diamond tool and the KDP

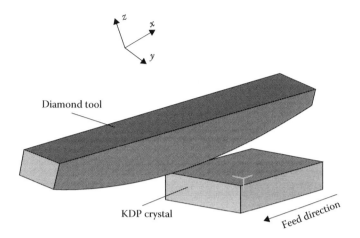

Figure 5.5 The FE cutting simulation model.

crystal are meshed with four-node tetrahedral elements. The diamond
tool rotates around the center of rotation *O* with a constant cutting velocity and has no displacement in direction *z*, as presented in Figure 5.5.
When the current cutting simulation is finished, the workpiece moves
once in direction *x* with a certain distance, which is identical to tool feed
rate, and then, the next run cutting simulation initiates. In addition, the
diamond tool is assumed to be a rigid body, whereas the KDP crystal is
regarded as a deformable body. Before simulation, the nodes on the bottom surface and the rightmost surface of KDP crystal are fixed in *y* and
z directions; that is, only one freedom in the feeding direction *x* is left.
The simulation is carried out with a diamond cutter tool, with −25° rake
angle, 8° clearance angle, and 5-mm tool nose radius under the following
processing parameters: a depth of cut of 15 μm, a feed rate of 60 μm/s, and
a spindle rotational speed of 300 r/min. The cutting force is extracted and
applied as the input force in the transfer function of machine tool system,
which is obtained from the FE model containing all relations considered
to be important to describe the mechanics phenomenon.

The cutting force over two revolutions is obtained from the cutting
simulation model, as shown in Figure 5.6. The values of cutting force are
around 0.8 and 1 in *y* and *z* directions, respectively. The cutting force in *x*
direction is nearly to zero.

5.3.1.3 Cutting path generation

Many interference factors are involved in the machining process, among
which cutting force is the most significant and unavoidable one. It has
a critical impact on the machining results. Although it is well known
that the relative vibration between the tool and the workpiece plays

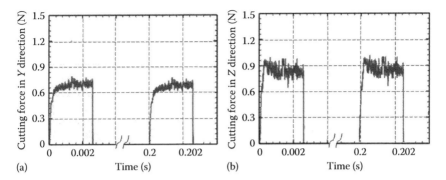

Figure 5.6 The simulated cutting force. (a) Cutting force in y direction and (b) cutting force in z direction.

an important role in the surface generation in single-point diamond fly cutting [36], most of the prior works have been focused on studying the relative tool-work vibration in a specific cutting path, while the changes of the cutting locus in the machining process is ignored, and the input force is always simplified as a constant force or a periodic one. However, in this study, to obtain the machining response in the whole surface of the workpiece, the cutting force changing with the cutting path is considered.

Figure 5.7a shows the schematic diagram of the fly-cutting machining process; the cutter rotates with a large fly-cutting head, and the KDP crystal is fed by a hydrostatic slide. Figure 5.7b shows the fly-cutting machining path on the workpiece. The variable δ is used to toggle the intermittent contact of the workpiece and the tool and is defined as unity during contact and as zero the remainder of the time. The duty ratio means the fraction of the total time per revolution in which the tool is cutting [37]. It can be found that the arc-shaped path of the rotating tool generates different radii of the workpiece with the slide feeding, so that the loading on the tool is changing along the feeding direction. Therefore, the cutting force is varied within the cutting time. There are three typical parts along the feed direction: part A′, B′, and C′, as presented in Figure 5.7b.

In part A′, the arc length that the cutter tool generates on the workpiece increases with the feeding. Therefore, the duty ratio is increasing in this part. In part B′, the arc length that the cutter tool generates on the workpiece is unchanged, and the period of cutting force is steady. In part C′, the arc length that the cutter tool generates on the workpiece decreases with the feeding. The duty ratio decreases until the cutter tool cuts out of the workpiece completely, while the tool cuts two times on the workpiece per revolution in this part.

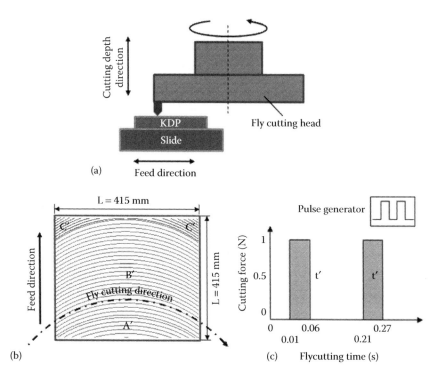

Figure 5.7 Fly-cutting machining. (a) Schematic diagram of the fly-cutting machining process, (b) the fly-cutting machining path, and (c) cutting force profile. (Reproduced with permission from Chen, W. et al., *Chin. J. Mech. Eng.*, 29(6), 1090–1095, 2016.)

According to Figure 5.7b, the arc length of tool cutting locus on the workpiece can be described as three parts, which correspond to parts A′, B′, and C′, respectively, as expressed by Equation 5.22.

$$L' = \begin{cases} 2R\arccos\left(\dfrac{R-ft}{R}\right) & , \quad t \le \sqrt{R^2 - \dfrac{L^2}{4}}\Big/f \\[3mm] 2R\arcsin\dfrac{L}{2R} & , \quad \sqrt{R^2 - \dfrac{L^2}{4}}\Big/f < t < L/f \quad (5.22) \\[3mm] 2R\left(\arcsin\dfrac{L}{2R} - \arccos\dfrac{R-(ft-L)}{R}\right) & , \quad t \ge L/f \end{cases}$$

where:
  L' notes the cutting length on the workpiece
  R notes the radius of the cutter head

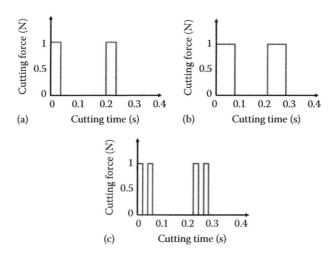

Figure 5.8 Cutting force of the three typical parts (a) A′, (b) B′, and (c) C′.

f notes the feed rate
L notes the length of the workpiece
t notes the cutting time

The amplitude of the cutting force obtained from the FE simulation is combined with the cutting path, and then, the cutting force in the whole machining path is obtained. Figure 5.8 shows the cutting force of the three typical parts, A′, B′, and C′.

5.3.2 Establishment of state space model based on finite element model

5.3.2.1 Configuration of the fly-cutting machine tool

In this study, the KDP crystals are machined by a homemade ultra-precision fly-cutting machine tool [38,39]. Figure 5.9 shows the configuration of this machine tool. It adopts a vertical architecture; a bridge supports a vertical-axis spindle and fly cutter over a horizontal-axis slide. Mounted to the horizontal slide is a vacuum chuck that fixes the workpiece by vacuum power. The main spindle is supported by externally pressurized ultraprecision cylindrical air bearings and is driven by a direct current (DC) motor, which can be rotated at 400 r/min. A large disk is fixed on the bottom of the principal axis, and the diamond tool is fixed at the edge of the big disk. The worktable is supported by a hydrostatic slide, which is driven by a linear motor with excellent slow feeding performance.

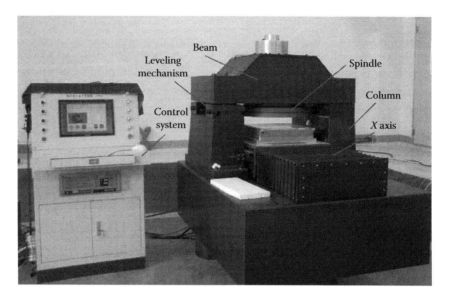

Figure 5.9 The configuration of the fly-cutting machine tool.

5.3.2.2   *Finite element modeling and state space establishment*

The FE model of the machine tool is built to establish the state space model of the machine tool. The joint characteristics of the machine tool, such as the bolt joint and the bearing connection, have great impact on the dynamic performance [16,40–41]. Therefore, the modeling approach of the junction directly determines the accuracy of the whole model of the machine tool. In order to make the FE model more accurately, both the fixed and unfixed joints are considered. A novel FE model of aerostatic bearing is proposed in [42], as shown in Figure 5.10, which links the theoretical study of the fluid film and the engineering analysis of the spindle shaft.

Based on the pressure distribution in the area of the gas film, the dynamic modeling approach proposed here is essentially different from the traditional modeling methods for aerostatic bearings. As outlined in Figure 5.11, the procedure for this approach can be described by the following key steps:

Step 1. Extract the aerostatic bearing structure parameters from the spindle structure, such as the bearing dimensions, clearance, and diameter of the orifice.

Step 2. Distribute the FEs of the aerostatic bearing, based on the structural dimensions.

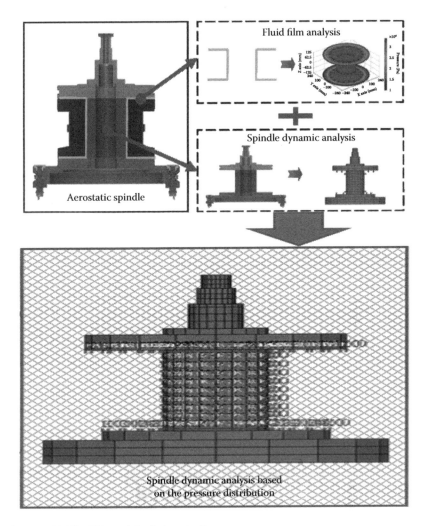

Figure 5.10 The FE model of air spindle.

Step 3. Calculate the pressure distribution in the bearing, based on the principle of flow equilibrium and the FE theory.

Step 4. Establish the equivalent spring element group, based on the pressure distribution of each element.

To accurately obtain the static and dynamic characteristics of the aerostatic bearing, it is necessary to calculate the theoretical pressure distribution of the gas film. Generally, air is used as the working medium in an

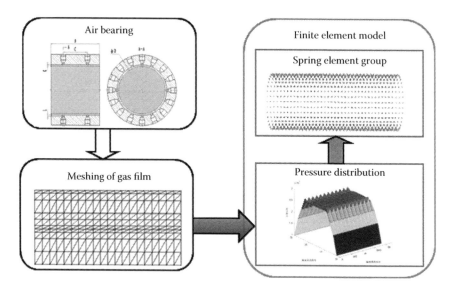

Figure 5.11 Outline of the dynamic modeling approach for the aerostatic bearing.

aerostatic spindle and can be considered a Newtonian fluid. The general form of the Reynolds equation can be written as follows [43]:

$$\frac{\partial}{\partial x}\left(h^3 p \frac{\partial p}{\partial x}\right) + \frac{\partial}{\partial y}\left(h^3 p \frac{\partial p}{\partial y}\right) = 12\frac{\partial(\rho h)}{\partial t} + 6\left[\frac{\partial}{\partial x}ph(u_1 + u_2)\right.$$
$$\left. + \frac{\partial}{\partial y}ph(v_1 + v_2)\right]$$

(5.23)

where:

u_1 and u_2 are the relative velocities of the two moving parts in the x-axis direction

v_1 and v_2 are the relative velocities of the two moving parts in the y-axis direction

h is the thickness of the gas film

To simplify the calculation, the nondimensional form of the parameters in Equation 5.23 is given as:

$$p = p_0\bar{p}, \ h = h_m\bar{h}, \ x = l\bar{x}, y = l\bar{y}, t = \frac{tl}{V}$$

(5.24)

where:

p_0 is the gas supply pressure

h_m is the thickness of gas film under the condition of flow equilibrium

V is the linear velocity of the gas film near the shaft
l is the selected reference length, such as the bearing width, perimeter, and other parameters

Equation 5.23 can then be simplified as:

$$\frac{\partial}{\partial x}\left(\bar{h}^3\frac{\partial \bar{p}^2}{\partial x}\right)+\frac{\partial}{\partial y}\left(\bar{h}^3\frac{\partial \bar{p}^2}{\partial y}\right)+\bar{Q}\delta_i = \Lambda_x\frac{\partial\left(\overline{hp}\right)}{\partial x}+\Lambda_y\frac{\partial\left(\overline{hp}\right)}{\partial y} \tag{5.25}$$

where:
δ_i is the Kronecker delta, taken as 1 at the orifice and 0 for other parts
Λ_x and Λ_y are dimensionless bearing numbers
Q is the flow rate factor for the gas mass flow into the orifice, referred to as the flow factor

For the bearings in steady-state operation, the first part on the right side of Equation 5.23 is equal to zero. Equation 5.25 can then be used to calculate the gas consumption and pressure distribution of the aerostatic bearing.

5.3.2.3 The stiffness equivalence principle based on the pressure distribution

A triangular element is generally used to divide the gas film in the bearing. The shape of the element is shown in Figure 5.12, and the nodes in the elements are marked in a counter-clockwise order. The interpolation of the function $f(z, x)$ must satisfy the following conditions: $f(z, x)$ is continuous in an element and between any adjacent elements; it can depict equal conditions of gas film pressure within an element; and the gradient of the gas film pressure is continuous in the field of every element.

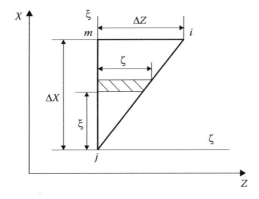

Figure 5.12 Triangular element.

The simplest interpolation function satisfying the above-mentioned constraints is:

$$p = A + Bz + Cx \tag{5.26}$$

In the case of each node, the function of pressure is written as:

$$p_i = A + Bz_i + Cx_i$$
$$p_j = A + Bz_j + Cx_j \tag{5.27}$$
$$p_m = A + Bz_m + Cx_m$$

where:
    z_i and x_i are the coordinates of node i, z_j
    x_j are the coordinates of node j
    z_m and x_m are the coordinates of node m

The coefficients A, B, and C can be obtained using Equation 5.28:

$$A = \frac{1}{2\Delta e}\left(a_i p_i + a_j p_j + a_m p_m\right)$$

$$B = \frac{1}{2\Delta e}\left(b_i p_i + b_j p_j + b_m p_m\right) \tag{5.28}$$

$$C = \frac{1}{2\Delta e}\left(c_i p_i + c_j p_j + c_m p_m\right)$$

where:
    Δe is the area of the FE
    $a_i = z_j x_m - x_m x_j$
    $b_i = x_j - x_m$
    $c_i = z_m - z_j$
    $a_j = z_m x_i - z_i z_m$
    $b_j = x_m - x_i$
    $c_j = z_i - z_m$
    $a_m = z_i x_j - z_j z_i$
    $b_m = x_i - x_j$
    $c_m = z_j - z_i$

By substituting Equation 5.28 into Equation 5.26, the square of the pressure function is obtained:

$$f = \frac{1}{2\Delta e}\left[\left(a_i + b_i z + c_i x\right)f_i + \left(a_j + b_j z + c_j x\right)f_j + \left(a_m + b_m z + c_m x\right)f_m\right] \tag{5.29}$$

Suppose:

$$N_i = \frac{1}{2\Delta e}\left(a_i + b_i z + c_i x\right)$$

$$N_j = \frac{1}{2\Delta e}\left(a_j + b_j z + c_j x\right) \tag{5.30}$$

$$N_m = \frac{1}{2\Delta e}\left(a_m + b_m z + c_m x\right)$$

and:

$$\Delta e = \frac{1}{2}\begin{vmatrix} 1 & X_i & Z_i \\ 1 & X_j & Z_j \\ 1 & X_m & Z_m \end{vmatrix} \tag{5.31}$$

$$N^{eT} = \begin{bmatrix} N_i \\ N_j \\ N_m \end{bmatrix}, p^e = \begin{bmatrix} p_i \\ p_j \\ p_m \end{bmatrix} \tag{5.32}$$

Then:

$$p = N^e p^e \tag{5.33}$$

Equations 5.5 and 5.8 are necessary as interpolation functions. They depict the relationship between the air pressure at any point within the FE and those at the element's nodes. The carrying capacity of the FE is:

$$\overline{W}_e = \int_{x_i}^{x_m}\left(\int_{z_i}^{z+\varsigma}\sqrt{A + Bz + Cx}\,dz\right)dx \tag{5.34}$$

The stiffness of the FE can be expressed as:

$$K_e = \frac{\partial \overline{W}_e}{\partial h} \tag{5.35}$$

As presented in Figure 5.13, node n belongs to elements 1, 2, 3, 4, 5, and 6, so the stiffness of the node n can be expressed as:

$$k_n = \frac{\sum_{t=1}^{6} K_t}{6} \tag{5.36}$$

Taking the adjacent elements 4 and 7, for example, the stiffness for each node of the two elements can be obtained using Equation 5.36.

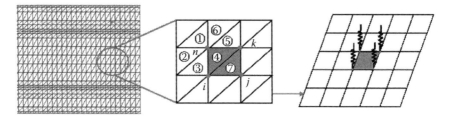

Figure 5.13 Meshing principle for the modeling method based on the pressure distribution.

The stiffness of each node is assigned to the real constant K of the corresponding spring element. It will generate an equivalent spring element group according to the pressure distribution. With this equivalence, the pressure distribution of the bearing is copied to the spindle shaft; therefore, the direct corresponding relationship (DCR) between the fluid film and the spindle dynamic performance can be established by the FE method. To analyze and substitute the pressure distribution conveniently, a general program for pressure distribution was developed. Combined with the pressure conditions, the program bridges the gap between the fluid film analysis and the spindle dynamic performance prediction.

5.3.2.4 Finite element modeling of the air spindle

A triangular element is used to separate the aerostatic bearings. Figure 5.14 shows the FE distribution of the bearings. The boundary conditions are as follows: the supply pressure in the orifices is 0.5 MPa, the atmospheric boundary condition is 0.1 MPa, and the coupling boundaries in the axial and radial bearings have the same pressure. The pressure distribution of the gas film is calculated based on the principle of flow equilibrium and the FE theory.

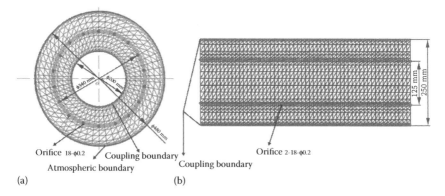

Figure 5.14 Finite element distribution of the bearing. (a) Finite element distribution of the axial bearing. (b) Finite element distribution of the radial bearing.

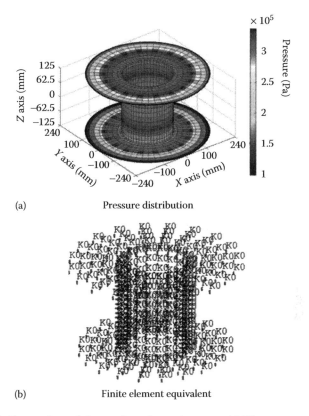

(a)   **Pressure distribution**

(b)   **Finite element equivalent**

Figure 5.15 Generation of the spring element group. (a) The pressure distributions of the gas film. (b) The spring element group.

The pressure on each element of the bearing is given in Figure 5.15a; it shows that the orifice elements have the highest pressure. The pressure of the elements decreases with increasing distance from the orifice elements, and we have atmospheric pressure at the atmospheric boundary. The spring element (Combin14) group automatically generated in accordance with Equation 5.13 is shown in Figure 5.15b. It can be seen that the pressure of each node in the gas film is equivalent, and an alternative, to the spring element.

5.4 Finite element modeling of the machine tool

The modeling method of the junction in this article is used as follows [44]: a spring element group generated by codes is introduced to represent the aerostatic and hydrostatic bearing and the axial stiffness of the linear motors, instead of the four or six elements in traditional modeling methods, to improve the accuracy of simulation. The Conta173 and targe170 elements are applied to the contact pairs of the adjustment pad and spherical pad, respectively. At the same time, the Prets179 element is used to simulate the bolt joint, which can exert the preload by node K. The finite element analysis (FEA) model of the whole machine is shown in Figure 5.16. The 3D-integrated modeling aims to accurately evaluate and optimize the dynamic performance of the overall machine. There are 17,000 elements and approximately 100,000 DOFs in this model.

Modal analyses are carried out on the FE model. The results in Figure 5.17 show that the first mode of the machine tool is the column swing from side to side (230 Hz; Figure 5.17a), the second mode is the turning of the beam (240 Hz; Figure 5.17b), and the third mode is the spindle swing (270 Hz; Figure 5.17d). Higher-order vibration shapes from the fourth to the sixth modes in Figure 5.17e, f, and h are regarded as combined shapes and are difficult to describe. It shows that the spindle and the slide have higher stiffness than the column, and the whole machine has good dynamic stiffness.

Then, a state space model is created to generate low-order models of complicated systems by defining the DOFs required for the desired frequency response, according to the results of modal analysis. In this study, the cutting force applies only at the node located at the tool tip in the x, y, and z directions, and the output is the corresponding displacement of the tool tip, caused by the cutting force. Therefore, only the DOFs of one node

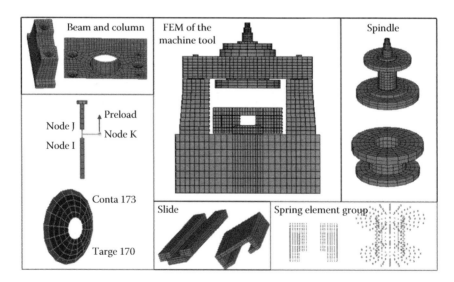

Figure 5.16 The FE model of the fly-cutting machine tool. (Reproduced with permission from Liang et al., *Int. J. Adv. Manuf. Technol.*, 70, 1915–1921, 2014.)

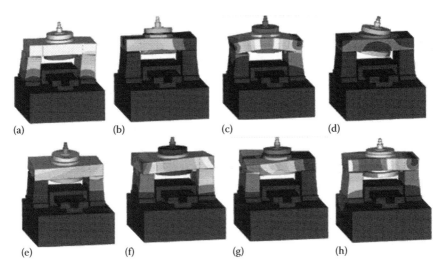

Figure 5.17 Dynamic modes of the machine tool: (a–h) 1st to 8th order modes vibration of the machine tool.

of the tool tip are required for the state space model. It makes the degree numbers required for the state space model to reduce from 100,000 to 3, which is approximately 33,000 times smaller than the original FE model.

For calculating the cutting force response by using with the MATLAB function *"lsim,"* a time vector *"t"* and input vector *"u"* are defined. The *"t"* and *"u"* are generated by the control system.

Figure 5.18 shows the results of the tool tip response calculated by the integrated method and the FE method, respectively. It can be found that

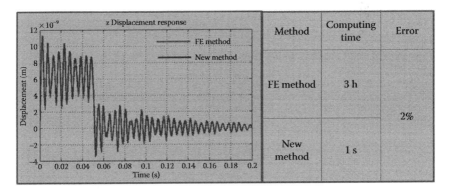

Figure 5.18 Tool tip response comparison between the FE method and the integration method. (Reproduced with permission from Liang et al., *Int. J. Adv. Manuf. Technol.*, 70, 1915–1921, 2014.)

the computational error is less than 2% between the two methods. The FE method takes 3 hours for the transient response under cutting force, whereas the integrated method takes only 1 second; all of the simulation are carried out with the same computer configuration (i5-2300@2.8GB). It shows that the integrated method improves the computational efficiency significantly.

5.5   Simulation of interaction between the machining process and the machine tool structures

Figure 5.19 shows the flow chart of the IMPMTS of the KDP crystal fly-cutting machining. The transfer function is obtained from the FE model. Then, the cutting force inputs to the transfer function of the machine tool, and the output is the tool tip vibration under the cutting force. The tool tip displacement is coupled with the cutting path generated by the motion control system, and the machined surface can be generated by this simulation model.

Figure 5.20a–c shows the typical machining process corresponding to parts A, B, and C, respectively. It presents the cutting force over two revolutions of the fly cutter and the tool tip response with the cutting

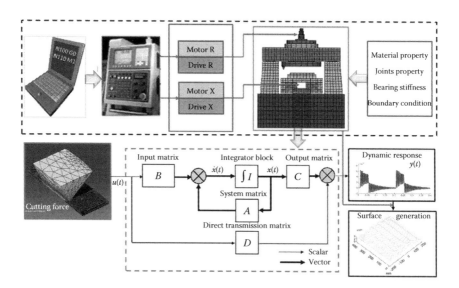

Figure 5.19 Flow chart of the IMPMTS of the KDP crystal fly-cutting machining. (Reproduced with permission from Liang et al., *Int. J. Adv. Manuf. Technol.*, 70, 1915–1921, 2014.)

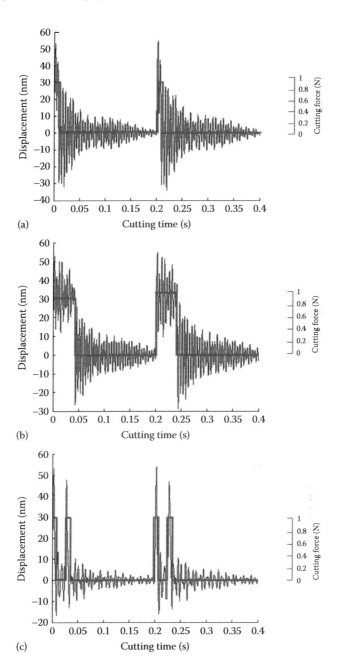

Figure 5.20 Typical cutting force response of (a) part A, (b) part B, and (c) part C.

force effect. It can be found that when the cutter comes into the cutting region, the cutter force changes from zero to 1 N suddenly. The impact effect will cause a large displacement of the tool tip first, leading to the high parts on the workpiece surface. Then, the tool tip will oscillate under the constant cutting force.

In part A', the cutting path is so short that only the impact effect occurs; therefore, the tool tip has a larger displacement in this area, leading to a high part on this part. In part B', the cutting path is long; in the whole cutting path, the tool tip has large displacement first, leading to a high part on cutting into place. It will then oscillate under the constant cutting force, forming the waviness along the cutting path on the machined surface. In part C', the cutter tool cuts the workpiece two times in one revolution. Each time, the cutting path is so short that only the impact effect occurs, leading to the high parts on these parts.

In the surface generation model, the tool tip response of each revolution is coupled with the cutting path to generate the machined surface. The surface generated by the integrated dynamic simulation model is shown in Figure 5.21. It can be noted that the waviness and the profile of the machined surface are well simulated.

The experimental results are examined by a 3D rough surface tester, Wyko RST-Plus (Veeco Metrology Group, Santa Barbara, California, United States), which has a 500-mm vertical measurement range and 3-nm vertical resolution. The measurement results with only tip, tilt, and piston removed are shown in Figure 5.22. It can be found that

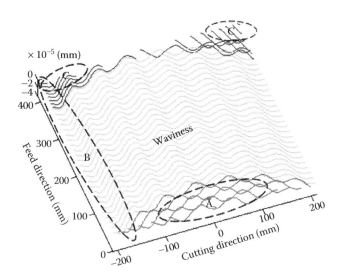

Figure 5.21 The surface generation by the proposed simulation method.

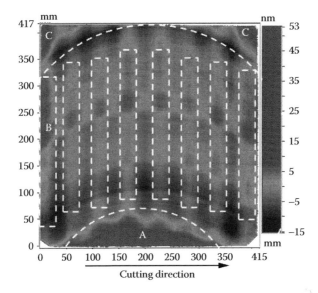

Figure 5.22 The tested result of the machined surface.

there are high parts in A, B, and C, and the waviness is also obtained in this simulation. The simulation results agree well with the experiments, which validate the theoretical models and analysis very well and provide the evidence of the approach being effective to fly-cutting machining.

5.6   Concluding remarks

This chapter presents an integrated method that can realize the modeling and simulation of the interactions between manufacturing processes and machine tool structures rapidly, without sacrificing computational accuracy. The main conclusions in this chapter are summarized as follows:

1. A simple machine tool dynamic analysis model is established by integrating the state space and FE methods. Based on this method, the dynamic response analysis of the machine tool is achieved rapidly, and hence, the computational efficiency is improved significantly.
2. The contour profile simulation method of the machined surface was proposed by integrating the whole machining path generated by the control system and the dynamic responses of the machine tool at each machining path. Using this method, one can achieve the surface simulation in short time and with high accuracy.

3. This method was used for an ultraprecision fly-cutting machine tool dynamic analysis; the result shows that it can reduce the computational time, with high accuracy. The bridge between the MPMT structure was established, which is useful for machining results prediction and machine tool design.

References

1. Chen JS, Hsu W Y. Dynamic and compliant characteristics of a Cartesian-guided tripod machine[J]. *Journal of Manufacturing Science and Engineering*, 2006, 128(2): 494–502.
2. Slocum AH. *Precision Machine Design*[M]. Englewood Cliffs, NJ: Prentice Hall, 1992.
3. Takasu S, Msduda M, Nishiguchi T. Influence of study vibration with small amplitude upon surface roughness in diamond machining[J]. *CIRP Annals-Manufacturing Technology*, 1985, 34(1): 463–467.
4. Brecher C, Esser M, Witt S. Interaction of manufacturing process and machine tool[J]. *CIRP Annals-Manufacturing Technology*, 2009, 58: 588–607.
5. Brecher C, Bäumler S, Guralnik A. Machine tool dynamics-advances in metrological investigation, modeling and simulation techniques, optimization of process stability[C]. *Proceedings of the 5th Manufacturing Engineering Society International Conference*, Zaragoza, 2013: 1–9.
6. Denkena B, Hollmann F. *Process Machine Interactions*[M]. London, UK: Springer, 2013.
7. Aurich JC, Biermann D, Blum H et al. Modelling and simulation of process: machine interaction in grinding[J]. *Production Engineering—Research and Development*, 2009, 3: 111–120.
8. Cao YZ, Altintas Y. Modeling of spindle-bearing and machine tool systems for virtual simulation of milling operations[J]. *International Journal of Machine Tools & Manufacture*, 2007, 47(9): 1342–1350.
9. Jiang SY, Zheng SF. Dynamic design of a high-speed motorized spindle-bearing system[J]. *Journal of Mechanical Design*, 2010, 132(2): 034501.
10. Cao Y, Altintas Y. A general method for the modeling of spindle bearing systems[J]. *Journal of Mechanical Design*, 2004, 126(6): 1089–1104.
11. Stefan B, Pawe G. Identication of the dynamic models of machine tool supporting systems. Part I: an algorithm of the method[J]. *Journal of Vibration and Control*, 2006, 12(3): 257–277.
12. Marsh ER, Arneson D, Doren MV et al. Interferometric measurement of workpiece flatness in ultra precision fly cutting[J]. *Sensor Review*, 2006, 26(3): 209–217.
13. Chen W, Lu L, Yang K et al. An experimental and theoretical investigation into multimode machine tool vibration with surface generation in flycutting[J]. *Proceedings of the Institution of Mechanical Engineers, Part B: Journal of Engineering Manufacture*, 2016, 230(2): 381–386.
14. Linc Y, Hung JP, Lot L. Effect of preload of linear guides on dynamic characteristics of a vertical column–spindle system[J]. *International Journal of Machine Tools & Manufacture*, 2010, 50(8): 741–746.

15. Huo D, Cheng K. A dynamics-driven approach to the design of precision machine tools for micro-manufacturing and its implementation perspectives[J]. *Proceedings of the Institution of Mechanical Engineers Proceedings of the Institute of Mechanical Engineers Part B: Journal of Engineering Manufacture*, 2008, 222(1): 1–13.

16. Lee SW, Mayor R, Ni J. Dynamic analysis of a Mesoscale machine tool[J]. *Journal of Manufacturing Science and Engineering*, 2006, 128(1): 194–203.

17. Altintas Y, Brecher C, Weck M et al. Virtual machine tool[J]. *CIRP Annals-Manufacturing Technology*, 2005, 54(2): 115–138.

18. Piendl S, Aurich J. FEM simulation of chip formation coupled with process dynamics[J]. *Journal of Manufacture and Engineering*, 2006, 220(10): 1597–1604.

19. Chaloux L. Part fixturing for diamond machining. *Proceedings of SPIE*, 1984; 508: 109–111.

20. Haynam CA, Wegner PJ, Auerbach JM et al. National Ignition Facility laser performance status. *Applied Optics*, 2007, 46: 3276–3303.

21. Burkhart SC, Bliss E, Nicola PD et al. National ignition facility system alignment. *Applied Optics*, 2011, 50: 1136–1157.

22. Yu HW, Jing F, Wei XF et al. Status of prototype of SG-III high-power solid-state laser. *Proceedings of SPIE*, 2008, 7131: 7131121–7131126.

23. Negres RA, Kucheyev SO, Mange PD et al. Decomposition of KH_2PO_4 crystals during laser-induced breakdown. *Applied Physics Letters* 2005, 86:171107.

24. Moses EI, Meier WR. The National Ignition Facility and the golden age of high energy density science. *IEEE Transactions on Plasma Science*, 2008, 36: 802–808.

25. Auerbach JM, Eimerl D, Milam D et al. Perturbation theory for electric-field amplitude and phase ripple transfer in frequency doubling and tripling. *Applied Optics*, 1997, 36: 606–612.

26. Montesanti RC, Locke SF, Thompson SL et al. Vertical-axis diamond flycutting machine for producing flat half-meter-scale optics. Optical Fabrication and Testing, OSA Technical Digest (Optical Society of America), 2000; paper OWA2.

27. Liang YC, Chen WQ, Bai QS et al. Design and dynamic optimization of an ultraprecision diamond flycutting machine tool for large KDP crystal machining. *International Journal of Advanced Manufacturing Technology*, 2013, 69:237–244.

28. Wang W, Li KY, Wang J et al. Analysis on dependence of phase matching angle on temperature in KDP crystal. *Optics & Laser Technology*, 2011, 43: 683–686.

29. Li KY, Jia HT, Wang CC et al. Theory and experiment analysis of factors affecting THG efficiency for the TIL prototype laser facility. *Optik*, 2009, 120: 1–8.

30. Auerbach JM, Wegner PJ, Couture SA et al. Modeling of frequency doubling and tripling with measured crystal spatial refractive index nonuniformities. *Applied Optics*, 2001, 40: 1404–1411.

31. Chen W, Liu H, Sun Y et al. A novel simulation method for interaction of machining process and machine tool structure[J]. *The International Journal of Advanced Manufacturing Technology*, 2016, 9: 1–8.

32. Oliver WC, Pharr GM. An improved technique for determining hardness and elastic modulus using load and displacement sensing indentation experiments. *Journal of Materials Research* 1992, 7(6):1564–1583.
33. Lin ZC, Lai WL, Lin HY et al. The study of ultraprecision machining and residual stress for NiP alloy with different cutting speeds and depth of cut. *Journal of Materials Processing Technology*, 2000, 97:200–210.
34. Lo S-P, Lin YY. An investigation of sticking behavior on the chip–tool interface using thermo-elastic–plastic finite element method. *Journal of Materials Processing Technology*, 2002, 121:285–292.
35. Dao M, Chollacoop N, Van KJ et al. Computational modeling of the forward and reverse problems in instrumented sharp indentation. *Acta Mater*, 2001, 49:3899–3918.
36. Wang H, To S, Chan CY et al. A theoretical and experimental investigation of the tool-tip vibration and its influence upon surface generation in single-point diamond turning. *Journal of Machine Tools and Manufacture*, 2010, 50:241–252.
37. Chen W, Huo D, Xie W et al. Integrated simulation method for interaction between manufacturing process and machine tool[J]. *Chinese Journal of Mechanical Engineering*, 2016, 29(6): 1090–1095.
38. Chen W, Liang Y, Sun Y et al. Design philosophy of an ultra-precision fly cutting machine tool for KDP crystal machining and its implementation on the structure design. *International Journal of Advanced Manufacturing Technology*, 2014, 70: 429–438.
39. Liang Y, Chen W, Bai Q et al. Design and dynamic optimization of an ultra-precision diamond fly cutting machine tool for large KDP crystal machining. *International Journal of Advanced Manufacturing Technology* 2013, 69: 237–244.
40. Kono D, Lorenzer T, Weikert S, Wegener K. Evaluation of modeling approaches for machine tool design. *Precision Engineering*, 2010, 34(3):399–407.
41. Yigit AS, Ulsoy AG. Dynamic stiffness evaluation for reconfigurable machine tools including weakly non-linear joint characteristics. *Proceedings of the Institution of Mechanical Engineers. Part B: Journal of Engineering Manufacturing*, 2002, 216(1):87–101.
42. Chen W, Liang Y, Sun Y et al. A novel dynamic modeling method for aerostatic spindle based on pressure distribution[J]. *Journal of Vibration and Control*, 2014, 1077546314523030.
43. Rowe WB. *Hydrostatic, Aerostatic and Hybrid Bearing Design*. Oxford, UK: Butterworth-Heinemann, 2012.
44. Liang Y, Chen W, Sun Y et al. A mechanical structure-based design method and its implementation on a fly-cutting machine tool design. *International Journal of Advanced Manufacturing Technology*, 2014, 70:1915–1921.

chapter six

Large-scale molecular dynamics simulations of nanomachining

Stefan J. Eder, Ulrike Cihak-Bayr, and Davide Bianchi

Contents

6.1 Introduction 141
6.2 Classical molecular dynamics in a nutshell 143
6.3 Atomistic simulation of nanomachining 146
    6.3.1 Preparing the model 146
    6.3.2 External constraints, boundary conditions, and
        simulation procedure 151
    6.3.3 Removing heat 152
    6.3.4 Dynamically identifying removed matter 153
    6.3.5 Determining the area of contact 156
    6.3.6 Evaluating the workpiece topography 157
6.4 System visualization 158
    6.4.1 Grain orientation 158
    6.4.2 Atomic displacement 160
    6.4.3 Temperature 162
6.5 Example: Grinding polycrystalline ferrite 162
6.6 Summary 172
Acknowledgment 173
References 173

6.1 Introduction

Machining of workpieces can massively change the microstructure close to the surface, so that the material properties there differ significantly from bulk properties. The latter are usually well known and documented in data sheets, but details about the mechanical and physical properties close to the surface usually are not documented. Especially, their reactions to machining processes or the interactions between abrasives and the substrate in a tribological contact are difficult to observe. On the other hand, these phenomena in the vicinity of the surface can and often will determine the wear resistance and friction behavior or the effectiveness of the given sets of machining process parameters.

In science as well as in technology, atomistic modeling and simulation play an important role as tools for discovery [1]. The method of molecular dynamics (MD) simulations was first applied in the 1950s [2], but its introduction to the field of tribology and nanomachining did not come until the late 1980s [3–5]. It allows the modeling of the behavior of solids and lubricants in systems where the surfaces come in such proximity that the resulting gap is of the order of a nanometer. In this case, the molecular/atomistic nature of matter can no longer be ignored, and continuum mechanics fails to correctly reproduce how the systems behave. Contrary to other simulation methods, MD simulation does not depend or rely on any semi-empirical constitutive material laws, so all reactions of the atoms within the simulated system are purely based on their physical interactions. They are not controlled by assumptions, measured values of chemical/physical material properties (such as stacking fault energies), or interactions between individual grains (such as grain boundary mobilities). Nowadays, MD is a well-established tool for studying many processes at nanoscale, be it friction and wear [6–8] or cutting, grinding, and machining [9,10]. With powerful and flexible MD codes abounding and often freely available over the internet [11–14], this basic numerical toolkit can be used by anyone with access to sufficient computing capacity. However, the art of scientific computing often lies in sensible model preparation as well as in the subsequent data analysis and interpretation.

Notable efforts of simulating scratching, cutting, or polishing atomistically are dedicated to the understanding of removal of a single nanoscale chip from a monocrystalline or amorphous flat surface [15–17] or from an isolated roughness feature [18–20] and to the study of some of the occurring crystallographic processes.

Molecular dynamics is especially interesting for tribology, as it can show physical reactions of lubricant molecules with surface atoms of the two counterbodies in relative motion. It can also simulate direct asperity–asperity interactions and their effect on the microstructure beneath. Fundamental plasticity behavior in a complex contact situation, namely a rough surface in contact with multiple abrasives, resulting in a variety of rake angles and indentation depths, may thus be observed and identified, even taking into account intergranular interactions or lattice orientations.

Machining is often blamed for causing work-hardened layers and phase transformations, especially when using blunt tools [21]. The impact of such layers on crack initiation and wear resistance is described in the literature, but their formation and an identification or a quantification of the main factors that determine the microstructural changes is neither well understood nor characterized in detail. Analytical studies have shown that for mild abrasive wear, for example, in high-end finishing processes or high-gloss polishing, and for steady-state wear regimes, the first nanometers beneath the surface change due to the tribological contact [22,23]. We therefore consider large-scale polycrystalline MD simulation as a promising candidate for studying the related phenomena. The crystal plasticity behavior of the substrate determines the wear rate, and the amount of energy is transformed into plastic deformation or other plasticity's processes due to friction at the contact. As the lateral sizes of experimentally observed surface grains are in the nanometric range, a polycrystalline MD model is able to adequately represent the uppermost workpiece layer close to the surface of a tribological contact.

The approach to atomistically treating machining processes discussed in this chapter deliberately sacrifices some of the physical detail found in other work, in order to explicitly simulate a system considerably more complex but closer to realistic nanotechnological workpieces. This added complexity includes, but is not limited to, polycrystalline workpiece microstructures, rough workpiece surface topographies, advanced abrasive geometries, random abrasive orientation and lateral distribution of multiple abrasives, realistic thermal conductivities for metallic workpieces, build-up of removed matter between abrasives, and the formation of polycrystalline chips.

6.2 Classical molecular dynamics in a nutshell

In classical MD simulations, atoms are treated as discrete particles characterized by properties such as their position, velocity, mass, and charge [24]. By integrating Newton's equations of motion for a set of these particles, the time development of the system can be followed. The forces acting on the particles are calculated as the negative gradient of their total energy, which is determined by the potentials governing the particle interactions. In this section, we will briefly introduce some basic concepts required for understanding the MD simulations in this chapter. For further information, the reader is referred to textbooks that address the subject matter in a more comprehensive manner [25–27].

The physics of an MD simulation is described by the potentials that model how the particles interact. The right choice of potentials used for a particular MD simulation is therefore of crucial importance. Potentials come in many classes; each one specialized in its own field of applications. Metals, for example, need to be treated differently than organic molecules or noble gases. To understand how such an interaction potential works,

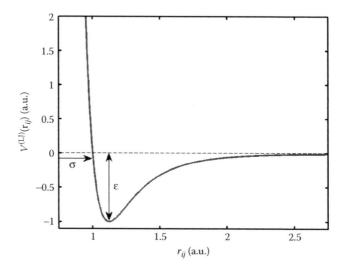

Figure 6.1 The Lennard–Jones potential for $\varepsilon = 1$ and $\sigma = 1$.

we will first discuss the Lennard–Jones (LJ) potential, shown in Figure 6.1, which is a simple but useful two-body (pairwise) potential that was introduced in the early 1930s [28]:

$$V^{(LJ)}(r_{ij}) = 4\varepsilon \left[\left(\frac{\sigma}{r_{ij}} \right)^{12} - \left(\frac{\sigma}{r_{ij}} \right)^{6} \right] \tag{6.1}$$

The term with the exponent 6 represents the van der Waals potential and models the dispersive dipole–dipole interactions between the particles. The other term may be interpreted as one that mimics the Pauli repulsion of electrons; however, the choice of the exponent 12 is not based on any physical law but mainly on numerical simplicity, since the term can then be easily calculated as the negative square of the van der Waals contribution. The energy parameter ε stands for the depth of the potential well, so higher values cause tighter binding and harder materials. The other parameter, σ, denotes the zero-crossing of the LJ potential function and is proportional to the equilibrium distance $2^{1/6}\sigma = 1.12\sigma$ between the two particles. The potential decreases toward zero rapidly. So, in order to save a considerable amount of computation time, it makes sense to introduce a cut-off radius r_c, which is usually of the order of 3σ, beyond which all interactions are neglected. The small jump in the resulting potential, which causes a spike in the gradient of the potential, and therefore in the interparticle force, can be remedied by shifting the entire function by $-V^{(LJ)}(r_c)$.

The LJ potential cannot adequately reproduce the behavior of metals, since it does not use information about the electronic structure. However, explicitly considering electrons would require a quantum mechanical treatment of the system, which is much more complex and would reduce the maximum number of atoms to ~ 1000. To overcome this difficulty, in the mid-1980s, several groups, most notably Daw and Baskes [29]; Finnis and Sinclair [30]; and Ercolessi, Parrinello, and Tosatti [31], developed potentials based on the general concept of density. As the local surrounding of an atom becomes denser, bonds become weaker; therefore, the relationship between cohesive energy and coordination should not be linear anymore, as is the case of pairwise potentials [32]. This changes the form of the attractive part of the potential, whereas the part modeling the repulsion between atomic cores may remain pairwise. In the approach by Daw and Baskes, the embedded atom method (EAM), an atom is viewed as an impurity embedded in a host of other atoms [29], and the total potential energy at the location R reads:

$$V^{(EAM)}(R) = \sum_i F_i \left(\sum_{j \neq i} \rho_j(r_{ij}) \right) + \frac{1}{2} \sum_i \sum_{j \neq i} \phi(r_{ij}) \qquad (6.2)$$

where:
   F_i is the embedding energy as a function of the sum over the electronic densities ρ_j (at the position of, but excluding the contribution of, atom i)
   ϕ is the repulsive pairwise potential

Although F_i is a multibody contribution, since it depends on all atoms within atom i's vicinity, the total energy is still a simple function of the positions of the atoms and therefore relatively straightforward to calculate. The general functional form in Equation 6.2 also applies to the Finnis–Sinclair potential and the glue model by the other groups mentioned earlier. It is the method leading to the functions F_i, ρ_j, and ϕ that vary greatly between the three approaches.

   In order to calculate the trajectories of the interacting particles in an MD simulation, their equations of motion must be integrated using a time integration algorithm. Time is discretized into finite time steps Δt, and by knowing the configuration space of the system at a given time t, the algorithm calculates the configuration space at a time $t + \Delta t$. The iteration of this scheme then yields the system's time development. There exists an abundance of time integration algorithms, an overview of which can be found in References 25 and 27. The Verlet algorithm has come to be one of the most widely used time integrators in MD, since it was first applied on a computer to calculate the phase diagram of argon in 1967 by Verlet [33]. In its velocity formulation, it explicitly produces the position R_i, the acceleration $\ddot{R}_i$, and the velocity $\dot{R}_i$ of particle i at the time $t + \Delta t$,

$$R_i(t + \Delta t) \quad = \quad R_i(t) + \dot{R}_i(t)\Delta t + \frac{1}{2}\ddot{R}_i(t)\Delta t^2$$

$$\ddot{R}_i(t + \Delta t) \quad = \quad -\frac{1}{M_i}\nabla_{R_i} V\left(R(t + \Delta t)\right) \qquad (6.3)$$

$$\dot{R}_i(t + \Delta t) \quad = \quad \dot{R}_i(t) + \frac{1}{2}\left[\ddot{R}_i(t) + \ddot{R}_i(t + \Delta t)\right]\Delta t$$

while being numerically more stable than the basic formulation.

6.3 Atomistic simulation of nanomachining

In this section, we will go through the technical details of carrying out an atomistic simulation of a nanomachining process based on [34–38]. All MD simulations discussed in this chapter are carried out with the open-source code LAMMPS [39]. We will start with the preparation of the poly-crystalline workpiece with a rough surface, followed by the counterbody consisting of hard, abrasive particles. We will then discuss how to set the boundary conditions and the kinematic constraints in a physically mean-ingful way. The matter of removing the heat introduced into the system by the machining process in a fashion consistent with the heat conductivity of a metal warrants a separate section. Finally, we will explain how sev-eral quantities of interest, some of which are, in principle, measurable via experiments, can be extracted from the large amount of data produced in such a simulation.

6.3.1 Preparing the model

The polycrystalline ferritic workpiece model used for the nanomachining simulations is built, starting out from a Voronoi construction produced, using MATLAB® [40]. For the first steps, only the physical size of the cuboid simulation box, the desired number of grains, and the lattice constant of iron are necessary. Although the final system will be two-dimensional (2D) periodic in the lateral x and y dimensions, with a fixed lower z bound-ary and a free surface at the upper z boundary, at this point, the entire system is assumed fully three-dimensional (3D) periodic for better ease of construction. To achieve this periodicity, the randomized locations of the Voronoi nodes defining the grains are replicated 26 times, so that the original system is padded in every direction (including diagonals) with

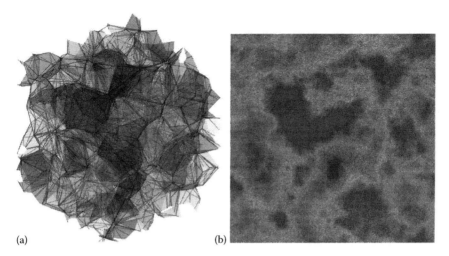

(a) (b)

Figure 6.2 (a) The initial 3D Voronoi construction that serves as the basis for the isotropic polycrystalline MD model of the workpiece. (b) Top view of the random, fractal, Gaussian surface, with topographic shading (dark = low/high, light = mid).

identical copies of itself. This $3 \times 3 \times 3$ superbox serves as the basis for the Voronoi construction. All Voronoi cells whose nodes lie outside of the original box are removed afterward, resulting in a 3D periodic representation of a polycrystalline structure (see Figure 6.2a). Note that the (purely mathematical) Voronoi construction will have features in its grain boundary structure that would be thermodynamically impossible. These artifacts will disappear during the dynamic heat treatment, carried out later in the model construction. The smaller the number of grains in the system (and consequently the larger the average grain size), the higher the likelihood of a Voronoi cell having an interface with itself across a periodic box boundary. This behavior is even more probable if the system box has an increased aspect ratio. In order to minimize the occurrence of the subsequent artifacts of artificial grain boundaries (with a grain boundary angle of zero), it makes sense to begin with a (near) cubic box shape and discard the parts of the system that are no longer required later. Each one of the resulting Voronoi cells is filled with a randomly oriented bcc Fe lattice, making sure that grains straddling a system box boundary are properly folded into the box to comply with the periodic boundary conditions. As a standard procedure, it should be checked if the misorientation angle distribution function reflects the theoretical curve calculated by MacKenzie [41] reasonably well and if the grain size distribution can be fitted to a one-parameter gamma distribution [42].

There are several ways to provide our system with a surface to be machined in the simulation. We can either cleave the polycrystal normal to the z axis, leading to an atomically flat surface, or we can remove all grains intersecting a given plane normal to the z axis, so that the resulting surface consists of intact grains, which produce a very rough surface with a non-Gaussian topographic distribution. As many naturally occurring surfaces feature fractal nature down to the nanoscale [43], we can also construct an isotropic, 2D-periodic, fractal workpiece surface topography, as was done in [37,44]. The power spectral density (PSD) of a fractal object possesses in a given frequency band a power law decay. We build the PSD by choosing a nominal fractal dimension, associated with the decay exponent, the desired root-mean-squared (RMS) roughness, and the typical lateral size of the roughness features, which determines the lower frequency cut-off of the PSD of the topographic height $z_{topo}(x, y)$ [45–47]. The 2D periodicity is enforced via the inverse Fourier transform of the Fourier spectrum associated with that PSD; see [37] for details. The required parameters cannot be set arbitrarily. The fractal dimension of natural surfaces lies within the range of $2-2.3$ [48], and the RMS roughness and the typical lateral roughness feature size have to be chosen, so that the average topographical slopes do not exceed a reasonably chosen maximum value. This limits either the number of relevant asperities that can be accommodated on the surface or the manageable roughness. Once the topographic height has been expressed as a function of the lateral system dimensions x and y, all atoms of the polycrystal constructed earlier whose z coordinate fulfills the inequality $z > z_{topo}(x, y)$ are discarded from the system, which leaves us with a crude, initial version of our workpiece; see Figure 6.2b for a topographically shaded top view. When choosing the average workpiece thickness, onto which the roughness is superimposed, it should be made sure that no grain contributing to the surface should extend to the (rigid) lower edge of the box, as this would hinder possible grain rotations and thus lead to artifacts.

As the Voronoi cells are filled with iron atoms, all the way up to their boundaries, as long as the atomic centers lie within the cell, it is likely that the atoms near the grain boundaries come into such close proximity with those of the neighboring grain that the repulsive forces would disintegrate the entire system within a few time steps of MD simulation. This issue can easily be resolved with an energy minimization requiring only a dozen or so iterations. In order to relax the grain boundaries and remove any thermodynamically unstable configurations, a dynamical MD run simulating a heat treatment is carried out. To reduce strong surface oscillations due to sudden thermal expansion, it is usually best to ramp the temperature from 0 K to the desired annealing temperature, which should be chosen with the melting temperature of the workpiece material and thermally activated recrystallization processes as well as possible phase transitions in mind. The thermostat controlling the temperature ramp should act on all atoms of the system to prevent steep temperature gradients and to ensure that the entire system is at the desired temperature at the end of the ramp. The annealing temperature is then held constant for a duration depending on the available computation power, but at least several hundred picoseconds (ps). During this period, it is advantageous to keep only the workpiece base coupled to the thermostat, so that the main part of the system can evolve without thermostat interference. Finally, the system temperature is ramped down to the desired simulation temperature and kept there for another several hundred ps, with the thermostat configured equivalently to before. At the end of the heat treatment, all thermodynamically impossible artifacts in the grain structure introduced through the Voronoi construction and the the surface introduced at an arbitrary position in the polycrystal without consideration of the local grain boundaries should have disappeared.

We now turn to designing the counterbody (or tool), consisting of numerous hard, abrasive particles that will machine the workpiece surface. While it would be possible to build abrasives that correctly represent the crystal structure and the hardness of, for example, diamond or SiC, we will consider our abrasives rigid and consisting of a simple bcc lattice, so that their defining properties are their size, shape, and orientation. We first construct their general geometry by cleaving a cube-shaped crystal along the six {1 0 0} and the eight {1 1 1} families of crystallographic planes in a way that they resemble typical abrasive nanoparticles [49,50]. The large, plate-shaped, abrasive particle shown in Figure 6.3 will be used in the application example later, surrounded by other geometries that can be produced by cleaving along crystallographic planes.

Particles of a given geometry are produced in several sizes, so that they can be mixed according to a Gaussian size-distribution extrapolated from the experimental data [51,52]; see Figure 6.4a. Now that it is known how many particles of all sizes are required, these are randomly distributed over the lateral simulation box dimensions. Care has to be taken that particles cannot overlap, especially close to the periodic box boundaries. The abrasives are rotated to produce a set of random orientations, and then, all atoms that come to lie outside of the simulation box are periodically mapped back inside. Figure 6.4b shows the result of the randomized lateral placement and orientation of 60 plate-shaped abrasives superimposed on the workpiece topography produced earlier. The generated set of abrasives can now be placed several angstroms above the heat-treated workpiece to produce the initial state of the nanomachining simulation model; see Figure 6.5.

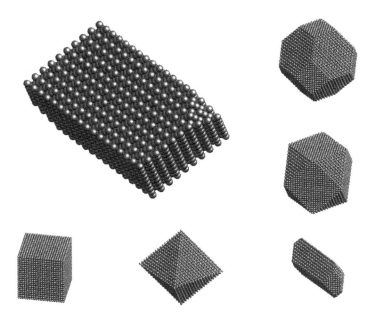

Figure 6.3 Six examples of abrasive particle geometries obtained by cleaving bcc crystals along {1 0 0} and {1 1 1} planes. The large particle in the top left is the plate-shaped type used in the examples throughout this chapter. The other types (counterclockwise from left) are cubic, octahedral, rod-shaped, cubo-octahedral, and truncated octahedral.

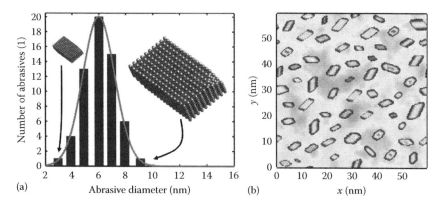

Figure 6.4 Gaussian size distribution (a) and random lateral placement and orientation (b) of 60 plate-shaped, abrasive particles.

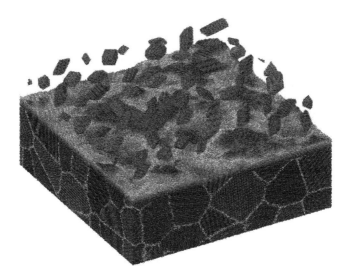

Figure 6.5 Fully assembled system consisting of a rough, polycrystalline work-piece about to be machined by 60 plate-shaped, hard, abrasive particles. Shading is according to a grayscale version of the hybrid scheme proposed in Eder et al. (Data from Eder et al., *Computer Physics Communications*, 212, 100–112, 2017. With permission.), where the surface has topographic (dark = low/high, light = mid) and the bulk crystallographic (dark = grains and white = grain boundaries) shading. The abrasives are shown in mid-gray.

6.3.2 External constraints, boundary conditions, and simulation procedure

In this section, we will explain how the model shown in Figure 6.5 will be handled numerically during the nanomachining simulations of 5 ns with a time step of 2 fs. A layer of 3Å thickness located at the lower base of the workpiece is kept rigid to act as a fixed anchor. As mentioned earlier, grains that contribute to this rigid layer should not come in direct contact with the abrasives, as their ability to rotate is impeded. The interactions between the iron atoms are governed by a state-of-the-art Finnis–Sinclair potential [53]. The workpiece and the abrasive particles interact via an LJ potential, with parameters similar to [16,54] ($\varepsilon_{LJ} = 0.125$ eV, $\sigma_{LJ} = 0.2203$ nm, and cut-off radius $r_c = 1$ nm), leading to slight adhesion between the two. The abrasives themselves do not wear.

When simulating two-body contact sliding, which would correspond to nanogrinding or fixed-grain nanolapping, the relative positions of the abrasives do not change during the simulations; that is, all abrasives are treated as one counterbody. They are pushed down onto the workpiece at a constant normal force in $-z$ direction that, divided by the cross-section of the simulation box, yields the normal pressure σ_z. The particles are moved laterally across the workpiece at a machining speed of 80 m/s and a slight angle (several degrees) with the x-axis. This angle ensures that each abrasive leaving the periodic simulation box to the right re-enters it at a different y coordinate at every pass, so that the particles do not scratch in their own grooves. During the simulations, the shear stress σ_x is calculated by summing up all the forces in x direction that are exerted on the lower rigid layer of atoms by all other atoms and then dividing by the lateral cross-section of the simulation box.

If desired, three-body contact sliding conditions may also be imposed without the necessity of an explicit counterbody. These conditions would correspond to nanopolishing or loose-grain nanolapping and require additional degrees of freedom. Here, the abrasive particles can rotate freely, and the relative positions of the abrasives are no longer fixed, which have several implications. As the particles can now engage in direct contact, the respective particle–particle interactions must be set. These might be assumed purely repulsive, which can be accomplished by setting $r_c = 2^{1/6}\sigma_{LJ}$ and shifting the potential by $+\varepsilon_{LJ}$ to ensure a continuous transition of the potential at r_c. However, care should be taken that the repulsion is not too hard (ε_{LJ} too high and/or σ_{LJ} too small), since this can quickly destroy the numerical stability of the simulations. The possibility of direct particle–particle contact may lead to interlocking and clustering of abrasives, with subsequent two-body contact sliding of the respective clusters. Another consequence of the additional degrees of freedom is that the y component of the particle movement is no longer explicitly controlled, which prohibits the velocity vector from being at a slight angle with the x axis, like in two-body contact sliding. This means that it is much more likely that the abrasives progressively dig deeper and deeper in their own grooves, which may be considered an artifact of small system size, but could also be a feature reflecting agglomeration of abrasive particles or wear particles, which is known to occur in real systems.

6.3.3   Removing heat

The friction and deformation processes occurring in the interface between the workpiece and the abrasives can produce large amounts of heat, especially at the high relative velocities typical of MD simulations. This heat needs to be removed from the system in a physically meaningful way. At the lowest level, this implies that the thermostat to which the system is coupled should dissipate the energy locally introduced into the system according to a thermodynamic ensemble. In our simulations, we employ a Langevin thermostat [55,56], acting only in y direction, that is, the direction orthogonal to sliding and normal pressure. In order not to overly interfere with the processes occurring in and near the interface, it would make sense to place the thermostatted region as far away from the interface as possible, for example, right above the rigid base of the workpiece. However, classical MD simulations do not feature any explicit information about the electrons, and the Finnis–Sinclair interaction potential governing how the iron behaves can only reproduce the phononic contribution to metallic heat conductivity. If we would thermostat only the workpiece base, we would be neglecting the dominating electronic contribution to heat conductivity, leading to huge temperature gradients within the workpiece.

There exist only few methods to correctly handle heat conduction in metals. A quasi-static two-temperature method has been applied in the modeling of laser annealing of voids [57,58], whereas another one uses dynamic coarse graining [59], effectively implementing a multiscale approach. An *ad hoc* technique featuring a coupling scheme to continuum has been used to examine frictional heating during sliding by solving the heat equation and imposing a thermal conductivity [60].

We have adopted an intelligent thermostatting approach put forward by [61,62], which assumes that the electrons of the metal can be seen as an implicit heat sink permeating the solid, so that coupling the thermostat to the entire workpiece is physically justified. What remains to be found, though, to put this approach into practice, is the time constant for the electron–phonon coupling in the modeled material. An estimation can be made via the Sommerfeld theory of metals [62], leading to a coupling time of approximately 0.5 ps for iron. However, such a strong coupling of the workpiece atoms to the Langevin thermostat effectively turns off heat conduction altogether, as all energy introduced into the system is removed almost instantaneously. In [38], we obtained the coupling time for the thermostat that best reproduces the macroscopic thermal conductivity of iron [63] in our particular system. This was done by equating the heat rate density from Fourier's law with the product of average shear stress and sliding velocity and comparing that value with the product of the experimental heat conductivity and the equilibrium temperature gradient in the workpiece for several coupling times. The best match was produced for an electron–phonon coupling time of 3.5 ps, which is in good agreement with other work featuring iron system components [58,64]. The electron–phonon coupling approach outlined previously also has the added benefit that the chips detached from the surface during the nanomachining process are allowed to cool off, which would otherwise have to be accomplished by, for example, including an explicit cooling fluid.

6.3.4   Dynamically identifying removed matter

All atoms that are abraded from the workpiece by the abrasive particles remain stuck to them in some way or another until the end of the simulation. As no atom is ever completely removed from the system, the question arises how to quantify the matter that has been removed from the workpiece. One relatively simple indicator for differentiating between workpiece and the abraded chips is the atomic advection velocity [34], that is, the nonthermal velocity component of each atom. Based on this quantity, each atom falls into one of three categories. All atoms moving at more than 90% of the imposed sliding velocity $v^{(abr)}$ can be safely considered removed matter stuck to an abrasive particle, and all atoms moving slower than 10% of the sliding velocity are considered stationary and therefore constitute the workpiece; the remaining atoms have velocities in between and are located within the shear zone. The simplified sketch in Figure 6.6 gives an overview of the discussed atomic categories. Although this approach may seem somewhat arbitrary, it is highly effective and can be justified from a crystallographic point of view. By calculating time-averaged radial distribution functions of iron for the removed matter, the workpiece, and the shear zone, it can be shown that the lattices of the first two categories reflect the thermalized bcc structure very well, whereas the latter features a strongly disturbed lattice structure due to the occurring shear [34].

For large systems with several million atoms, calculating each atom's advection velocity at every time step in postprocessing is not feasible, but fortunately, the necessary filtering procedures can be implemented directly into LAMMPS, so that the abraded chips can be identified on the fly [37]. By averaging the x component of the momentary atomic velocity

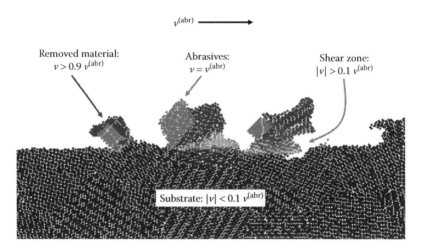

Figure 6.6 How to determine which atoms are currently considered removed material (dark, attached to abrasives), substrate (dark, at bottom), or within the shear zone (light, in between), depending on the atomic advection velocity v. The abrasives move at a constant speed of $v^{(abr)}$.

over several thousand consecutive time steps, the thermal vibrations are sufficiently suppressed, so that only the advective component remains. Logical vectors with one entry per atom can then be defined to count the number of atoms for which the conditions discussed previously apply. As soon as the number of atoms contributing to the removed matter is known for a given time step, the corresponding wear volume can be easily calculated by multiplying with a constant per-atom volume (for the case of Fe, it is calculated simply by the lattice constant at a reasonable average temperature cubed, divided by the number of atoms per bcc cell, $(a_{Fe}(\langle T \rangle))^3/2 \approx 11.6\text{Å}^3$). As the workpiece is laterally periodic, it may be more meaningful to specify the average wear depth h_w, which is the wear volume divided by the (constant) lateral cross-section of the simulation box A_{nom}.

In simulations of multibody abrasion, it may be of interest to know how much matter is removed by the individual abrasive particles, for example, for tailoring an optimized particle size distribution or selecting beneficial abrasive orientations based on this knowledge. This abrasive/chip affiliation usually depends on the abrasive particle size, geometry, orientation, relative position, and initial point of contact with the workpiece [36]. While it may be a rather simple task for a human to identify which chips of removed matter were caused by which abrasive particles, it is surprisingly difficult for a clustering algorithm that has no additional information about the data it is handling. We have therefore resorted to a partly knowledge-based iterative approach to affiliate the chips of removed matter with the respective abrasives. Since the abrasives themselves do not wear, the problem can be substantially simplified by once searching for all surface atoms of the abrasives and noting their atomic indices, thus greatly reducing the computational effort of handling the counterbody. We then iterate over all atoms previously identified as removed matter and check which ones lie closer than a distance criterion that includes up to the third-next neighbors in the radial distribution function. Larger distance criteria certainly lead to faster convergence but have the downside of falsely affiliating atoms as soon as the chips of removed matter come closer to each other and nearby other abrasive particles. Once an atom fulfills the distance criterion for a given abrasive, it is affiliated with it, so that at the next iteration, it will be considered a part of the counterbody. This algorithm leads to an iterative growth of affiliation indexing. Since machining takes place predominantly in $+x$ direction, it is much more likely that removed chips lie to the right of their causing abrasives. This knowledge about the process may be incorporated into the clustering procedure by searching only for unaffiliated chip atoms whose x coordinates are greater than those of the counterbody. This directional search hinders artificial *backwards affiliation* to preceding abrasives that have come in contact with a chip. Furthermore, care must be taken that the

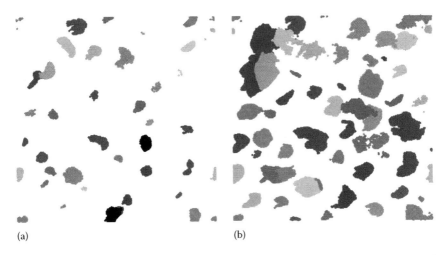

(a)   (b)

Figure 6.7 Affiliating the chips of removed matter with the abrasives that caused them at normal pressures of 0.1 GPa (a) and 0.4 GPa (b) using a partly knowledge-based clustering algorithm. Different shades represent different abrasives.

periodic boundaries are properly handled, that is, the (up to four) parts of a chip straddling one or both boundaries are not falsely identified as multiple chips and therefore possibly also falsely affiliated; see the dark black particle in Figure 6.7a. At higher normal pressures or for certain abrasive particle configurations, chips of removed matter may start coalescing. At this point, even a human interpreter of the data may have trouble deciding which atoms belong to which abrasive; see Figure 6.7b. To a degree, this might be resolved by no longer treating the time steps separately but rather by retaining the affiliation from the preceding time step. However, this would come at considerably higher computational cost, as postprocessing may no longer be parallelized.

6.3.5   Determining the area of contact

As the contact area between the abrasives and the workpiece can be correlated with both the friction occurring due to the machining process and the volume of removed material, its exact knowledge allows one to optimize machining and wear processes with respect to energy efficiency and material removal rate. Several groups have given a great deal of thought to the atomistic contact area between two bodies and attempted to calculate it in a physically meaningful manner [65–72], as continuum methods fail at this length scale [73]. Yet, the issue of what constitutes the physical, mechanical, or chemical contact between counterbodies is still disputed. For simplicity, we adopt an approach based on the number of atoms in contact, determined by a contact distance criterion that includes nearest and second-nearest neighbors of the bcc ion lattice [34,74]. We consider the counterbody as consisting of the abrasives plus the chips of removed matter and then identify all atoms of the substrate and the shear zone that fulfill the contact distance criterion. We then assume a constant per-atom area of contact to convert the number of atoms in contact with an areal expression. If chip affiliation clustering has already been carried out, as described earlier, one can simultaneously determine which part of the contact area is caused by which abrasive.

6.3.6   Evaluating the workpiece topography

The basis for determining the workpiece topography was laid by the time-resolved identification of the substrate atoms in Section 6.3.4. While the evaluation of the substrate surface could, in principle, be done on the fly, the necessary periodic redefinition of the static group of substrate atoms within LAMMPS is awkward and slows down the simulations considerably. We therefore consider the postprocessing approach from [37], outlined later, which is computationally more efficient. The substrate atoms constituting the surface must be identified and mapped to a regular mesh, which then allows surface visualizations, quantitative operations (difference images, etc.), and the calculation of surface texture parameters.

Our approach to identifying the surface atoms is aided by the evaluation of the time-averaged centrosymmetry (CS) parameter [75] for each atom. While this quantity may not be the first choice to distinguish between point defects and stacking faults in systems with temperatures beyond 500 K, it can safely distinguish between bulk and surface atoms. However, this only works where the substrate actually has a free surface but not where it is in direct contact with an abrasive and/or a chip of removed matter. We therefore produce two subsets of all atoms of a given time step. All atoms that have been identified as substrate atoms and have a CS parameter greater than 18 can be safely considered surface atoms in a bcc lattice [37]. This subset consists of several thousand atoms. For the case where a surface atom cannot be automatically identified, we produce a reduced subset of substrate atoms that lie closer to the surface than the lowest point of the shear zone, which consists of up to several million atoms.

We now construct a regular mesh covering the lateral extent of the system with a resolution coarser than the lattice constant of Fe to reduce oversampling effects but fine enough to provide several thousand data points for smooth surface histograms. The meshed surface is then determined by scanning over all mesh elements and searching in each one for the atom from the smaller subset with the maximum topographic value or, if the latter is unavailable, from the much larger subset, noting the respective $z(x,y;t)$ value for each element. From this height distribution, we can then calculate the arithmetic mean height, the RMS roughness, the skewness [35], and other relevant texture quantities derived from Abbot–Firestone bearing area curves such as the core, peak, and valley depth parameters [76].

6.4   System visualization

In addition to some of the global quantities discussed earlier, which may be studied as functions of time and load, some quantities related to the microstructural development of the workpiece require space-resolved analysis and appropriate visualization. In this respect, it is beneficial to produce images in a style that experimentalists are accustomed to, that is, similar to electron microscopy and the related material imaging techniques. Computer tomographs are highly informative visualizations of sections through the workpiece that can be colored according to various properties described in the following subsections.

6.4.1   Grain orientation

For a meaningful analysis of the microstructural development of the poly-crystalline workpiece, it is critical to uniquely identify the orientation of every single grain during a machining simulation. The basic approach is to emulate the visualization style of electron backscatter diffraction (EBSD) used in conjunction with scanning electron microscopy for the estimation of the grain orientations. In EBSD, some of the accelerated electrons entering the sample may scatter back and be diffracted according to the Bragg condition. The regular lattice planes of the sample will then produce the so-called Kikuchi patterns, which are parts of the two diffraction cones per lattice planes with given Miller indices. Based on these Kikuchi line distances and angles, the lattice structure and crystal orientation can be deduced. The latter can serve as a basis for coloring the individual scanning points, so clusters of points with identical colors constitute single grains.

The MD simulation provides the position of all the atoms; hence, we need a suitable method for determining the orientation. We use a polyhedral template matching (PTM) [77] algorithm implemented in the open visualization tool (OVITO) [78] for determining the grain orientation. The OVITO is an open-source visualization and analysis software for atomistic simulation data. It has served in a growing number of computational simulation studies as a useful tool to analyze, understand, and illustrate simulation results. The PTM classifies structures according to the topology of the local atomic environment, without any ambiguity in the classification, and with greater reliability than, efor example, common neighbor analysis in the presence of thermal fluctuations. The PTM does not rely exclusively on the closest neighbors for determining the crystal-line structure, but also on the Voronoi cell generated by these atoms. Each crystalline pattern, that is, sc, bcc, fcc, and so on, produces a quasi-unique Voronoi cell. The PTM selects only those near neighbors whose Voronoi cell overlaps a given template. As the matching condition also includes the rotation of the template, the orientation of the lattice surrounding a given atom is an output of the PTM.

A time-dependent analysis of systems consisting of several millions of atoms is inconvenient by using the graphical user interface of OVITO, where one interactively calculates the orientations of the grains at each time frame. To perform this operation efficiently, it is beneficial to use the Python scripting mode of OVITO, making use of the NumPy library [79]. This approach allows automation and the possibility to distribute the ori-entation analysis over a cluster by performing the calculations for indi-vidual time steps in parallel.

Orientations are provided as misorientation with the sample frame and are defined as the transformation necessary for rotating an object from frame A to frame B. The minimum number of rotations necessary for such a transformation is three. Because of crystal symmetry, the num-ber of possible combinations to reorient the crystal axes are multiple; for example, in Euler space, a cubic lattice has 24 representations of the same misorientation, which will make the rotated object indistinguishable. For this reason, the orientations are commonly given, so that they lie in the fundamental zone, which is defined as the minimum amount of orienta-tion space required to describe all the orientations [80]. In practice, it is the smallest set of rotations necessary to move from frame A to frame B. In the fundamental zone, each orientation can be described as one unique point, commonly known as disorientation. We chose the standard stereo-graphic triangle (SST) for representing the grain disorientation. For cubic structures, the SST is the area inside the stereographic projection of the (001), (111), and (011) axes on the orientation sphere. An orientation point in the SST represents the inclination of the (001) axis of the cubic lattice with respect to the delimiting axes.

Orientation can be calculated using several methods: as sequence of rotations of the object around an axis of a reference frame, for exam-ple, Euler angles, or as rotation of an object around an axis, for example, Rodrigues vectors and quaternions. Euler angles have two main advan-tages: they require minimum storage information, as only the minimum amounts of data, that is, three values, have to be saved, and direct coloring is possible, as each triad of angles can be associated with an RGB color. The major drawback is that the axes of rotation are usually codependent, which can generate the so-called Gimbal lock, where two axes of rotation degenerate into one, thus becoming indistinguishable and creating color-ing artifacts. Both Rodrigues vectors and quaternions express the same

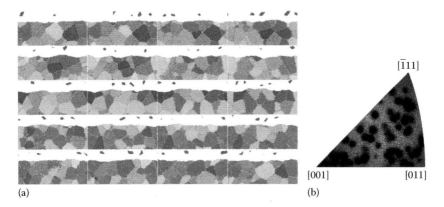

Figure 6.8 Substrate tomographs with EBSD-IPF grain orientation shading of the initial system configuration (a). Abrasives are mid-gray. In the IPF triangle legend in (b), the individual grain orientations within the workpiece are superimposed as black clusters.

type of rotation mechanism. As Rodrigues vectors store the minimum information of three values, some of the symmetry operations can cause infinities. By contrast, quaternions are four-dimensional (4D) vectors that, despite providing redundant information, prevent the disadvantages of both other methods. The PTM employed here uses quaternions for the calculation of the orientations of the crystal lattice.

Once the grain orientations are available for each atom in the workpiece, they are processed using MTEX [81,82], a free MATLAB toolbox for analyzing and modeling crystallographic textures by means of EBSD or pole figure data. Figure 6.8a shows 20 substrate tomographs of the workpiece as an example. We use the inverse pole figure (IPF) coloring scheme in the fundamental zone of the associated symmetry group. The coloring scheme is the stereographic projection of the cylindrical hue–saturation–value (HSV) scheme in the fundamental zone; see Figure 6.8b. As is common in this scheme, the hue (H) represents the angle between the (001) axis of crystals oriented parallel to the z axis of the sample substrate and the (001) axis of other crystals in the substrate, the saturation (S) is the length of such a projection, and the value (V) is chosen to be 1, so that the colors are always bright and easy to read.

6.4.2   Atomic displacement

When searching for phenomena such as grain rotation, it is practical to visualize the displacement of the individual workpiece atoms. A stable approach uses the information from three subsequent time steps by calculating the displacement vector of a given atom between the first step and the third step and placing that vector at the position of the atom at the step in between. Of course, it would be possible to display the resulting vectors with their actual lengths, but as the advective displacements close to the machining interface are some orders of magnitude larger than the purely thermal displacements further into the workpiece, these images would not be readable. It is therefore beneficial to rescale all the vectors to a length that is the same for all atoms and appropriate for visualization purposes and to color them according to their original lengths. As the most interesting displacements for grain rotations in the workpiece correspond to velocities of the order of several meters per second (compared with the machining velocity of 80 m/s), it makes sense to define a maximum velocity of, say, 8 m/s, so that the velocity differences ranging from 0 to 8 m/s are well resolved. Any atom with an (advection) velocity beyond that value would, according to our definition from Section 6.3.4, not be considered part of the workpiece. As the tomographic slices are all normal to the y axis of the simulation box, it only makes sense to show the x and z components of each displacement vector, which are sufficient for identifying sliding planes or vortices emerging within grains; see the example in Figure 6.9. For a better overview, the vectorial images are superimposed with the atoms that have a CS parameter [75] greater than 6, shown in black, so that the grain boundaries and surfaces are marked, which allow a comparison with the EBSD-IPF visualizations described earlier.

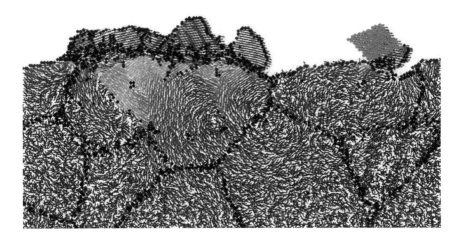

Figure 6.9 Exemplary atomic displacement tomograph with normalized vector lengths. The shading corresponds to atomic drift velocities ranging from 0 m/s to 8 m/s to resolve the *slow* displacements within the workpiece (lightest shading = 4 m/s). Removed matter and shear zone have saturated to dark shading. Abrasives are mid-gray.

6.4.3 Temperature

If one intends to interpret the microstructural changes in the workpiece occurring as a consequence of the machining process, thermal visualization and analysis are necessary to put the observations into perspective. In principle, the workpiece temperature is a simple function of the momentary atomic velocities, which are readily available at each time step. However, as machining speeds may reach values of several tens of meters per second, where they cannot be neglected, compared with the thermal velocities of the atoms, care must be taken that these advective components do not contribute to the temperature. Furthermore, due to the possibly high thermal gradients occurring in the machining interface, it is desirable to obtain a space-resolved thermal analysis of the workpiece cross-sections. Temperature cannot be defined for a single atom, but only for an ensemble of atoms, so the spatial resolution of such a thermal analysis is limited to clusters of atoms large enough to allow a meaningful definition of temperature. With these two aspects in mind, a simple but effective approach to visualizing the temperature distribution is to define a control volume around each atom of the system and calculate its average velocity. If this volume was chosen large enough, say, a sphere with a radius of 1 nm, the thermal velocity fluctuations within the control volume will average out, so only the average advection velocity $\langle v \rangle$ of the atoms in the cluster remains. This advective component can be subtracted from every atom in the volume, so that a corrected *temperature* T_j of the jth atom

$$T_j = \frac{m}{3Nk_B} \left(\sum_i^N v_i - \langle v \rangle \right)^2 \tag{6.4}$$

can be defined. Here, N is the number of atoms in the control volume, m is the mass of an atom, and k_B is the Boltzmann constant.

6.5 Example: Grinding polycrystalline ferrite

In this section, we give an example as to how the simulation, analysis, and visualization approach discussed previously can be put into practice. Two parameters that are often changed in machining—whether it is grinding, polishing, or lapping—are the normal pressure and the machining velocity. As machining velocities of the order of several meters per second and below require considerable amounts of computational power, this discussion will focus on the variation of the normal pressure at constant velocity. Based on practical experience, higher normal loads are assumed to cause higher indentation depths of the abrasives and more wear, likely at the expense of surface smoothing. The relationship between the latter and

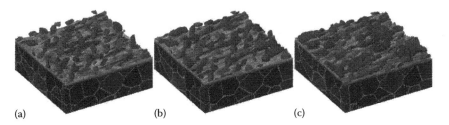

(a)   (b)   (c)

Figure 6.10 After 1 ns of nanomachining: (a) $\sigma_z = 0.1$ GPa, (b) $\sigma_z = 0.4$ GPa, and (c) $\sigma_z = 0.7$ GPa. Shading scheme identical to Figure 6.5.

the normal pressure is usually not quantified and often compensated by long grinding times. The resulting effects on the microstructural evolution of the surface layer are either neglected or not well understood.

In the present machining example illustrated in Figures 6.10 and 6.11, a variation of the normal pressure with highly nonglobular abrasives was carried out, which evidently cut deeper into the substrate than the more globular abrasives employed in the previous studies [37,38] and lead to more pronounced grooves. Figure 6.11 shows cross-sections through the substrate ground, with three different normal pressures acting on the abrasives. The images reveal that the average substrate height is markedly reduced for loads increasing from 0.1 GPa to 0.4 GPa and 0.7 GPa. So, the increase of normal load results in higher wear of the substrate, which is also obvious by the large chip-like wear particles for the highest load. Figure 6.12(a, c) depicts the evolution of the mean wear depth h_w and the arithmetic mean height z_{subst} as a function of grinding time. It is evident that hardly any wear or workpiece height reduction occurs up to 0.3 GPa. For these small normal pressures, the wear particle volume, reflected by h_w, increases in the first nanoseconds and stays relatively constant for the rest of the simulation. In addition, the 0.4 GPa and 0.5 GPa variants show nearly constant h_w toward the end of the grinding process but are accompanied by a steady decrease in substrate height z_{subst}. For high loads, the wear particle volume becomes large, and the trend is rather unstable as parts of the wear particles do recrystallize back onto the substrate.

The atoms that are neither classified as wear particles nor as substrate constitute the shear zone, quantified as the mean shear zone thickness $h_{shear} = V_{shear} / A_{nom}$. Its evolution of time is given in Figure 6.12b, where the variants up to 0.4 GPa show constant shear zone thickness, but the high-load cases exhibit steadily increasing shear zones, along with steadily decreasing z_{subst} values. A combined analysis of h_w, h_{shear}, and z_{subst} indicates a change of the wear behavior at 0.5 GPa. At approximately 3 ns, the wear height stops increasing further and remains constant at the value that it has reached at that time, whereas the shear zone thickness

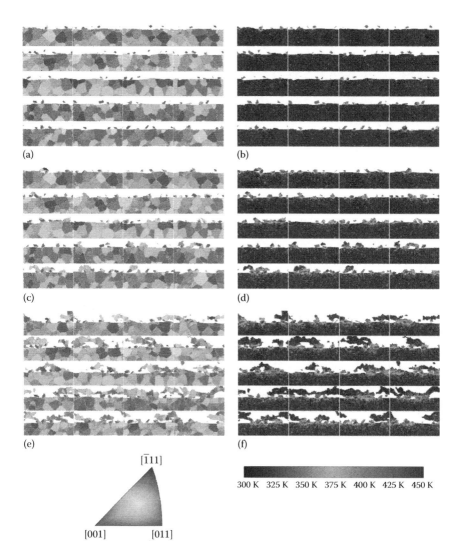

Figure 6.11 Substrate tomographs after 5 ns of grinding at 0.1 GPa (a and b), 0.4 GPa (c and d), and 0.7 GPa (e and f). Abrasives are mid-gray. (a,c,e) Shading according to grain orientation (EBSD-IPF standard, see legend below). (b,d,f) Shading according to temperature (see bar below, the removed matter in (f) is the hottest).

is relatively constant and low up to 3 ns, and then, it starts to increase steadily. This leads to a combined effect, resulting in a steadily decreasing substrate height z_{subst}, without any kinks. Furthermore, the h_{shear} discloses the reason for sudden drops of h_w. As parts of the previously generated wear particles recrystallize onto the substrate, their advection velocity

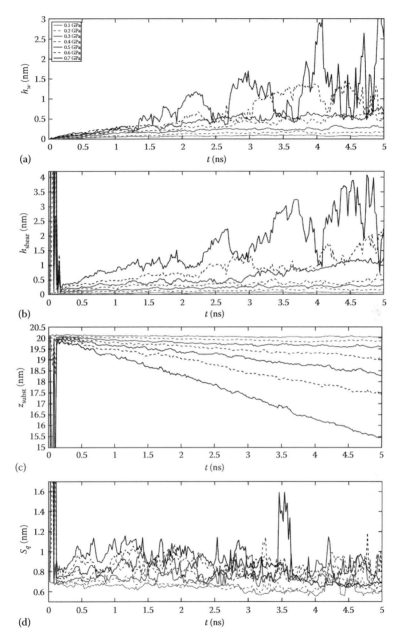

Figure 6.12 Mean wear depth h_w (a), mean shear zone thickness h_{shear} (b), arithmetic mean height z_{subst} (c), and root-mean-square roughness S_q (d) over time.

decreases, and the respective volume section is exposed to high shear stresses and strains. Consequently, peaks in the h_{shear} curve correlate with minima in the h_w curve and vice versa. The change in the initial behavior, that is, abrasion of material by formation of wear particles but hardly any shear zone up to approximately 0.5 GPa, to increasing shear zone formation is also supported by the final averaged values in Figure 6.13. The wear depth increases linearly up to 0.4 GPa, where the shear zone h_{shear} leaves the linear increase with further increasing load and proceeds progressively. z_{subst} and h_w seem to follow the same parabolic (yet mirrored) trend up to 0.4 GPa, but for higher pressures, the amount of wear cannot be increased with the same rate any more.

Topographic analysis of the grinding simulations is exemplarily shown by the surface roughness S_q value in Figure 6.12d. For the sharp-edged abrasives employed in this example, there is, in general, no smoothing of the surface for high loads, and longer grinding times do not lead to improvement of the surface quality but rather reveal some erratic trends. For normal pressures between 0.3 GPa and 0.6 GPa, the surface becomes rougher with increasing normal pressure, and for the highest loads, the surface roughness does not seem to be determined by the chosen normal pressure. The final roughness values in Figure 6.13e show only a weak correlation with normal grinding pressures.

Apart from a slight reduction of z_{subst}, visible in Figure 6.11a, grinding at 0.1 GPa does not change the microstructure. Only some small grains situated directly at the surface are abraded completely, whereas the larger grains are unaffected and merely reduced in height. Grain boundary migration cannot be observed. Figure 6.12 reveals that the wear particles are created in the first 0.5 ns and travel with the abrasives they adhere to, throughout the rest of the simulation. Rarely, parts of the wear particles touch the substrate again and recrystallize onto the surface or transfer shear to an asperity, as can be seen in the last slice no. 20 in Figure 6.11a and b. In the temperature gradients in Figure 6.11b, d, and f, the friction zones are visible as temperatures greater than 300 K. Furthermore, the simulation reveals that at 0.1 GPa, the majority of the 60 abrasives do not take part in the removal of material. The possibility for an abrasive to effectively abrade matter is determined by its relative orientation to the asperities and its rake angle, not by its size.

When increasing the normal pressure, the microstructural changes become more numerous, and the size and shape of the wear particles change considerably. At 0.4 GPa, shown in Figure 6.11c, the wear particles are much larger, and even chip-like shapes are created, which can only be attributed to the plate shape of the abrasives, as the system is simulated without any ambient medium covering the surface. Occasionally, wear particles become trapped between two abrasives, thus forming substantially larger particles; see the tomographic slices

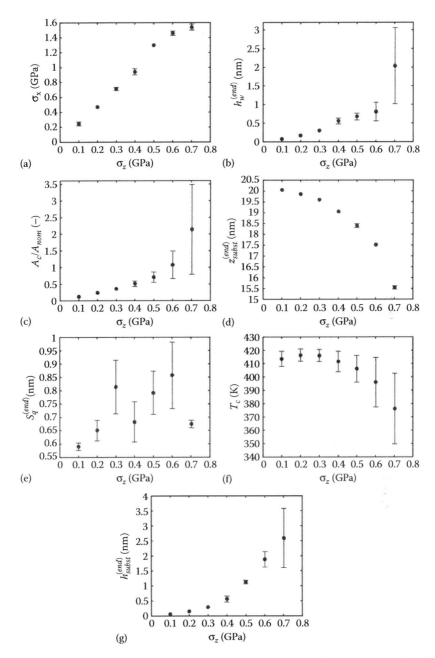

Figure 6.13 Mean shear stress σ_x (a), final wear depth h_w (b), mean normalized real contact area A_c/A_{nom} (c), final arithmetic mean height z_{subst} (d), final root-mean-square roughness S_q (e), mean contact temperature T_c (f), and final mean shear zone thickness h_{shear} (g) over normal pressure σ_z.

nos. 17 and 18 in Figure 6.11c. The corresponding temperature plot in panel (d) shows much more extended zones with elevated temperatures compared with the 0.1 GPa simulation in panel (b). Yet, the temperature within the substrate is as low as that at 0.1 GPa, due to effective cooling by using the thermostatting scheme described earlier. The same can be observed within the wear particles themselves. Previously formed particles can be distinguished from the newly formed ones by the temperature and its gradient.

At the highest applied pressure of 0.7 GPa, the wear particles form extremely long chips; see Figure 6.11e. As the plate-shaped abrasives' surface planes effectively transport the wear particles away from the surface, the wear particles end up some distance above the surface and seem to be stable. Only small proportions recrystallize onto the substrate, which happens at the locations of asperities or rims of formerly created grooves. Owing to the large wear particles and their elongated shape, the contact area A_c becomes enormous and even exceeds the nominal contact area A_{nom} by far. An A_c/A_{nom} ratio greater than 1 is a direct result of the fact that elongated wear particles cover rims of formerly formed grooves, which increase substantially in number and height with normal pressure; see Figure 6.13c and Figure 6.10 for the 3D rim structure. In addition, A_c is subjected to fluctuations in the steady-state regime of high load grinding; thus, the error becomes magnitudes larger than that for the low loads.

The contact area A_c is the zone where heat is generated due to the friction at the interface between the abrasives (as well as the wear particles traveling along with them) and the substrate. The effect of this friction energy on the microstructure depends significantly on the temperature gradients that evolve in the substrate and how stable they are as a function of time, which is effectively taken care of by applying electron–phonon coupling. This approach enables a cooling of the wear particles, so they can form multiple grains within, which is necessary for the formation of chip-shaped wear particles. Wear particles that remain hot are more likely to form a single lattice, where any new atoms will perfectly crystallize. Thanks to electron–phonon thermostatting, the frictional heat is quickly transported into the substrate as well as into the wear particles. In accordance with Fourier's law, higher friction energies at higher normal pressures cause higher temperature gradients in the workpiece, which can be seen in the temperature cross-sections in Figure 6.11b, d, and f for the three load cases. This is a direct result of the chosen thermostatting scheme. Surprisingly, the average contact temperature T_c does not increase as the normal pressure is increased, as shown in Figure 6.12f. This discrepancy can be understood if the detailed structure of the wear particles is taken into consideration. As the load is increased, the wear particles grow substantially larger and also grow sideways. Slice no. 15 of the 0.4 GPa variant shown in Figure 6.11c and d cuts through an edgewise end of a giant wear particle. This section of the wear particle has been formed in previous steps, so it has already had some time to cool down. Still traveling with the abrasive visible in slice no. 17, it creates a large contact area A_c, which heats up directly beneath the abrasive, where it exerts normal pressure and shear onto the substrate but not beneath the edgewise ends, like the one visible in slice no. 15. There, the already partially cooled section of the wear particle simply slides across the substrate, without producing much heat. Beneath such wear particles or sections of wear particles, the shear zone is thin. By contrast, it is thicker in slice no. 11, where the abrasive is plowing deeply into a rim and heat is generated locally as high normal stresses act together with shear. In slice no. 15 of the 0.4 GPa simulation, there are examples of both cases.

The microstructure in the substrate, although nearly unaffected when ground with 0.1 GPa, undergoes modifications at higher loads. At 0.5 GPa, slice no. 15 displays a highly fine-grained structure beneath the wear particle, and in slice no. 9, even finer near-surface grains can be observed for the 0.5 GPa variant; see Figure 6.14. In these near-surface locations, it is difficult to judge where the interface between abrasives and substrate is actually located. Possibly, the grains are created during crystallization of the wear particles onto the surface, as the relative velocity decreases, or they are formed as a result of dislocation-dominated hardening processes. Therefore, advection velocity and atomic displacement plots of slice no. 15 were produced for the 0.4 GPa, the 0.5 GPa, and the 0.6 GPa grinding processes; see the two central rows in Figure 6.15.

The detailed velocity plots not only reveal the location of the interface but also the lattice rotations that are taking place during the grinding process. At 0.5 GPa, for example, the substrate undergoes surface grain refinement, which is not caused by an increase in dislocation densities. Grain formation is rather provoked by a partial lattice rotation,

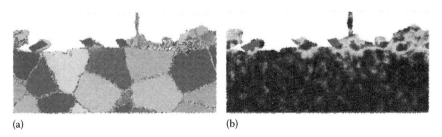

(a)   (b)

Figure 6.14 Detail tomographs of slice no. 9 located at $y = 28.5$ nm after 5 ns of machining at 0.5 GPa. Abrasives are mid-gray. (a) EBSD-IPF grain orientation shading (see SST legend in Figure 6.11), and (b) temperature shading (dark = 300 K/450 K and light = 375 K).

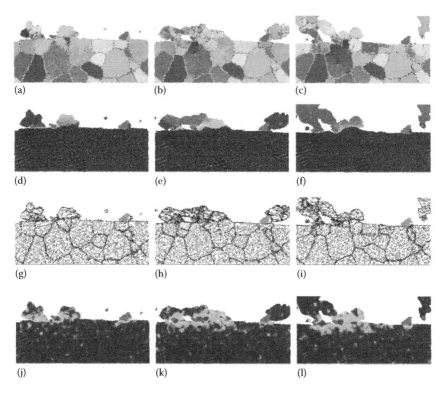

Figure 6.15 Detail tomographs of slice no. 15 located at $y = 46.5$ nm after 5 ns of machining. Abrasives are mid-gray. Left: 0.4 GPa, center: 0.5 GPa, and right: 0.6 GPa. (a–c) EBSD-IPF grain orientation shading (see SST legend in Figure 6.11), (d–f) advection velocity shading (dark: $\langle v_x \rangle = 0$ m/s or 80 m/s, light: $\langle v_x \rangle = 40$ m/s), (g–i) atomic displacement vector plots (arrow shading according to equivalent velocities ranging from 0 m/s to 8 m/s), and (j–l) temperature shading (dark $= 300$ K/450 K and light $= 375$ K).

which is a result of the local stress state that is a combination of a shear component and the normal pressure. Although the velocities within the substrate grains are small, and are thus shown as light gray arrows, the direction of the arrows is still distinctly different for the particular grains and shows the rotation direction that led to the formation of the grain boundary, shown in Figure 6.15h and i.

The plots in Figure 6.15 illustrate that the grain size structure at the surface is strongly dependent on the normal grinding pressure. Smaller grains are only generated at higher loads. Yet, the grain size appears to be unaffected, as long as no abrasive or wear particle is in contact with the surface. Thus, grain rotation is provoked only by external shear produced by the passing wear particles but not by dislocation density increase. The latter was checked with the help of the OVITO software.

The reason for no dislocation pile up can be found in the initial grain size, which is likely small enough to exhibit nanocrystalline plasticity behavior in parts of the workpiece. Temperature can be excluded as a main influencing factor in the present example, as hardly any frictional heat is stored near the surface, and thus, no annealing effect occurs in the substrate. The decrease in the grain size proves to be stable for multiple passes of abrasives, as the grain structure in Figure 6.15c is persistent and not the result of the abrasives that just passed the particular location.

Figure 6.15d–f gives some insight into the formation of the wear particles. Some previously formed wear particles that are still traveling with the abrasive are cooled down, like the one in panels (d and j: 0.4 GPa) and in (e and k: 0.5 GPa), and a large one is about to cool down. At 0.6 GPa in panels (f and l), a grinding rim has formed and the wear particle is following its shape, without incorporating new atoms, but forming a relatively thick shear zone (i). Thus, the temperature is high only in the active zone at the leading edge of the abrasive but not at the other side of the rim.

The increased occurrence of fine-grained structures at higher grinding pressures and with time changes the plastic response to the external loading of the abrasives. This is visible in the shear stress as a function of normal pressure in Figure 6.13a or as a function of normalized contact area in Figure 6.16a. Both curves initially increase linearly, but the increase saturates at high normal pressures. The average shear stress σ_x acting on the basal plane increases linearly with the normal pressure up to 0.6 GPa; see Figure 6.13a. At high normal pressures, the resistance against shear is reduced. As this cannot be attributed to high temperatures in the contact or the substrate due to the employed thermostatting scheme, the reason for this softening must be found in the microstructure evolving during grinding. A fine-grained structure can reduce the friction energy via grain boundary sliding and thus result in lower resistance to deformation and sliding, which are reflected in lower shear stresses. As long as the shear stress is

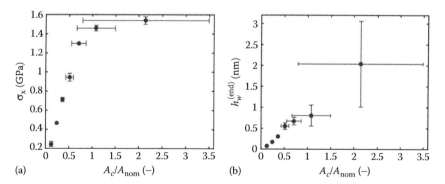

(a)

(b)

Figure 6.16 (a) Shear stress σ_x and (b) final wear depth $h_w^{(end)}$ over the normalized contact area A_c/A_{nom} with $A_{nom} = 3595\,nm^2$.

linearly dependent on the contact area, it follows the Bowden–Tabor law of kinetic friction, describing adhesive contacts [83,84].

The wear height h_w correlates linearly with the normalized contact area A_c/A_{nom}, as long as the contact area does not exceed the nominal area of the simulation box; see Figure 6.16b. For an A_c/A_{nom} ratio greater than 1, the errors increase massively and the wear height increases less than the real contact area, which becomes more and more complex due to wear particles wrapping around abrasives.

Finally, the presented example shows that the wear particle size and shape as well as the resulting microstructure within the workpiece are dramatically influenced by the abrasives' shape and the normal pressure. With increasing normal grinding pressure, the simulated microstructures feature more finer grains at the surface than in the initial condition. Locations where parts of large wear particles recrystallize onto the substrate are fine-grained, as can be seen in some slices at 0.7 GPa in Figure 6.11e and in the detailed view of slice no. 15 at 0.5 GPa and 0.6 GPa in Figure 6.15b and c. For pressures less than 0.4 GPa, the microstructures merely exhibit different extents of abrasion of small grains from the surface. The smaller the initial grains, the more unstable they are during grinding. While machining, grain growth toward the surface can be observed as long as the normal pressure is high enough, in this case greater than 0.3 GPa, which is not accompanied or assisted by higher temperatures. Thus, the microstructural changes such as grain boundary movement must be caused by thermomechanical driving forces. For even higher normal pressures, grain refinement beneath passing abrasives dominates. Again, the mechanical driving force for grain rotation determines the microstructural evolution and thus the shear stress. Neither the specific lattice orientation of the initial grains proved to be of major relevance, nor do the final microstructures show any preferred lattice orientations.

6.6   Summary

In this chapter, we gave an introduction on how to simulate a nanomachining process of a polycrystalline workpiece with multiple abrasive particles, using classical MD simulations. Special emphasis was put on the model preparation and the thermostatting procedure, as we felt that these aspects are often treated cursorily in the pertinent literature, which makes it nontrivial to assess and reproduce published results. A second focus was the system visualization, using workpiece tomography, allowing an interpretation of the atomistic simulations similar to analytical electron microscopy, even going beyond the data depth that may be extracted from SEM micrographs. As an application example of the described simulation and analysis approach, we discussed the influence of normal pressure variation on the high-speed machining process of a rough, nanocrystalline, ferritic workpiece ground by sharp-edged abrasives. Among the observable phenomena are the abrasion of grains, grain growth, the formation of fine-grained structures, as well as changes in the plasticity behavior at high loads that might be falsely attributed to the temperature by using conventional approaches. The methodology outlined in this chapter may be used as a tool for understanding and optimizing material removal processes, aiming at ultra-smooth surface topographies or strictly specified near-surface grain microstructures.

Acknowledgment

This work was funded by the Austrian COMET-Program (Project K2, XTribology, No. 849109) and carried out at the "Excellence Centre of Tribology."

References

1. U. Landman. Materials by numbers: computations as tools of discovery. *Proceedings of the National Academy of Sciences U.S.A.*, 102(19):6671–6678, 2005.
2. B. J. Alder and T. E. Wainwright. Phase transition for a hard sphere system. *Journal of Chemical Physics*, 27(5):1208, 1957.
3. U. Landman, W. D. Luedtke, and M. W. Ribarsky. Structural and dynamical consequences of interactions in interfacial systems. *Journal of Vacuum Science & Technology A: Vacuum, Surfaces, and Films*, 7(4):2829–2839, 1989.
4. P. A. Thompson and M. O. Robbins. Simulations of contact-line motion: Slip and the dynamic contact angle. *Physical Review Letters*, 63(7):766–769, 1989.
5. J. F. Belak and I. F. Stowers. A molecular dynamics model of the orthogonal cutting process. Technical report, Lawrence Livermore National Lab., Livermore, CA, 1990.
6. M. Müser. *Atomistic Simulations of Solid Friction.* In Peter Nielaba, Michel Mareschal, and Giovanni Ciccotti, editors, *Bridging Time Scales: Molecular Simulations for the Next Decade*, vol. 605, *Lecture Notes in Physics*, pages 289–317. Springer, Berlin, 2002.
7. J. David Schall, P. T. Mikulski, G. M. Chateauneuf, G. Gao, and J. A. Harrison. *Molecular Dynamics Simulations of Tribology.* In *Superlubricity*, pages 79–102. Elsevier, Amsterdam, 2007.
8. I. Szlufarska, M. Chandross, and R. W. Carpick. Recent advances in single-asperity nanotribology. *Journal of Physics D: Applied Physics*, 41(12):123001, 2008.
9. E. Brinksmeier, J. C. Aurich, E. Govekar, C. Heinzel, H.-W. Hoffmeister, F. Klocke, J. Peters, R. Rentsch, D. J. Stephenson, E. Uhlmann, et al. Advances in modeling and simulation of grinding processes. *CIRP Annals-Manufacturing Technology*, 55(2):667–696, 2006.
10. S. Goel, X. Luo, A. Agrawal, and R. L. Reuben. Diamond machining of silicon: a review of advances in molecular dynamics simulation. *International Journal of Machine Tools and Manufacture*, 88:131–164, 2015.
11. LAMMPS. http://lammps.sandia.gov/.

12. GROMACS. http://www.gromacs.org/.

13. Moldy. http://www.ccp5.ac.uk/moldy/moldy.html.

14. NAMD. http://www.ks.uiuc.edu/research/namd/.

15. R. Komanduri, N. Chandrasekaran, and L. M. Raff. MD simulation of indentation and scratching of single crystal aluminum. *Wear*, 240(1/2):113–143, 2000.

16. Q. X. Pei, C. Lu, H. P. Lee, and Y. W. Zhang. Study of materials deformation in nanometric cutting by large-scale molecular dynamics simulations. *Nanoscale Research Letters*, 4(5):444–451, 2009.

17. L. Si, D. Guo, J. Luo, and X. Lu. Monoatomic layer removal mechanism in chemical mechanical polishing process: A molecular dynamics study. *Journal of Applied Physics*, 107(6):064310, 2010.

18. J. B. Adams, L. G. Jr. Hector, D. J. Siegel, H. Yu, and J. Zhong. Adhesion, lubrication and wear on the atomic scale. *Surface and Interface Analysis*, 31(7):619–626, 2001.

19. P. M. Agrawal, L. M. Raff, S. Bukkapatnam, and R. Komanduri. Molecular dynamics investigations on polishing of a silicon wafer with a diamond abrasive. *Applied Physics A*, 100(1):89–104, 2010.

20. J. Zhong, R. Shakiba, and J. B. Adams. Molecular dynamics simulation of severe adhesive wear on a rough aluminum substrate. *Journal of Physics D: Applied Physics*, 46(5):055307, 2013.

21. G. M. Robinson, M. J. Jackson, and M. D. Whitfield. A review of machining theory and tool wear with a view to developing micro and nano machining processes. *Journal of Materials Science*, 42(6):2002–2015, 2007.

22. I. Zarudi and L. Zhang. Subsurface damage in single-crystal silicon due to grinding and polishing. *Journal of Materials Science Letters*, 15(7):586–587, 1996.

23. L. Zhang and I. Zarudi. Towards a deeper understanding of plastic deformation in mono-crystalline silicon. *International journal of mechanical sciences*, 43(9):1985–1996, 2001.

24. S. Eder. Molecular dynamics evidence of a three-term kinetic friction law for mixed- and boundary-lubricated nanotribological systems. PhD thesis, Vienna University of Technology, 2012.

25. M. P. Allen and D. J. Tildesley. *Computer Simulation of Liquids*. Oxford University Press, Oxford, 1987.

26. J. M. Haile. *Molecular Dynamics Simulation: Elementary Methods*. Wiley, March 1997.

27. M. Griebel, S. Knapek, and G. Zumbusch. *Numerical Simulation in Molecular Dynamics: Numerics, Algorithms, Parallelization, Applications*. 1st edition, Springer, Incorporated, 2007.

28. J. E. Lennard-Jones. Cohesion. *Proceedings of the Physical Society*, 43:461, 1931.

29. M. S. Daw and M. I. Baskes. Embedded-atom method: Derivation and application to impurities, surfaces, and other defects in metals. *Physical Review B*, 29(12):6443–6453, 1984.

30. M. W. Finnis and J. E. Sinclair. A simple empirical n-body potential for transition metals. *Philosophical Magazine A*, 50(1):45–55, 1984.

31. F. Ercolessi, M. Parrinello, and E. Tosatti. Simulation of gold in the glue model. *Philosophical Magazine A*, 58(1):213–226, 1988.

32. F. Ercolessi. A molecular dynamics primer. Spring College in Computational Physics, ICTP, Trieste, http://www.fisica.uniud.it/ ercolessi/md/md/, June 1997.

33. L. Verlet. Computer "experiments" on classical fluids. I. Thermodynamical properties of Lennard-Jones molecules. *Physical Review*, 159(1):98, 1967.

34. S. J. Eder, D. Bianchi, U. Cihak-Bayr, A. Vernes, and G. Betz. An analysis method for atomistic abrasion simulations featuring rough surfaces and multiple abrasive particles. *Computer Physics Communications*, 185:2456–2466, 2014.

35. S. J. Eder, U. Cihak-Bayr, A. Vernes, and G. Betz. Evolution of topography and material removal during nanoscale grinding. *Journal of Physics D: Applied Physics*, 48:465308, 2015.

36. S. J. Eder, U Cihak-Bayr, and D Bianchi. Single-asperity contributions to multi-asperity wear simulated with molecular dynamics. *IOP Conference Series: Materials Science and Engineering*, 119(1):012009, 2016.

37. S. J. Eder, D. Bianchi, U. Cihak-Bayr, and K. Gkagkas. Methods for atomistic abrasion simulations of laterally periodic polycrystalline substrates with fractal surfaces. *Computer Physics Communications*, 212:100–112, 2017.

38. S. J. Eder, U. Cihak-Bayr, D. Bianchi, G. Feldbauer, and G. Betz. Thermostat influence on the structural development and material removal during abrasion of nanocrystalline ferrite. *ACS Applied Materials & Interfaces*, 9(15):13713–13725, 2017.

39. S. J. Plimpton. Fast parallel algorithms for short-range molecular dynamics. *Journal of Computational Physics*, 117:1–19, 1995.

40. MATLAB, *MATLAB Release*, The MathWorks, Natick, MA, 2015a.

41. J. K. Mackenzie. Second paper on statistics associated with the random disorientation of cubes. *Biometrika*, 45(1/2):229–240, 1958.

42. K. Marthinsen. Comparative analysis of the size distributions of linear, planar, and spatial poisson voronoi cells. *Materials Characterization*, 36(2):53–63, 1996.

43. T. R. Thomas. *Rough surfaces*. World Scientific, 1998.

44. P. Spijker, G. Anciaux, and J.-F. Molinari. Relations between roughness, temperature and dry sliding friction at the atomic scale. *Tribology International*, 59:222–229, 2013.

45. M. V. Berry and J. H. Hannay. Topography of random surfaces. *Nature*, 273:573, 1978.

46. F. M. Borodich and D. A. Onishchenko. Similarity and fractality in the modelling of roughness by a multilevel profile with hierarchical structure. *International Journal of Solids and Structures*, 36(17):2585–2612, 1999.

47. R. F. Voss. *Fractals in Nature: From Characterization to Simulation*. Springer, 1988.

48. B. N. J. Persson. On the fractal dimension of rough surfaces. *Tribology Letters*, 54(1):99–106, 2014.

49. D. V. De Pellegrin, N. D. Corbin, G. Baldoni, and A. A. Torrance. Diamond particle shape: Its measurement and influence in abrasive wear. *Tribology International*, 42(1):160–168, 2009.

50. D. V. De Pellegrin and G. W. Stachowiak. Sharpness of abrasive particles and surfaces. *Wear*, 256(6):614–622, 2004.

51. M. Bielmann, U. Mahajan, and R. K. Singh. Effect of particle size during tungsten chemical mechanical polishing. *Electrochemical and Solid-State Letters*, 2(8):401–403, 1999.

52. J. Luo, D. Dornfeld et al. Effects of abrasive size distribution in chemical mechanical planarization: modeling and verification. *IEEE Transactions on Semiconductor Manufacturing*, 16(3):469–476, 2003.

53. M. I. Mendelev, S. Han, D. J. Srolovitz, G. J. Ackland, D. Y. Sun, and M. Asta. Development of new interatomic potentials appropriate for crystalline and liquid iron. *Philosophical Magazine*, 83:3977–3994, 2003.

54. K. Maekawa and A. Itoh. Friction and tool wear in nano-scale machining— A molecular dynamics approach. *Wear*, 188(1):115–122, 1995.

55. T. Schneider and E. Stoll. Molecular-dynamics study of a three-dimensional one-component model for distortive phase transitions. *Physical Review B*, 17(3):1302–1322, 1978.

56. B. Dünweg and W. Paul. Brownian dynamics simulations without Gaussian random numbers. *International Journal of Modern Physics C*, 2:817, 1991.

57. P.-H. Huang and H.-Y. Lai. Nucleation and propagation of dislocations during nanopore lattice mending by laser annealing: Modified continuum-atomistic modeling. *Physical Review B*, 77(12):125408, 2008.

58. C. Schäfer, H. M. Urbassek, and L. V. Zhigilei. Metal ablation by picosecond laser pulses: A hybrid simulation. *Physical Review B*, 66(11):115404, 2002.

59. X. Liu and S. Li. Nonequilibrium multiscale computational model. *The Journal of Chemical Physics*, 126(12):124105, 2007.

60. J. David Schall, C. W. Padgett, and D. W. Brenner. Ad hoc continuum-atomistic thermostat for modeling heat flow in molecular dynamics simulations. *Molecular Simulation*, 31(4):283–288, 2005.

61. A. Caro and M. Victoria. Ion-electron interaction in molecular-dynamics cascades. *Physical Review A*, 40(5):2287, 1989.

62. Q. Hou, M. Hou, L. Bardotti, B. Prével, P. Mélinon, and A. Perez. Deposition of Au_N clusters on Au (111) surfaces. I. Atomic-scale modeling. *Physical Review B*, 62(4):2825, 2000.

63. Y. S. Touloukian, R. W. Powell, C. Y. Ho, and P. G. Klemens. *Thermophysical Properties of Matter—The TPRC Data Series*. Vol. 1. *Thermal Conductivity—Metallic Elements and Alloys*. 1970.

64. C. Björkas and K. Nordlund. Assessment of the relation between ion beam mixing, electron–phonon coupling and damage production in Fe. *Nuclear Instruments and Methods in Physics Research Section B: Beam Interactions with Materials and Atoms*, 267(10):1830–1836, 2009.

65. R. W. Carpick, D. Frank Ogletree, and M. Salmeron. A general equation for fitting contact area and friction vs load measurements. *Journal of Colloid and Interface Science*, 211(2):395–400, 1999.

66. B. N. J. Persson. Contact mechanics for randomly rough surfaces. *Surface Science Reports*, 61(4):201–227, 2006.

67. G. Carbone and F. Bottiglione. Asperity contact theories: Do they predict linearity between contact area and load? *Journal of the Mechanics and Physics of Solids*, 56(8):2555–2572, 2008.

68. C. Yang and B.N.J. Persson. Contact mechanics: contact area and interfacial separation from small contact to full contact. *Journal of Physics: Condensed Matter*, 20(21):215214, 2008.

69. Y. Mo, K.T. Turner, and I. Szlufarska. Friction laws at the nanoscale. *Nature*, 457(7233):1116–1119, 2009.

70. S. Cheng and M.O. Robbins. Defining Contact at the Atomic Scale. *Tribology Letters*, 39(3):329–348, 2010.

71. N. Prodanov, W. B. Dapp, and M. H. Müser. On the contact area and mean gap of rough, elastic contacts: Dimensional analysis, numerical corrections, and reference data. *Tribology Letters*, 53(2):433–448, 2014.

72. M. Wolloch, G. Feldbauer, P. Mohn, J. Redinger, and A. Vernes. *Ab initio* calculation of the real contact area on the atomic scale. *Physical Review B,* 91(19):195436, 2015.

73. B. Luan and M. O. Robbins. The breakdown of continuum models for mechanical contacts. *Nature,* 435:929–932, 2005.

74. S. Eder, A. Vernes, and G. Betz. Methods and numerical aspects of nanoscopic contact area estimation in atomistic tribological simulations. *Computer Physics Communications,* 185:217–228, 2014.

75. C. L. Kelchner, S. J. Plimpton, and J. C. Hamilton. Dislocation nucleation and defect structure during surface indentation. *Physical Review B,* 58(17):11085–11088, 1998.

76. E. J. Abbott and F. A. Firestone. Specifying surface quality: A method based on accurate measurement and comparison. *Mechanical Engineering,* 55:569–572, 1933.

77. P. M. Larsen, S. Schmidt, and J. Schiøtz. Robust structural identification via polyhedral template matching. *Modelling and Simulation in Materials Science and Engineering,* 24(5):055007, 2016.

78. A. Stukowski. Visualization and analysis of atomistic simulation data with OVITO–the open visualization tool. *Modelling and Simulation in Materials Science and Engineering,* 18(1):015012, 2009.

79. S. van der Walt, S. Chris Colbert, and G. Varoquaux. The NumPy array: a structure for efficient numerical computation. *Computing in Science & Engineering,* 13(2):22–30, 2011.

80. G. S. Rohrer, D. M. Saylor, B. El Dasher, B. L. Adams, A. D. Rollett, and P. Wynblatt. The distribution of internal interfaces in polycrystals. *Zeitschrift für Metallkunde,* 95(4):197–214, 2004.

81. F. Bachmann, R. Hielscher, and H. Schaeben. *Texture Analysis with MTEX– Free and Open Source Software Toolbox.* In *Solid State Phenomena,* vol. 160, pages 63–68. Trans Tech Publ, 2010.

82. G. Nolze and R. Hielscher. Orientations–perfectly colored. *Journal of Applied Crystallography,* 49(5):1786–1802, 2016.

83. F. P. Bowden and D. Tabor. The area of contact between stationary and between moving surfaces. *Proceedings of the Royal Society of the London, Series A,* 169:391–413, 1939.

84. F. P. Bowden and D. Tabor. *The Friction and Lubrication of Solids.* Oxford University Press, Oxford, 1950.

chapter seven

Multiobjective optimization of support vector regression parameters by teaching-learning-based optimization for modeling of electric discharge machining responses

Ushasta Aich and Simul Banerjee

Contents

Nomenclature 180
7.1   Introduction... 180
7.2   Experiment... 182
7.3   Unified learning system development... 184
    7.3.1   Support vector machine... 184
    7.3.2   Multiobjective teaching-learning-based optimization........ 189
        7.3.2.1   Modifications and marching procedure................. 190
    7.3.3   Testing of unified learning system....................................... 199
7.4   Conclusion 206
Acknowledgment... 206
Appendix... 207
References.. 209

Nomenclature

ASR	Average surface roughness (μm)
b	Bias
C	Regularization parameter
cur	Current setting (A)
d	Training input space dimension
$f(\mathbf{x})$	Target function
iter_{max}	Maximum number of iterations
$K(x_i, \mathbf{x})$	Kernel function
$\text{MATE}_1, \text{MATE}_2$	Mean absolute training error in MRR, ASR
MRR	Material removal rate (mm^3/min)
n	Number of learners in class
N	Number of training data
rand	A pseudorandom number generated following standard uniform distribution within range (0,1)
rw_1, rw_2	Random weight factors
t_{off}	Pulse-off time (μs)
t_{on}	Pulse-on time (μs)
TF_{iter}	Teaching factor at iterth iteration
$\mathbf{w}$	Weight vector
$\mathbf{x}$	Training input vector
$\mathbf{y}$	Training output vector
$\bar{y}$	Mean of training output set
z	Number of attributes
E	Radius of loss insensitive hyper-tube
$\eta_i, \eta_i^*, \alpha_i, \alpha_i^*$	Lagrange multipliers
ξ_i, ξ_i^*	Slack variables
σ	Standard deviation of radial basis function (kernel function)
$\Phi(\mathbf{x})$	Feature space

7.1   Introduction

Mathematical modeling of any process would be a stepping stone for working in virtual environment. In virtual world, near-exact representation of process is necessary to freeze the procedure at the earliest in the preproduction stage. Representative model should be robust in nature. In case of inherent stochastic-type nontraditional manufacturing process such as electric discharge machining (EDM), model development and subsequent prediction of process outcomes with reasonable accuracy would become difficult. Advanced learning-based systems, being devoid of four problems—efficiency in training, efficiency in testing, overfitting and, algorithm parameter tuning—would be effective in such situation.

In the present study, experiments are carried out on EDM process in the semi-finishing and roughing zone, with different combinations of three significant process parameters—current (cur), pulse-on time (t_{on}), and pulse-off time (t_{off}). Material removal rate (MRR) and average surface roughness (ASR) are considered as two performance measures. In case of machining, rate of material removal determines the productivity of the process; that is, higher MRR results in higher productivity. Besides, to meet the specific functional aspects of product, quality must be maintained. One of the major surface quality measurements is the ASR. Therefore, at the product design and manufacturing stage, both of these performance measures—MRR and ASR—are to be predicted simultaneously.

As the EDM process is itself stochastic in nature, the predictions of outcomes become more challenging. Aich and Banerjee [1] built two independent explicit models of MRR and ASR in EDM through a supervised batch learning-based support vector machine (SVM) regression procedure. A meaningful physical significance of the insensitive zone of the learned system is to provide a space to allow the tolerances on uncontrollable variations in the EDM process. They outlined a way of setting all the internal structural parameters—regularization parameter (C), radius of loss insensitive hyper-tube (ε), and standard deviation of Gaussian radial basis function (σ) chosen as kernel function ($K(x_i, x)$)—for SVM learning, employing particle swarm optimization (PSO). However, the behavior of the developed models near the boundary of the experimental domain appears as not very accurate, and selection of internal parameters of PSO itself is a critical job to ensure smooth convergence toward global optimum.

Performances of the swarm-based optimization techniques, rather evolutionary algorithms, are affected by their own control parameter settings [2]. Unlike those probabilistic approaches, algorithm-specific parameter-less teaching-learning-based optimization (TLBO), introduced by Rao et al. [3], is proved to be more effective for complex-type multimodal high-dimensional nonlinear objective functions [3,4].

With the aid of TLBO, internal structural parameters of SVM are tuned by Aich and Banerjee [5] for developing two independent learning systems, one for each of these two individual process outcomes—MRR and ASR in the EDM process. Compared with their earlier work [1], selection of internal parameters of optimization technique is now no longer required. Thus, methodology of independent learning systems becomes user-friendly [5]; yet, multiple sets of optimum internal structural parameters of learning system do not permit the use of the methodology for concurrent prediction of multiple responses—MRR and ASR for the same set of input parameters. Therefore, a compact learning system is to be developed that could estimate multiple responses from a single set of internal parameters.

In the present work, development of a unified structure of SVM regression for predicting multiple responses is attempted. Unified learning is performed by simultaneous minimization of errors in estimation of MRR and ASR by modified TLBO. This development is an advancement of mathematical modeling toward the compact virtual data generator. In the proposed modified TLBO, combined rank method, an improvement in multiobjective optimization by TLBO, is suggested for simultaneous optimization of multiple objective functions and an optimum unique set of C, ε, and σ is obtained. With the optimum unique set of SVM internal structural parameters, C, ε, and σ, two separate sets of Lagrange multipliers, one for each of the MRR and ASR, are calculated on feeding respective training vectors. Subsequently, MRR and ASR are estimated from the calculated corresponding sets of Lagrange multipliers. It is to be noted that Aich and Banerjee [5] in their previous work generated two sets of Lagrange multipliers (each for MRR and ASR) from two independent sets of C, ε, and σ. The novelty of the present study lies in the development of such unification of SVM regression structures for concurrent prediction of conflicting-type multiple responses with the aid of modified TLBO. The basics of SVM and TLBO are introduced in Sections 7.3.1 and 7.3.2, respectively. This modification could be generalized for solving any such multiple objective functions in an efficient way. The proposed procedure may become a building block for expert system.

This chapter is comprised of three sections. A brief discussion on the EDM process, selection of machining and performance parameters, their levels and performance measurements are given in Section 7.2. In Section 7.3, detailed discussions on the steps involved in unified structure development, including SVM, TLBO, and the results obtained, are provided. Finally, conclusions for the present study are concisely listed in Section 7.4.

7.2 Experiment

One of the most sophisticated and precise nontraditional machining processes, EDM, is used in machining conductive material (resistivity should not exceed 100 ohm-cm), regardless of its hardness, toughness, and strength [6]. This process is generally employed for manufacturing complex surface geometry and integral angles in mold, die, aerospace, surgical components, and so on [7].

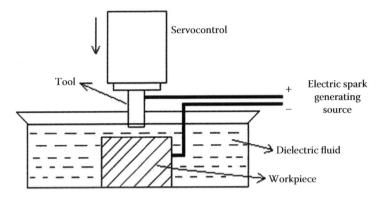

Figure 7.1 Schematic of electrical discharge machining process.

In EDM, material is eroded by series of spatially discrete and chaotic high-frequency electric discharges (sparks) of high power density between the tool electrode and the workpiece separated by a fine gap of dielectric fluid [8]. The working zone is completely immersed into dielectric fluid medium for enhancing electron flow in the gap, cooling after each spark, and easy flushing of eroded particles. Basic scheme of the EDM is shown in Figure 7.1.

Experiment is carried out on Tool Craft A25 EDM machine under open circuit voltage of 66 V operating with commercially available kerosene oil as dielectric medium [5] and different combinations of four levels of each of the three most dominating process parameters, namely current (cur), pulse-on time (t_{on}), and pulse-off time (t_{off}). Standard high-speed steel-cutting tool (C: 0.80%, W: 6%, Mo: 5%, Cr: 4%, and V: 2%) equivalent to grade M2 is chosen as the workpiece material (measured density 8006 kg/m³), and it is connected in reverse polarity. Electrolytic copper with density 8904 kg/m³ and cross-sectional diameter of 12 mm is used as tool material.

For working in the semi-finishing and roughing zone, based on the availability of the machine settings, levels of the input process parameters are chosen (Table 7.1).

Table 7.1 Process parameters and their levels

	Level 1	Level 2	Level 3	Level 4
Current setting (A)	6	9	12	15
Pulse-on time (µs)	50	100	150	200
Pulse-off time (µs)	50	100	150	200

Total 64 mutually exclusive combinations of four different levels of each of the three process parameters are set to the EDM machine, and corresponding process outcomes are noted. For determination of MRR, work sample weights are taken at standard measuring balance (AFCOSET—ER182A) of least count 0.01 mg before and after machining. Weight loss is then divided by the measured density of workpiece material, in order to convert it into volumetric term, and is further divided by the actual machining time to obtain the MRR in terms of mm^3/min. Centerline ASR values of the machined surface of workpiece along three mutually 120° apart directions are measured by the Taylor Hobson Precision Surtronis 3$^+$ Roughness Checker, with sample length of 4 mm and stylus tip radius of 5 μm. Mean of the three measured centerline ASR (Ra) values is considered as the representative ASR of the EDM machined surface. Fifteen percent of the total 64 unique treatments, that is, 10 sets, is chosen randomly and kept aside for testing purposes. Rest of the data sets are used for the training of SVM learning system.

7.3   Unified learning system development

For building a unified structure of SVM regression learning system that provides concurrent prediction of multiple responses, randomly 54 data sets are taken for training. Different sets of randomly chosen 54 data are taken, and same results are obtained. Here, results of learning system development with a typical set of randomly chosen 54 data are reported. Fitted learning systems are tested through rest of the 10 sets of data.

As prerequisites for this proposed unified learning systems development, brief discussions are given on SVM and TLBO. Modification of standard TLBO and steps for building unified learning system are presented in Section 7.3.2.1.

7.3.1   Support vector machine [5]

Different techniques such as multivariable regression analysis [9], response surface methodology [10], and artificial intelligence-based neural network [11] are rigorously used for modeling empirical data. Suffering from generalization of model estimation, overfitting might occur in artificial neural network (ANN). Besides, random variations in process outcomes are obvious in stochastic-type machining process. These random fluctuations in experimental results are to be absorbed with specified tolerance value for efficient predictions. Structural risk minimization-based [12] SVM, which is one of the most advanced supervised batch learning system, could be a smart way of capturing these fluctuations.

Suppose, a representative model is to be developed for a disjoint, independent, and identical distributed data set $\{(x_1, y_1), (x_2, y_2), \ldots (x_N, y_N)\}$

in d dimensional input space (i.e., $\mathbf{x} \in R^d$). Target function may be represented in the form [13]:

$$f(\mathbf{x}) = \langle \mathbf{w}, \mathbf{x} \rangle + b \qquad (7.1)$$

where $\langle\,,\,\rangle$ indicates dot product in vector space.

Nonlinearity in the relation between input and output patterns (Figure 7.2 [14]) is handled through mapping the high-dimensional input space to a feature space $\Phi(\mathbf{x})$ via kernel functions. So, optimal choice of weight factor $\mathbf{w}$ and threshold b (bias term) is a prerequisite of accurate modeling. Flatness of the model is controlled by minimizing Euclidean norm $||\mathbf{w}||$. Besides, empirical risk of training error should also be minimized [15]. So, regularized risk minimization problem for model developing can be written as follows:

$$R_{\text{reg}}(f) = \frac{||\mathbf{w}||^2}{2} + C\Sigma_{i=1(1)N}L(y_i, f(x_i)) \qquad (7.2)$$

Weight vector $\mathbf{w}$ and the bias term b can be estimated by optimizing this function, Equation 7.2, which minimizes not only empirical risks, but also reduces generalization error; that is, overfitting of model simultaneously. A loss function is to be introduced to penalize overfitting of model with training points.

A number of loss functions, namely quadratic loss function, Huber loss function, ε-insensitive loss function, and so on, are already developed for handling different types of problems [16]. In general, these loss functions are some modified measurements of distances of the points

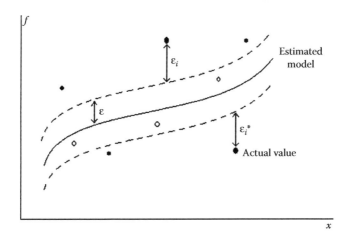

Figure 7.2 Nonlinear SVM regression model.

and their corresponding estimated values. Square values of the distances between actual points and corresponding estimated values are considered for assigning loss in quadratic loss function. Quadratic loss function corresponds to the conventional least squares error criterion. Huber loss function is the combination of linear and quadratic loss functions. This robust loss function exhibits optimal properties when the underlying distribution of the data is unknown. Still, these two loss functions—quadratic and Huber—will produce no sparseness in the support vectors. To address these issues, Vapnik [12] proposed ε-insensitive loss function as a trade-off between the robust loss function of Huber and one that enables sparsity within the support vectors. ε-Insensitive loss function (refer Figure 7.3) may be defined as [14]:

$$L(y_i, f(x_i)) = \left|y_{i,\,\text{experimental}}\quad f(x_i)\right| - \varepsilon, \quad \text{if}\left|y_{i,\,\text{experimental}} - f(x_i)\right| \geq \varepsilon$$
$$= 0, \qquad\qquad\qquad \text{if}\left|y_{i,\,\text{experimental}} - f(x_i)\right| < \varepsilon$$

(7.3)

In most of the model-building techniques, data are fitted with least training error calculation to estimate the unknown coefficient or weight vectors associated with training inputs. That is, all the data are tried to fit as close as possible to the deemed model. In SVM regression, an insensitive zone wrapped around the estimated function is defined. This insensitive zone is expected to capture the fluctuations within permissible tolerances specified by process outcomes. Thereby, radius of this hyper-tube directly controls the allowable complexity of the learning system. In nomenclature, the outliers around this tube are named as support vectors. Here, ε-insensitive loss function, refer to Equation 7.3, is considered to penalize overfitting of the system with training points.

As this radius of insensitive hyper-tube increases, model would become more flat, being unable to reveal the unseen nature of variation in the outcomes, whereas lower radius might make the model more complex.

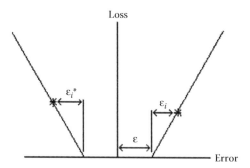

Figure 7.3 ε-Insensitive loss function.

Thus, a trade-off between complexity and flatness of the estimated model is required. Two positive slack variables, ξ_i and ξ_i^*, are introduced [12,13] to cope with infeasible constraints of the optimization problem. Hence, the constrained problem can be reformulated as:

$$\text{minimize:} \frac{\|\mathbf{w}\|^2}{2} + C\Sigma_{i=1(1)N} \left(\xi_i + \xi_i^*\right)$$

$$y_{i,\,exp} - \langle \mathbf{w}, \mathbf{x} \rangle_i - b \leq \varepsilon + \xi_i \tag{7.4}$$

$$\text{subject to:} \langle \mathbf{w}, \mathbf{x} \rangle_i + b - y_{i,\,exp} \leq \varepsilon + \xi_i^*$$

$$\xi_i, \xi_i^* \qquad \geq 0 \qquad\qquad i = 1(1)N$$

This problem can be efficiently solved by standard dualization principle, utilizing Lagrange multiplier. A dual set of variables is introduced for developing Lagrange function. It is found that this function has a saddle point with respect to both primal and dual variables at the solution. Lagrange function can be stated as:

$$L = \left(\frac{\|\mathbf{w}\|^2}{2}\right) + C\Sigma_{i=1(1)N} \left(\xi_i + \xi_i^*\right) - \Sigma_{i=1(1)N} \left(\eta_i \xi_i + \eta_i^* \xi_i^*\right)$$

$$- \Sigma_{i=1(1)N} \alpha_i \left(\varepsilon + \xi_i - y_i + \langle \mathbf{w}, \mathbf{x} \rangle_i + b\right) \tag{7.5}$$

$$- \Sigma_{i=1(1)N} \alpha_i^* \left(\varepsilon + \xi_i^* + y_i - \langle \mathbf{w}, \mathbf{x} \rangle_i - b\right)$$

where:
  L is the Lagrangian and
  η_i, η_i^*, α_i, α_i^* are Lagrange multipliers satisfying $\eta_i, \eta_i^*, \alpha_i, \alpha_i^* \geq 0$

So, partial derivatives of L with respect to W, b, ξ_i, and ξ_i^* will give the estimates of $\mathbf{w}$ and b.

Support vectors can be easily identified from the value of difference between Lagrange multipliers (α_i, α_i^*). Very small values (close to zero) indicate the points inside the insensitive hyper-tube, but nonzero values belong to support vector group [17]. The $\mathbf{w}$ can be calculated by [13]:

$$\mathbf{w} = \Sigma_{i=1(1)N} (\alpha_i - \alpha_i^*) \Phi(x_i) \tag{7.6}$$

The idea of kernel function $K(x_i, \mathbf{x})$ gives a way of addressing the curse of dimensionality [16]. It helps to enable the operations to be performed in the feature space ($\Phi(\mathbf{x})$) rather than potentially high-dimensional input space.

A number of kernel functions satisfying Mercer's condition were suggested by researchers [17,18]. Each of these functions has its own specialized applicability. Use of polynomial kernel function is a popular method for nonlinear modeling. The long-established multilayer perceptron with a single hidden layer has a valid kernel representation for certain values of the scale and offset parameters, whereas Fourier series kernel is probably not a good choice, because its regularization capability is poor, which is evident by consideration of its Fourier transform. Among different splines, specifically *b*-spline is also a popular choice for modeling because of its flexibility. Exponential radial basis function produces a piecewise linear solution, which can be attractive when discontinuities are acceptable. Apart from all these kernel function, Gaussian radial basis function has received significant attention, as this kernel is implicit with each support vector contributing one local Gaussian function centered at that data point.

Here, Gaussian radial basis function with σ standard deviation, Equation 7.7, is used for its better potentiality to handle higher-dimensional input space.

$$K\left(x_i, \mathbf{x}\right) = \exp\left(\frac{-\left\|x_i - \mathbf{x}\right\|^2}{2\sigma^2}\right) \tag{7.7}$$

Thus, representative model of the learning system, with optimum choice of the most significant structural parameters, C, ε, and σ, may be presented as [13]:

$$f(\mathbf{x}) = \Sigma_{i=1(1)N}\left(\alpha_i - \alpha_i{}^*\right)K(x_i, \mathbf{x}) + b$$

$$\left| \begin{array}{l} C_{\text{optimum}} \\[6pt] \varepsilon_{\text{optimum}} \\[6pt] \sigma_{\text{optimum}} \end{array} \right. \tag{7.8}$$

To get the benefit of this exclusive feature of SVM regression over other model development techniques, a regularization parameter is required to penalize the support vectors, whereas the points that lie inside the insensitive zone are considered to be of zero loss. Thereby, to fit the data with reasonable accuracy, internal structural parameters of SVM, namely the regularization parameter (C), which controls the penalty associated with support vector; radius of insensitive tube (ε); and standard deviation of Gaussian radial basis kernel function (σ), are to be properly tuned. Improper choice of SVM internal parameters may lead to underfitting or overfitting of the actual process [1,5]. Thus, for each set of input–output combination, an optimum set of SVM internal structural parameters, C, ε and σ, is expected.

Structure of SVM learning system should vary for each of the different input–output combinations. In case of multiple process outcomes of a manufacturing process with same settings in machine control parameters, there should be separate sets of optimum C, ε, and σ for each of the process outcomes. In the present study, a methodology is proposed to develop a unified structure of SVM regression learning system for concurrent prediction of multiple process outcomes of a manufacturing process. That is, an optimum unique set of structural parameters, C, ε, and σ, is searched to exist, instead of multiple sets of optimum C, ε, and σ corresponding to multiple responses [5]. Robust optimization techniques could be employed to tune these internal structural parameters. In this regard, algorithm-specific parameter-less TLBO would be a justified choice.

7.3.2 Multiobjective teaching-learning-based optimization

Compared with traditional deterministic approaches for optimization of multimodal, high-dimensional nonlinear large-scale engineering problems, metaheuristic algorithms exhibit more promising performances [2]. Natural phenomena-inspired trajectory- and population-based different algorithms are still suffering from the problem of tuning their own internal parameters [3,4]. Rao et al. [3] introduced an algorithm-specific parameter-less optimization technique that mimics the ideology of teaching-learning process, called teaching-learning-based optimization (TLBO). A class of learners is considered as the population of the optimization algorithm. In TLBO, different control variables and scores of the learners are analogous to different subjects offered to the learners and objective function values, respectively. Marching steps of TLBO to reach global optimum are broadly divided into two phases: teacher phase and learner phase.

In teacher phase, teacher always tries to pull forward the batch of learners, aiming to his/her own level. Gaining more knowledge from teacher helps the learners score better marks. Therefore, teacher gradually increases the mean score of the learners according to his/her own capability. Still, the knowledge dissemination by the teacher and acquiring of knowledge by learners are not always same for all teacher–learner combinations. Therefore, a teaching factor (TF) should play a typical role in this teacher phase. In the present study, adaptive TF [19], depending on the current performance level of the whole batch, is deployed instead of randomly selected integer between 1 and 2 [3]. This adaptive TF, calculated as a ratio of mean of the learners' value to teacher value of latest population [19], aids in converging the simulation with lesser time.

Gaining of knowledge by the learners is further enhanced through different scheme of interactions among themselves, namely group discussions, presentations, formal communications, and so on. These intralearner interactions are performed in the second phase; that is, learner phase. In this learner phase, each of the learners is randomly selected and compared with another randomly selected different learner. If the other learner has more knowledge than him/her, then the former learner gains some knowledge from the other one. By this way, scores of the learners are increased. Main steps involved in TLBO are briefly listed as follows:

Step 1: The learner having best score is identified and considered as teacher.

Step 2: Adaptive TF is calculated, and all learners are modified toward the teacher.

Step 3: All modified learners (two at a time) are randomly selected, and they upgrade themselves.

Step 4: Check the termination criterion. If satisfied, current teacher is declared as optimum setting; otherwise, repeat the steps with current upgraded learners till termination criterion is satisfied.

However, exploitation of the search space is done in teacher phase, whereas learner phase does the exploration. In every iteration, objective function values, that is, current learners' scores in each subject gradually move toward optimum zone. Still, no such guideline is found in the literature for simultaneous optimization of multiple objective functions.

In the present study, TLBO is modified by introducing a combined ranking method, with weight infected rank selection (wherever necessary) for simultaneous optimization of multiple objective functions, and thus is employed for tuning the internal structural parameters of SVM. Further, different modifications of termination criterion, initial population, and selection of teacher in case multiple learners achieve same score are suggested. These modifications and implementation of modified TLBO in searching of an optimum unique set of C, ε, and σ are expounded in the next section.

7.3.2.1 Modifications and marching procedure

Proper choice of searching ranges is a prerequisite for faster convergence of modified TLBO. In addition, objective function should be justifiably selected according to the proposed goal.

For better implementation of SVM methodology, it is suggested [18,20] to normalize the training input vectors within the range (0, 1). Thus, the chosen control parameters of the EDM process, that is, current, pulse-on time, and pulse-off time, are normalized within their corresponding experimental ranges.

$$x_{1,norm} = (cur - 6) / (15 - 6)$$

$$x_{2,norm} = (t_{on} - 50) / (200 - 50)$$

$$x_{3,norm} = (t_{off} 50) / (200 - 50) \tag{7.9}$$

$$MRR_{norm} = (MRR - 1.00) / (28.25 - 1.00)$$

$$ASR_{norm} = (ASR - 3.50) / (9.25 - 3.50)$$

Searching of C, ε, and σ should be robust in nature. Wide range of search spaces of C, ε, and σ may be a good choice, but irrelevant movements would take a lot of time to converge. Hence, searching range of these three parameters should be logically chosen. Aich and Banerjee [1] reported some experimental data-based techniques to choose these three ranges. For setting a searching range of C, upper end of six sigma range [18,20] of response values was considered. Near the boundary, there are some duplication errors due to further selection of a range, considering normal distribution over the upper end value of six sigma range. This seems to be erroneous in physical significance. Actually, the regularization parameter C should lie within the limit obtained from experimental values of corresponding response variable [5]. Therefore, range of experimental values might be a robust reasonable choice for searching the range of C.

$$\left(MRR_{exp}\right)_{min} \leq C_{MRR} \leq \left(MRR_{exp}\right)_{max}$$
$$\left(ASR_{exp}\right)_{min} \leq C_{ASR} \leq \left(ASR_{exp}\right)_{max} \tag{7.10}$$

Besides, searching ranges of ε and σ are chosen as [18,20]:

$$\varepsilon = \left(\frac{\bar{y}}{30}, \frac{\bar{y}}{10}\right); \sigma = \left[(0.1)^{1/z}, (0.5)^{1/z}\right] \tag{7.11}$$

Here, z indicates the number of most influencing attributes in the process. In EDM, these are three, namely current (cur), pulse-on time (t_{on}), and pulse-off time (t_{off}).

Searching ranges of C, ε, and σ are decided based on the experimental values of the respective response parameter. In the present study, as an optimum unique set of C, ε, and σ for both MRR and ASR is to be looked, searching ranges of C, ε, and σ for both MRR and ASR should be the same. Experimental values of MRR and ASR lie in different ranges. Therefore, they are normalized using Equation 7.9. Limits of searching ranges are revised based on these normalized response values. Combined searching range is obtained by union operation between these two individual searching ranges of C, ε, and σ. For example, based on normalized MRR and ASR, searching ranges of ε are calculated first by using Equation 7.11. For MRR, it is (0.0123, 0.0369), and for ASR, it is (0.0167, 0.0501). Performing union operation between these two ranges, combined search range is identified. Lower limit of combined search range of ε is estimated as maximum (0.0123, 0.0167) and upper limit as minimum (0.0369, 0.0501). Finally, combined search range of ε is decided as (0.0167, 0.0369). Similarly, searching ranges of C and σ are also identified. Optimum unique values of C, ε, and σ are to be searched within these combined searching ranges (Table 7.2).

Choice of different sets of internal structural parameters, C, ε, and σ, changes the values of Lagrange multipliers for each of MRR and ASR.

To build the best structure of the learning system for near-accurate predictions of responses, chance of generalization errors should be reduced in the learning process. Hence, internal parameters (C, ε, and σ) must be tuned in such a fashion so as to reduce the training errors in the learning process. Thereby, in this study, mean absolute training errors (MATE) in prediction of process responses, MRR ($MATE_1$) and ASR ($MATE_2$), are chosen as two objective functions.

$$MATE\ (\%) = \left(\frac{100}{N}\right) \Sigma_{i=1(1)N} \left[\frac{(|y_{i,exp} - y_{i,est}|)}{y_{i,exp}}\right] \qquad (7.12)$$

Training by experimental results with proper internal structural parameters of SVM regression (C, ε, and σ) is necessary to get near-exact representation of the process. The three internal structural parameters should be optimally tuned for each individual output–input combination. Thus, for multiple responses of a process with same input control parameters, separate sets of optimum C, ε, and σ for different responses are expected. In the present work, a methodology is proposed to build a unique structure of SVM regression for predicting multiple responses, that is, to search an optimum unique set of C, ε, and σ, instead of separate sets of C, ε, and σ for the responses. In the proposed steps, simultaneous minimization of $MATE_1$ and $MATE_2$ is carried out for selection of an optimum unique

Table 7.2 Searching ranges of SVM internal structural parameters—C, ε, and σ

SVM internal parameters	Material removal rate	Average surface roughness	Combined
C	(0.0000, 1.0000)	(0.0000, 1.0000)	(0.0000, 1.0000)
ε	(0.0123, 0.0369)	(0.0167, 0.0501)	(0.0167, 0.0369)
σ	(0.4642, 0.7937)	(0.4642, 0.7937)	(0.4642, 0.7937)

set of SVM's internal structural parameters C, ε, and σ. The TLBO with certain modifications is employed for this tuning operation. During simulation process, different sets of C, ε, and σ reshape the learning system. With same training input vectors and a particular set of C, ε, and σ, two different sets of Lagrange multipliers for two responses are calculated using corresponding individual training output vectors. Subsequently, with these two sets of Lagrange multipliers, normalized MRR and normalized ASR are predicted. These predicted values are denormalized with the help of Equation 7.9, and corresponding MATEs are evaluated using Equation 7.12, based on denormalized MRR and ASR. Finally, two different sets of Lagrange multipliers are calculated from the simulated optimum unique set of C, ε, and σ. When training errors become stable at their achievable minimum value, with corresponding set of Lagrange multipliers, MRR and ASR are estimated separately using Equation 7.8. Here, this multiobjective optimization is performed by algorithm-specific parameter-less TLBO.

Within the estimated searching ranges (refer Table 7.2), modified TLBO is applied for simultaneous minimization of MATE_1 and MATE_2. Although TLBO is a parameter-less optimization technique, still, to get this benefit in optimizing any nonlinear high-dimensional objective functions, termination criteria should be logically defined. In most of the optimization techniques, a termination criterion is defined by the maximum number of iterations or change in objective function value below a predefined margin. When optimizing a new objective function, it is very difficult to know earlier the required number of iteration to meet a certain target. Even to attain certain accuracy, change in objective function values may vary due to their different scale ranges. In some cases, attainable optimum objective function value is difficult to predict earlier. As such, a general termination criterion is required to propose for population-based searching techniques. A general meaningful criterion is suggested based on spread of population relative to searching range in different dimension; that is, spread-range (SR) ratio [5] is defined as a ratio of standard deviation of population to span of searching range expressed in %:

$$\text{SR ratio (\%)} = 100 \times \frac{\left(\text{standard deviation of population}\right)}{\left(\text{span of searching range}\right)} \quad (7.13)$$

Thereby, simulation will be stopped when this SR ratio along each of the input parameters' dimensions simultaneously goes down below a predefined limit. Here, this limit is chosen as 1%; that is, searching operation would be flagged off when SR ratio along C, ε, and σ dimensions simultaneously drops below 1%.

Metaheuristic techniques march to the global optimum with some randomly generated probabilistic logical movements. Whatever might be the termination criterion that is considered, if simulation is stopped by watching that the specific user-defined measurement just reaches below a certain value in any iteration, then it may be prematured. Simulation should be allowed for a few more iterations to finally freeze down below that specified limit. In the present work, SR ratio of latest population, that is, learner in each direction, C, ε, and σ, is used as termination criterion, and simulation is terminated when SR ratio values along all dimensions (C, ε, and σ) satisfy the termination criterion, that is, go below 1% in last consecutive five iterations.

In case of population-based optimization technique, a widely spread initial population must be ensured for better exploration in the whole range. As discussed earlier, a latest population-based termination criterion, that is, SR ratio of the latest population along each dimension, is considered. Therefore, initial SR ratio of the population must have a high value along all dimensions to ensure proper exploration of the search space. In the present work, considering initial SR ratio as at least 40% along each of the three directions, a set of 20 learners is randomly generated within specified search space (Table 7.2). For maintaining the repeatability of the simulation steps, initial learners are given in Table A1.

In each step of iteration, with different set of learners, that is, set of C, ε, and σ, shape of learning system changes. Teacher of any iteration should be selected as that set of C, ε, and σ having lowest training error (MATE) value. When optimizing multiple objective functions, the same set of C, ε, and σ might not give minimum value for both the objective functions. To overcome this difficulty, here, a ranking method is proposed. In a typical iteration, at first, rank the learners separately according to the objective functions' values, for example, that set of learners gets two sets of ranks, $rank_1$ based on $MATE_1$ and $rank_2$ based on $MATE_2$. These two rank matrices are element-wise multiplied to get a combined rank for all the current-set learners. Say, a learner that is a set of C, ε, and σ gets two ranks—4 and 17. These two rank values are multiplied, that is, $17 \times 4 = 68$. Similarly, combined ranks of other learners are calculated. These combined rank values always lie between 1 and (number of learners)^2. According to this combined rank matrix, the best learner is marked and set as teacher for subsequent teaching purpose. Here, objective function values are not multiplied at all; combined rank values are obtained only by element-wise multiplication of $rank_1$ and $rank_2$ matrices.

In most of the published studies of optimization algorithms [21], it is reported that the best one of the latest population works as guide for next iteration. The best one is chosen with either minimum or maximum objective function value. However, if multiple best settings in the population with same minimum or maximum objective function value are found, then confusion will come to choose only one among all those best settings. In case of selecting the current teacher, a combined rank method is already proposed in the last paragraph. If multiple learners give same best combined rank value, it would become difficult to choose only one among those learners. Improper choice may guide the following iterations in a wrong way and finally might be trapped inside any local optimum. A weight-combining method is reported by Aich and Banerjee [5]. Although their method is applied on learners having same objective function value, here, similar approach is taken on learners who give same combined rank value. In the present work, a weighted combination of all those learners is to be evoluted, such that new evoluted learner must give both $MATE_1$ & $MATE_2$ lower (higher for maximization) than either of the $MATE_1$ & $MATE_2$ corresponding to the learner having second-best combined rank at current population, applicable only for the first iteration, or the minimum $MATE_1$ & $MATE_2$ gained at immediate last iteration. For example, in case of simultaneous minimization of bivariable two objective functions within search space. ([0, 20], [0, 20]), at any iteration, one learner (9.7, 13.5) gets two ranks as 12 and 3 and another learner (18.2, 7.3) gets two ranks as 4 and 9. Therefore, two learners have same combined rank, $12 \times 3 = 4 \times 9 = 36$. At last iteration, minimum $MATE_1$ and minimum $MATE_2$ were 53.92 and 24.73, respectively. Now, one must choose the teacher among these two learners for next iteration. No such clear guidance is reported till now to choose the right one among these two. Here, a weighted combination of these two learners along their respective dimensions is calculated. Randomly, two weights (rw_1, rw_2) are generated between (0, 1), such that $rw_1 + rw_2 = 1$. A new learner is evoluted as ($rw_1 \times 9.7 + rw_2 \times 18.2$, $rw_1 \times 13.5 + rw_2 \times 7.3$). For $rw_1 = 0.4$ and $rw_2 = 0.6$, new learner would be (14.80, 9.78), which gives $MATE_1$ as 19.66 and $MATE_2$ as 10.29. New evoluted learner gives both $MATE_1$ and $MATE_2$ less than the minimum $MATE_1$ (53.92) and $MATE_2$ (24.73) gained at last iteration. In case of first iteration, comparison would be done, with $MATE_1$ and $MATE_2$ corresponding to the learner having second-best combined rank at current population. Thus, learner (14.80, 9.78) would be the teacher for next iteration; otherwise, the steps are repeated with another random set of weights (rw_1, rw_2), until the above said condition is fulfilled. However, there is no need to update current population with this evoluted teacher. This proposition is expected to be effective to avoid ambiguity to choose the right teacher at any iteration.

Adapting all the above said modifications, steps of modified TLBO algorithm used for searching an optimum unique set of C, ε, and σ by simultaneously minimizing MATEs (Equation 7.12) in the estimation of both the responses MRR ($MATE_1$) and ASR ($MATE_2$) are discussed below.

Step 1: Normalize the control parameters, cur, t_{on}, and t_{off}, and process responses, MRR and ASR, using Equation 7.9. The MATEs (Equation 7.12) in the estimation of MRR ($MATE_1$) and ASR ($MATE_2$) are separately considered as two objective functions. Set $n = 20$ and $iter_{max} = 250$.

Step 2: Calculate two searching ranges of C, ε, and σ, based on normalized MRR and ASR separately. Do the union operation between these two searching ranges, and get the combined searching ranges of C, ε, and σ (refer Table 7.2).

Step 3: Set iter = 1 and termination criterion as SR ratio along all three dimensions <1% in consecutive five iterations. Randomly (following uniform distribution), generate n set of learners (Table A1), with SR ratio along each dimension >40% within search space.

Step 4: With normalized training input and output vectors, for each of the current set of n learners, two different sets of Lagrange multipliers for normalized MRR and normalized ASR are calculated separately. With these Lagrange multipliers, normalized MRR and normalized ASR are estimated, corresponding to the current set of learners. These estimated normalized MRR and normalized ASR are denormalized with the help of Equation 7.9, and corresponding $MATE_1$ and $MATE_2$ are calculated.

Step 5: Rank all n learners with respect to their corresponding $MATE_1$ and $MATE_2$; store these two sets of ranks in $rank_1$ and $rank_2$ matrices, respectively. Get combined rank by element-wise multiplication of $rank_1$ and $rank_2$ matrices. If multiple learners have same best combined rank, go to step 6; otherwise, learner having the best combined rank is selected as current $teacher_{iter}$; then, go to step 7.

Step 6: Learners having same best combined rank are identified. Make a weighted combination of those identified learners, such that the new evoluted learner must give both $MATE_1$ and $MATE_2$ lower than either the $MATE_1$ and $MATE_2$ corresponding to the learner having the second-best combined rank at current population (applicable only for first iteration) or the minimum $MATE_1$ and $MATE_2$ gained at immediate last iteration. The new evoluted learner is selected as current $teacher_{iter}$.

Step 7: Find out mean of current all n learners and estimate the adapted TF as:

$$TF_{iter} = \frac{current\ mean_{iter}}{current\ teacher_{iter}} \tag{7.14}$$

Step 8: Calculate SR ratio along all three dimensions—C, ε, and σ. If termination criterion is satisfied, stop simulation and declare the latest teacher as the optimum unique set of C, ε, and σ; otherwise, go to step 9.

Step 9: If iter = iter$_{max}$, then go to step 3 and restart the simulation with higher iter$_{max}$; otherwise, set $t = 1$ and go to step 10.

Step 10: Calculate the new tth learner taught by the current teacher$_{iter}$ following the relation

new learner$_t$ = current learner$_t$ + rand

$$\times \left(\text{current teacher}_{iter} - \text{TF}_{iter} \times \text{current mean}_{iter} \right) \quad (7.15)$$

$$t = 1(1)n$$

Step 11: If MATE$_{1, \text{new } t}$ < MATE$_{1, t}$ and MATE$_{2, \text{new } t}$ < MATE$_{2, t}$, then replace tth learner of current population by new tth learner and go to step 12; otherwise, tth learner of current population is kept unaltered. Then, go to step 12.

Step 12: If $t = n$, then set $k = 1$ and go to step 13; otherwise, set $t = t + 1$ and go to step 10.

Step 13: Select random integer r between 1 and n, except k.

Step 14: If MATE$_{1, k}$ < MATE$_{1, r}$ and MATE$_{2, k}$ < MATE$_{2, r}$, then calculate new kth learner sharing knowledge with current rth learner, using Equation 7.16, and go to step 16; otherwise, go to step 15.

new learner$_k$ = current learner$_k$ + rand

$$\times \left(\text{current learner}_k - \text{current learner}_r \right) \quad (7.16)$$

Step 15: If MATE$_{1, k}$ > MATE$_{1, r}$ and MATE$_{2, k}$ > MATE$_{2, r}$, then calculate new kth learner sharing knowledge with current rth learner, using Equation 7.17, and go to step 16. Otherwise, kth learner of current population is kept unaltered; then, go to step 16.

new learner$_k$ = current learner$_k$ + rand

$$\times \left(\text{current learner}_r - \text{current learner}_k \right) \quad (7.17)$$

Step 16: If $k = n$, replace the learners of current population by corresponding new learners; set iter = iter + 1; and go to step 4. Otherwise, set $k = k + 1$ and go to step 13.

Therefore, latest teacher is selected as unique set of C, ε, and σ. With this set of C, ε, and σ, two separate sets of Lagrange multipliers are calculated using the corresponding normalized training output vectors of MRR and ASR. The unified structure of SVM regression learning system of the EDM process could be represented by Equation 7.8. Prediction of normalized MRR and normalized ASR could be done separately by pouring their respective set Lagrange multipliers into this Equation 7.8. Predicted normalized MRR and ASR are denormalized and subsequently finally achieved training errors—$MATE_1$ and $MATE_2$—are estimated.

Now, using the above said TLBO algorithm adapted with all discussed modifications, training errors in the prediction of MRR ($MATE_1$) and ASR ($MATE_2$) are minimized simultaneously for different settings of C, ε, and σ within combined searching ranges (refer to Table 7.2). As the simulation marches, with different values of C, ε, and σ, the shape of the learning system gets modified and, consequently, training errors are changed. Finally, the optimum unique set of C, ε, and σ within the specified searching ranges (Table 7.2) with achievable minimum MATE—$MATE_1$ and $MATE_2$—is found and reported in Table 7.3. Modified TLBO algorithm is coded in MATLAB R2012a and LibSVM command line functions are used for the SVM learning process.

Optimum unique value of C is shifted toward the upper end of search space. This indicates the complexity of the model, which is in favor of the stochastic behavior of the EDM process. The random fluctuations could be controlled by proper choice of ε. Here, lower value of ε indicates that the learning system could be able to absorb the random variations adequately. Besides, small σ value claims that the unified learning system is stable and generalized by entrapping the oscillatory patterns in outputs outside the insensitive zones.

With this simulated optimum unique set of C, ε, and σ (listed in Table 7.3), two sets of Lagrange multipliers (α_i, α_i^*) for normalized MRR and for ASR are calculated separately (Table A2). Representative models of the developed unified structure of SVM regression learning system are given by Equation 7.18.

Table 7.3 Results of tuning internal structural parameters (C, ε, and σ) of SVM for unified learning

Response	Optimum unique SVM internal parameters for normalized responses			Simulation time (s)	No. of support vectors	Bias	Performance	
	C	ε	σ				MATE (%)	r^2
MRR	1.0000	0.0167	0.4642	871.0905	37	0	6.50	0.9855
ASR					49	0	3.31	0.9527

$$f(\mathbf{x}) = \Sigma_{i=1(1)N} \left(\alpha_i - \alpha_i^* \right)_j K\left(x_i, \mathbf{x} \right) + b \left| \begin{array}{rcl} C &=& 1.0000 \\ \varepsilon &=& 0.0167 \\ \sigma &=& 0.4642 \end{array} \right. \tag{7.18}$$

with $j = 1$ for normalized MRR, $j = 2$ for normalized ASR, and

$$K(x_i, \mathbf{x}) = \exp\left(-\frac{\|x_i - \mathbf{x}\|^2}{2\sigma^2} \right) \Bigg|_{\sigma = 0.4642}$$

Marching steps for searching an optimum unique set of C, ε, and σ in simultaneous estimation of MRR and ASR are given in the following flow-chart (Figure 7.4). Corresponding to current teacher$_{iter}$, of each iteration, MATE$_1$ and MATE$_2$ are calculated, and their gradual decaying patterns are represented in Figures 7.5 and 7.6. Observing the components of SR ratio of current population at the end of each iteration, influence of three internal structural parameters—C, ε, and σ—on simultaneous minimization could be understood (refer to Figure 7.7).

In case of minimizing MATEs, MATE$_1$ and MATE$_2$, relative to C and ε, the effect of σ is marginally lower, as SR ratio for σ decreases at a faster rate relative to C and ε (Figure 7.8). After a few iterations, absence of irregular fluctuations of SR ratios along all three dimensions indicates the convergence of simulation procedure toward global optimum in a smooth way.

7.3.3   Testing of unified learning system

Unified learning system of MRR and ASR (Equation 7.18) is tested with 10 disjoint data sets obtained from separate follow-up experimental runs. For testing purpose, testing input vectors are normalized using Equation 7.9. Sets of Lagrange multipliers of normalized MRR and normalized ASR (refer Table A2) are separately fed to the learning system (Equation 7.18), and testing output vectors—normalized MRR and normalized ASR—are estimated separately. These estimated normalized outputs are denormalized with the help of Equation 7.9. Steps for concurrent estimation of MRR and ASR in testing are shown in Figure 7.8. The absolute errors in prediction, with corresponding experimental values, are calculated and presented in Tables 7.4 and 7.5.

Mean absolute testing errors (Tables 7.4 and 7.5) for both MRR (3.51%) and ASR (3.37%) indicate the practical adequacy of the developed unified

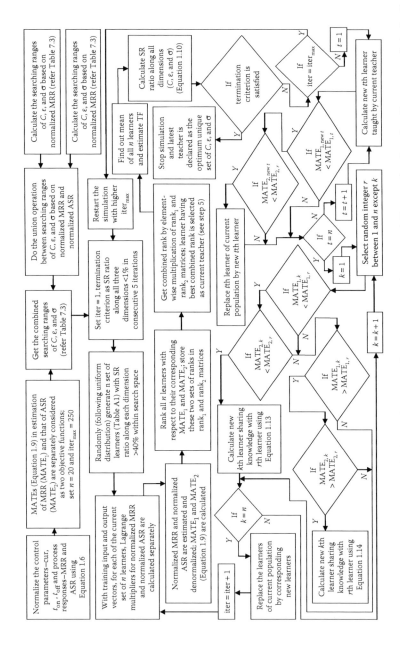

Figure 7.4 Sequence diagram of modified TLBO to search optimum unique set of C, ε, and σ by simultaneous minimization of $MATE_1$ and $MATE_2$.

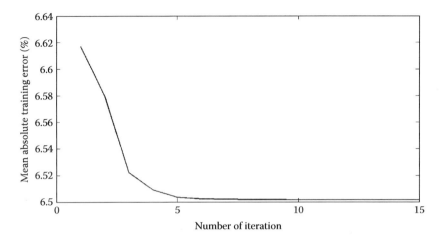

Figure 7.5 Changes in MATE in the estimation of MRR (MATE$_1$).

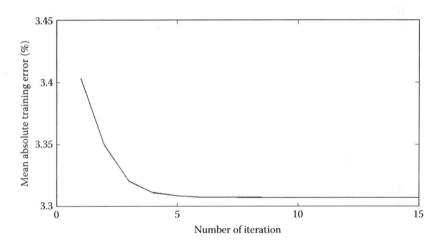

Figure 7.6 Changes in MATE in the estimation of ASR (MATE$_2$).

structure of SVM regression learning system for prediction of MRR and ASR in the EDM process within their experimental ranges.

To depict the effects of different process parameters (current, pulse-on time, and pulse-off time) on responses, surface plots for MRR and ASR are generated using Equation 7.18 and subsequent denormalization (Figures 7.9 through 7.14).

For both MRR and ASR, current shows a strong positive influence, whereas the other two control parameters—pulse-on time and pulse-off

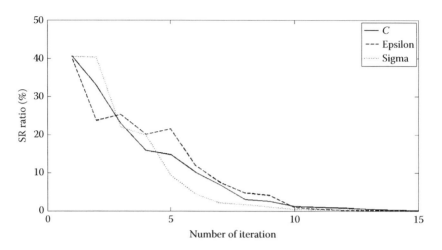

Figure 7.7 Change of SR ratio along C, ε, and σ during simultaneous minimization of $MATE_1$ and $MATE_2$.

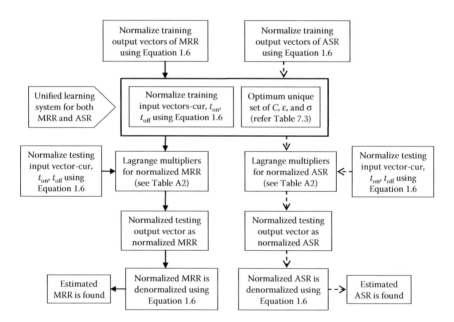

Figure 7.8 Steps for concurrent estimation of MRR and ASR from unified structure of SVM regression learning system.

Table 7.4 Testing of estimated MRR

	Control parameters			Material removal rate		
S. no.	Current (A)	Pulse-on time (µs)	Pulse-off time (µs)	Experimental (mm³/min)	Estimated (mm³/min)	Absolute error (%)
1	6	100	50	5.48126	5.51254	0.57
2	6	200	150	4.56557	4.79086	4.93
3	9	100	100	9.13364	8.60628	5.77
4	9	150	50	13.50951	12.70528	5.95
5	9	200	100	10.48887	10.45321	0.34
6	12	50	100	9.46479	9.93234	4.94
7	12	100	50	19.36570	18.94783	2.16
8	12	150	200	11.36906	11.43744	0.60
9	15	100	100	18.06487	16.80110	7.00
10	15	150	50	24.95816	25.65435	2.79
Mean absolute testing error (%)						3.51

Table 7.5 Testing of estimated ASR

	Control parameters			Average surface roughness		
S. no.	Current (A)	Pulse-on time (µs)	Pulse-off time (µs)	Experimental (µm)	Estimated (µm)	Absolute error (%)
1	6	100	50	4.61	4.59	0.43
2	6	200	150	4.21	4.31	2.38
3	9	100	100	5.79	6.00	3.63
4	9	150	50	6.95	6.40	7.91
5	9	200	100	6.22	6.20	0.32
6	12	50	100	5.88	5.78	1.70
7	12	100	50	7.44	6.90	7.26
8	12	150	200	7.35	7.55	2.72
9	15	100	100	7.48	7.26	2.94
10	15	150	50	8.49	8.12	4.36
Mean absolute testing error (%)						3.37

time—are found to be not so effective with their changes. At lower values of current, effects of both pulse-on time and pulse-off time on MRR and ASR are almost insignificant. In the higher zone of current values, higher MRR could be obtained by increasing pulse-on time or by lowering pulse-off time. Although variation of pulse-off time does not show significant change in ASR, even at higher values of current, with increase of pulse-on time, ASR is found to be increased at upper zone of current space.

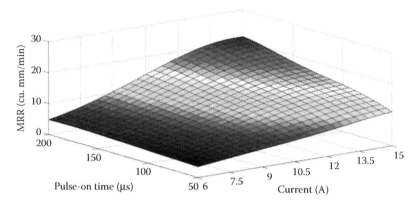

Figure 7.9 Effect of current and pulse-on time on MRR at pulse-off time 125 μs.

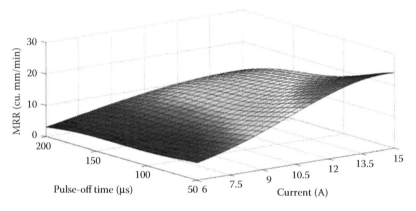

Figure 7.10 Effect of current and pulse-off time on MRR at pulse-on time 125 μs.

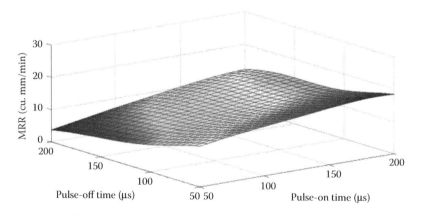

Figure 7.11 Effect of pulse-on time and pulse-off time on MRR at current 10.5 A.

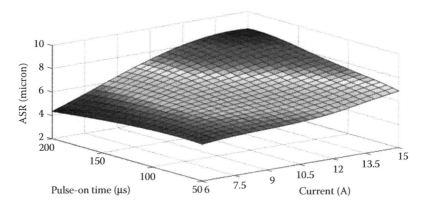

Figure 7.12 Effect of current and pulse-on time on ASR at pulse-off time 125 μs.

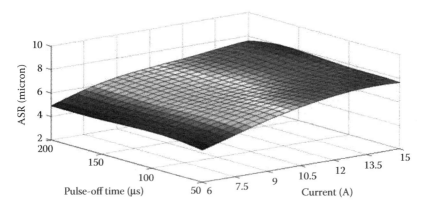

Figure 7.13 Effect of current and pulse-off time on ASR at pulse-on time 125 μs.

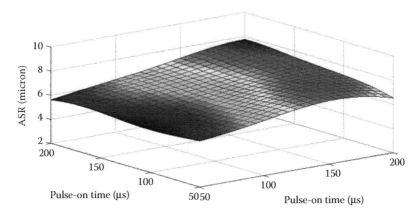

Figure 7.14 Effect of pulse-on time and pulse-off time on ASR at current 10.5 A.

The higher coefficient of determinations (r^2s) (refer to Table 7.3) indicates the strong correlations between experimental and estimated responses (denormalized).

Although the proposed methodology is described with the aid of experimental results of MRR and ASR in the EDM process, the suggested steps could be easily adopted for other representative model development problems in a generalized way.

7.4 Conclusion

In the present work, a simple methodology is proposed to develop a unified structure of SVM regression-based learning system of MRR and ASR in the EDM process, with internal structural parameters tuned by modified TLBO.

- Modification over standard TLBO—combined rank method is introduced for simultaneous minimization of MATEs in the estimation of MRR and ASR.
- In combined rank method, learners of current population are combined with weight vectors instead of objective functions ($MATE_1$ and $MATE_2$) to reserve the independent impacts of objective functions.
- An optimum single set of SVM internal structural parameters—C, ε, and σ—for both the MRR and the ASR is obtained, instead of two separate sets of C, ε, and σ for two individual responses.
- Optimum unique set of C, ε, and σ, generates two sets of Lagrange multipliers on feeding the corresponding normalized process outcomes separately to the developed unified learning system for estimation of MRR and ASR, respectively.

The proposed way for developing unified structure of SVM regression learning system for concurrent prediction of multiple outcomes could be a guideline for building compact data generator in virtual world of any such process. Simultaneous handling of multiple objective functions, without affecting their individual influences, as suggested in the present study, would be a stepping stone to multiobjective optimization.

Acknowledgment

This work is supported by the Council of Scientific and Industrial Research, Human Resource Development Group, India (File no. 09/096 (0833)/2015-EMR-I).

Appendix

Table A.1 Initial learner population for searching optimum unique set of C, ε, and σ by modified TLBO

Learner no.	C	ε	σ
1	0.0562	0.0169	0.6413
2	0.9825	0.0168	0.4735
3	0.9739	0.0329	0.4921
4	0.3274	0.0168	0.7914
5	0.8495	0.0352	0.7303
6	0.2533	0.0182	0.4764
7	0.9156	0.0261	0.7927
8	0.8193	0.0365	0.4666
9	0.0469	0.0351	0.6027
10	0.0624	0.0234	0.7503
11	0.9999	0.0335	0.7616
12	0.0312	0.0187	0.4731
13	0.7384	0.0366	0.7820
14	0.1198	0.0369	0.7528
15	0.8958	0.0183	0.7443
16	0.0899	0.0232	0.7526
17	0.1775	0.0176	0.7766
18	0.0691	0.0175	0.5132
19	0.7849	0.0178	0.4744
20	0.9833	0.0262	0.6943
Initial SR ratio (%)	40.64	40.18	40.45

Table A.2 Difference of Lagrange multipliers (α_i, α_i^*) for normalized MRR and normalized ASR

Sl. no.	Training input vector (cur, t_{on}, t_{off})	Difference of Lagrange multipliers for normalized MRR	Difference of Lagrange multipliers for normalized ASR
1	(6, 50, 50)	−0.000000002796	−0.764554207673[b]
2	(6, 50, 100)	−0.000000009512	0.884361167229[b]
3	(6, 50, 150)	0.000000000122	0.510197961001[b]
4	(6, 50, 200)	−0.000000016877	−0.999999994898[b]
5	(6, 100, 100)	−0.432080569555[a]	−0.999999998139[b]
6	(6, 100, 150)	0.000000032870	0.154408781906[b]
7	(6, 100, 200)	−0.000000005594	0.710914971657[b]

(Continued)

Table A.2 **(Continued)** Difference of Lagrange multipliers (α_i, α_i^*)
for normalized MRR and normalized ASR

Sl. no.	Training input vector (cur, t_{on}, t_{off})	Difference of Lagrange multipliers for normalized MRR	Difference of Lagrange multipliers for normalized ASR
8	(6, 150, 50)	0.127462045052[a]	0.037989657879[b]
9	(6, 150, 100)	0.141377784440[a]	0.999999995288[b]
10	(6, 150, 150)	0.464622968706[a]	−0.999999995293[b]
11	(6, 150, 200)	−0.217389039837[a]	−0.551957179210[b]
12	(6, 200, 50)	−0.009938496912[a]	0.000000001114
13	(6, 200, 100)	−0.000000004138	−0.135215960250[b]
14	(6, 200, 200)	−0.430589900521[a]	0.077156773467[b]
15	(9, 50, 50)	0.355462431335[a]	0.999999994509[b]
16	(9, 50, 100)	0.273609069455[a]	−0.295041059654[b]
17	(9, 50, 150)	−0.000000002954	−0.999999996557[b]
18	(9, 50, 200)	0.000000002403	0.999999996087[b]
19	(9, 100, 50)	−0.000000007705	−0.090219565910[b]
20	(9, 100, 150)	0.217241099393[a]	0.999999996909[b]
21	(9, 100, 200)	−0.000000001207	−0.289849585477[b]
22	(9, 150, 100)	−0.420062202030[a]	0.999999997362[b]
23	(9, 150, 150)	0.000000002994	−0.436752900699[b]
24	(9, 150, 200)	−0.138829893909[a]	0.999999998078[b]
25	(9, 200, 50)	−0.000000029292	−0.875041277073[b]
26	(9, 200, 150)	−0.273962234136[a]	0.000000001502
27	(9, 200, 200)	0.999999998834[a]	0.321212248902[b]
28	(12, 50, 50)	−0.332688993668[a]	−0.301423445472[b]
29	(12, 50, 150)	−0.725705107150[a]	−0.000000010391
30	(12, 50, 200)	0.071476440548[a]	0.079249244142[b]
31	(12, 100, 100)	−0.469673124348[a]	−0.914066753160[b]
32	(12, 100, 150)	0.898274485156[a]	0.770128415857[b]
33	(12, 100, 200)	0.000000001649	−0.999999996900[b]
34	(12, 150, 50)	−0.000000013432	0.999999998585[b]
35	(12, 150, 100)	0.999999995452[a]	−0.917033628344[b]
36	(12, 150, 150)	−0.999999992952[a]	−0.999999996259[b]
37	(12, 200, 50)	0.538360015461[a]	−0.086447265691[b]
38	(12, 200, 100)	−0.999999998048[a]	0.999999998030[b]
39	(12, 200, 150)	0.999999993285[a]	−0.000000009813
40	(12, 200, 200)	−0.574857959601[a]	−0.094448806725[b]
41	(15, 50, 50)	0.920212678063[a]	0.999999997745[b]
42	(15, 50, 100)	−0.999999996271[a]	−0.828828033849[b]

(Continued)

*Table A.2 (**Continued**)* Difference of Lagrange multipliers (α_i, α_i^*) for normalized MRR and normalized ASR

Sl. no.	Training input vector (cur, t_{on}, t_{off})	Difference of Lagrange multipliers for normalized MRR	Difference of Lagrange multipliers for normalized ASR
43	(15, 50, 150)	0.776567167323[a]	0.999999998710[b]
44	(15, 50, 200)	−0.000000001486	−0.451792235691[b]
45	(15, 100, 50)	0.495825151268[a]	−0.486546465507[b]
46	(15, 100, 150)	0.999999998702[a]	0.913875910423[b]
47	(15, 100, 200)	−0.969917192846[a]	0.000000007371
48	(15, 150, 100)	−0.999999998601[a]	0.999999996940[b]
49	(15, 150, 150)	−0.568482595264[a]	−0.999999998590[b]
50	(15, 150, 200)	0.999999902894[a]	0.616622070391[b]
51	(15, 200, 50)	0.846108281196[a]	0.587529799685[b]
52	(15, 200, 100)	0.000000000960	−0.999999998437[b]
53	(15, 200, 150)	0.999999996666[a]	0.832877246791[b]
54	(15, 200, 200)	−0.515946613189[a]	0.417561755751[b]

[a] indicates support vectors for normalized MRR.
[b] indicates support vectors for normalized ASR.

References

1. Aich, U. and Banerjee, S. (2014). Modelling of EDM responses by support vector machine regression with parameters selected by particle swarm optimization. *Applied Mathematical Modeling, 38*:2800–2818.
2. Rao, R. V. and Patel, V. (2012). An elitist teaching-learning-based optimization algorithm for solving complex constrained optimization problems. *International Journal of Industrial Engineering Computations, 3*:535–560.
3. Rao, R. V., Savsani, V. J. and Vakharia, D. P. (2012). Teaching-Learning-Based Optimization: An optimization method for continuous non-linear large scale problems. *Information Sciences, 183*:1–15.
4. Rao, R. V., Savsani, V. J. and Vakharia, D. P. (2011). Teaching-learning-based optimization: A novel method for constrained mechanical design optimization problems. *Computer-Aided Design, 43*:303–315.
5. Aich, U. and Banerjee, S. (2016). Application of teaching learning based optimization procedure for the development of SVM learned EDM process and its pseudo Pareto optimization. *Applied Soft Computing, 39*:64–83.
6. Konig, W., Dauw, D. F., Levy, G. and Panten, U. (1988). EDM future steps towards the machining of ceramics. *Annals of CIRP, 37*:623–631.
7. Ho, K. H. and Newman, S. T. (2003). State of the art electrical discharge machining (EDM). *International Journal of Machine Tools and Manufacture, 43*:1287–1300.
8. Kunieda, M., Lauwers, B., Rajurkar, K. P. and Schumacher, B. M. (2005). Advancing EDM through fundamental insight of the process. *Annals of the CIRP, 54*:599–622.

9. Liao, Y. S., Huang, J. T. and Su, H. C. (1997). A study on the machining-parameters optimization of wire electrical discharge machining, *Journal of Materials Processing Technology*, 71:487–493.

10. Banerjee, S., Mahapatro, D. and Dubey, S. (2009). Some study on electrical discharge machining of ({WC+TiC+TaC/NbC}-Co) cemented carbide. *International Journal of Advanced Manufacturing Technology*, 43:1177–1188.

11. Panda, D. K. and Bhoi, R. K. (2005). Artificial neural network prediction of material removal rate in electrodischarge machining. *Materials and Manufacturing Processes*, 20:645–672.

12. Vapnik, V. (1995). *The Nature of Statistical learning Theory.* Springer, New York.

13. Smola, A. J. and Scholkopf, B. (2004). A tutorial on support vector regression. *Statistics and Computing*, 14:199–222.

14. Cristianini, N. and Taylor, J. S. (2000). *An Introduction to Support Vector Machines and other Kernel-based-Learning Methods.* Cambridge University Press, Cambridge.

15. Sapankevych, N.I. and Sankar, R. (2009). Time series predictionusing support vectormachines: A survey. *IEEE Computational Intelligence Magazine*, 4:24–438.

16. Gunn, S. R. (1998). Support vector machines for classification and regression. Technical report, University of Southampton.

17. Yu, P. S., Chen, S. T. and Chang, I. F. (2006). Support vector regression for real-time flood stage forecasting. *Journal of Hydrology*, 328:704–716.

18. Levis, A. A. and Papageorgiou, L. G. (2005). Customer demand forecasting via support vector regression analysis. *Chemical Engineering research and Design*, 83:1009–1018.

19. Rao, R. V. and Patel, V. (2013). Multi-objective optimization of two stage thermoelectric cooler using a modified teaching-learning-based optimization algorithm. *Engineering Applications of Artificial Intelligence*, 26:430–445.

20. Cherkassky, V. and Ma, Y. (2004). Practical selection of SVM parameters and noise estimation for SVM regression. *Neural Networks*, 17:113–126.

21. Rao, S. S. (2009). *Engineering Optimization Theory and Practice.* John Wiley & Sons, Hoboken, NJ.

chapter eight

Modeling of grind-hardening

Angelos P. Markopoulos, Emmanouil L. Papazoglou,
Nikolaos E. Karkalos, and Dimitrios E. Manolakos

Contents

8.1 Introduction 211
8.2 Modeling of grinding wheel, forces, and heat partition 215
    8.2.1 Modeling of grinding wheel topology 215
    8.2.2 Process forces... 217
        8.2.2.1 Slip force calculation.. 217
        8.2.2.2 Cutting force 218
    8.2.3 Production and partition of heat 219
8.3 Analytical modeling... 221
8.4 Numerical solution using ANSYS... 223
    8.4.1 Description of general details of the model......................... 223
    8.4.2 Verification of the model... 226
    8.4.3 Thermomechanical material properties 226
8.5 Results and discussion... 228
    8.5.1 Analytical model.. 228
    8.5.2 Results obtained from ANSYS simulations 232
    8.5.3 Method's accuracy test 236
    8.5.4 Effect of grinding fluid use 237
        8.5.4.1 Grinding fluid model.. 237
        8.5.4.2 Results of grinding fluid use in
                  grinding-hardening experiments............................ 238
Nomenclature 241
References.. 243

8.1 Introduction

High-precision production of steel components usually includes a hardening process, in order to alter the structure of the components' surface. Conventional thermal processing methods are characterized by increased energy consumption, and they involve the use of pollutant solvents—salts.

At the same time, it is usual for the hardening process to be conducted away from industries and production lines, leading to additional energy, environmental, and financial cost for the fabrication of final products. Furthermore, the need for cleaning of products before the thermal process occurs requires large amounts of water.

Grind-hardening or grinding-hardening process is considered as a hybrid process, which can be employed for the simultaneous surface hardening and grinding of metal components, thus reducing the number of steps of the total manufacturing procedure, along with the disadvantages, which are related to each of those steps. Specifically, this process can be conducted in the same machine tool, using the same setup. Grind-hardening process is based on the control of the amount of heat produced by the process itself, which leads to a localized heating of the workpiece and the subsequent increase of surface roughness through a martensitic transformation.

Before introducing the methodology followed in this study, it is considered noteworthy to discuss some significant previous works on grind-hardening process, most of which were conducted during the past decade.

At first, a concise and global presentation of this manufacturing process, leading to a sufficient level of understanding of grind-hardening, was given by Brockhoff and Brinksmeier [1]. Focusing primarily on the needs of industry for technoeconomic-efficient surface thermal processing of workpieces, they proposed the utilization of heat amount produced during grinding for conducting a hardening process at the same time. Initially, they conducted studies on the heat produced during grinding and mainly the amount absorbed by the workpiece, with a view to avoid undesired thermomechanically induced phenomena such as surface cracks or regions with different hardness value. Required power during the process, the temperature field in the workpiece, and the partition of heat were able to be computed, in correlation with process parameters and material properties by conducting experiments and through analytical models. The next step proposed by the authors was the utilization of produced heat in the framework of a hybrid manufacturing process.

From the aforementioned work, valuable information concerning the effect of process conditions to the outcome of the hardening process can be extracted. In the first place, it is pointed out that an increase of depth of cut with constant feed speed can lead not only to larger forces and power but also to a reduction of specific power, due to an increase in the contact length. Specific energy, absorbed by the workpiece, is increased with an increase in depth of cut, reaching a maximum of $e_w = 150$ J/mm^2 for a depth of cut $a_e = 1$ mm. By increasing feed speed, cutting forces and power are increased. However, due to the reduction of processing time, the amount of specific energy absorbed by the workpiece is reduced. More specifically, for an increase in feed speed from $u_w = 0.00167$ m/s to

$u_w = 0.0833$ m/s, a reduction of specific energy from $e_w = 1150$ J/mm² to $e_w = 25$ J/mm² is observed. Consequently, when using low feed speeds, a high heat partition ratio toward the workpiece and lower power are observed, whereas when using high feed speeds, the opposite situation may occur. Workpiece material is also an important process parameter, as maximum hardness value depends on carbon content and percentages of other alloying materials. Quenched steels can be processed to achieve larger hardness penetration depths (HPDs) in comparison with annealed steels, due to the more favorable distribution of carbides. Furthermore, the use of cutting fluid is not suggested, as a large amount of heat is dissipated away, except for the case of workpieces with small volume, for which cooling is not sufficient.

In the work of Nguyen et al. [2], the use of liquid nitrogen as a cooling medium is proposed and its advantages are discussed. It is accentuated that grind-hardening process can induce several undesired effects to the workpiece, such as reduced surface quality, low dimensional accuracy, residual stresses, unfavorable workpiece material microstructure alterations, and intense surface corrosion. Liquid nitrogen is already being used during welding to avoid corrosion and also as a cryogenic process, so that the remaining austenite from classical thermal process is transformed to martensite. Taking this into consideration, the authors proposed the use of liquid nitrogen as a cooling medium, given also its capability to lead to high rates of cooling. They used AISI 1045 steel as workpiece material (with an orthogonal parallelepiped geometry) and the following process parameters: $u_s = 23$ m/s, $u_w = 0.0067$ m/s, and $a_e = 0.02$ mm. It must be noted that these conditions are considered as rather low feed speed and depth of cut values, respectively. The authors observed an improved surface quality and almost barely noticeable oxidation marks in comparison with surfaces processed in air, which also have very rough surfaces. Moreover, they managed to achieve very high hardness, with a maximum value of $HV_{(500gr)} = 1100$, whereas without the use of liquid nitrogen, the maximum hardness achieved by grinding-hardening was $HV_{(500gr)} = 750$ and, when they employed a conventional cutting fluid, no increase in hardness was observed. On the other hand, using liquid nitrogen as a cooling medium, HPD is reduced significantly.

In another work, Nguyen and Zhang [3] used a 3D finite element method (FEM) computational model to compute the temperature field in the workpiece during grind-hardening and HPD; they investigated the possibility of using liquid nitrogen as a cooling medium. AISI 1045 steel workpieces of orthogonal parallelepiped geometry were again employed, and selected process parameters were $u_s = 25$ m/sec, $u_w = 0.01667$ m/sec, and $a_e = 0.05$ mm and 0.1 mm. Investigations were conducted both for grinding-hardening in air and with the use of liquid nitrogen. As it was also observed in their earlier study, relatively low values of feed speed and depth of cut were chosen. Their results indicate that a deviation of 7.8%–12.9% in comparison with the experimental temperature values was noted for the FEM model, but it was able to predict HPD with a deviation of 4.2%–7.9%.

Han et al. [4] stated that although in several cases of industrial practice such as the fabrication of crankshafts in automotive industries, nonquenched steel components are currently more preferable, the need of components with high surface hardness is still present in industrial applications, and, in this manner, alternative methods of surface hardness, such as high- or medium-frequency induction hardening, are tested in industrial applications. In the current framework of demands from the industry for efficient surface hardness processes, grind-hardening process was studied for 400Cr steel workpieces of orthogonal parallelepiped geometry. The process parameters were $u_s < 40$ m/s, $u_w = 0.005$–0.025 m/s, and $a_e = 0.05$–0.25 mm, and the process was conducted under dry conditions. The results indicated a reduction in HPD when feed speed was increased, whereas increasing depth of cut up to a point led to an increase of HPD, but after that, a decrease of HPD was observed. Maximum surface hardness value was measured to be 700 HV, 2.8 times higher than the initial hardness value. Finally, they also stated that greatest improvement in components' hardness can be observed in nonquenched steels.

Zhang et al. [5] conducted a computational study of grind-hardening process and compared their results with experimental ones. They chose AISI 1020 as work material; orthogonal parallelepiped workpieces and process parameters were selected as: $u_s = 19.6$ m/s, $u_w = 0.01$–0.05 m/s, and $a_e = 0.1$–0.3 mm, while the process was conducted under dry conditions. The computed maximum temperatures are similar to the experimental ones. Moreover, it was found that maximum surface hardness increased from 220 HV to 520–660 HV with increasing depth of cut; this effect can be justified by the uneven distribution of ferrite and perlite in the workpiece. The HPD was calculated with a deviation of 7%, and based on these results, HPD's correlation with depth of cut and feed speed was established. More specifically, it was found that HPD's correlation with feed speed can be considered nonlinear. For some of the process parameters' combinations, no hardening was observed, due to the low temperatures.

Liu et al. [6] investigated the effect of various parameters on the development of zones with different microstructure inside the machined workpiece. More specifically, they conducted experiments on AISI 1060 steel workpieces with orthogonal parallelepiped geometry, and process parameters were: $u_s = 26.3$ m/sec, $u_w = 0.008$ m/sec, and $a_e = 0.2$–0.5 mm under dry conditions. They observed two distinct zones within the HPD: a fully hardened zone and a transitional zone. Hardness within the fully hardened zone was between 750 HV and 780 HV, 1.4 times higher than that of common quenched steels and 2.4 times higher than that of common annealed steels. The HPD was found to increase almost linearly with respect to the cutting depth, and almost no difference was attributed to the initial condition of workpiece, as quenched steel workpieces had slightly higher HPD than annealed ones.

Apart from grind-hardening process for orthogonal parallelepiped workpieces, investigation for grinding-hardening of cylindrical workpieces was also conducted [7–10]. In the present study, it is intended that only orthogonal parallelepiped workpieces will be studied, and no further details on particularities of grinding-hardening process for cylindrical workpieces will be discussed. Other noteworthy works on grind-hardening are the work conducted by Alonso et al. [11] and Salonitis [12], concerning residual stresses developed on workpieces during grind-hardening.

In this study, the focus is set on the study of the grind-hardening process for relatively high feed speed and depths of cut. In the relevant literature, as aforementioned, the majority of researchers have used feed speeds less than 1 m/min. So, it is intended to study grind-hardening process for a different range of feed speeds with both numerical and analytical models, which will be developed with a view to investigate this previously not studied area. AISI O1 and AISI D2 steel types are employed for experimental work, as these materials have rarely been investigated in the relevant literature for grind-hardening process, and so, it is an excellent opportunity to conduct an intriguing research on the processing of these two materials. The experimental results are used for the analytical and numerical modeling of grind-hardening.

8.2   Modeling of grinding wheel, forces, and heat partition

8.2.1   Modeling of grinding wheel topology

The volumetric concentration of pores can be evaluated from an empirical formula, according to the grinding wheel type, proposed by Malkin and Guo [13], as:

$$V_p = \frac{1}{100} \cdot \left(45 + \frac{S^{-2n}}{1.5} \right) \tag{8.1}$$

Similarly, the volumetric concentration of grains can be also calculated from an empirical formula, as:

$$V_g = \frac{2}{100} (32 - S) \tag{8.2}$$

Finally, the volumetric concentration of the bonding material can be calculated from the values of volumetric concentration of the two other constituent materials of the grinding wheel:

$$V_b = 1 - V_g - V_P \qquad (8.3)$$

The ratio of static to active grains depends on factors such as elasticity of grinding wheel, wheel and workpiece deformation, and processing time. In the model proposed in this study, an expression correlating the volumetric concentration of bonding material and the ratio of static to active grains is employed. This expression was derived by the work conducted by Hou and Komanduri [14]. A *normalization coefficient*, which can relate a reference ratio Φ_{ref} to the properties of each grinding wheel, is also calculated.

$$\text{N.F.} = 20.535 \cdot V_b - 0.217 \qquad (8.4)$$

$$\Phi_a = \Phi_{ref}(\text{N.F.}) \qquad (8.5)$$

$$\Phi_a = \frac{n_a}{n_s} \qquad (8.6)$$

The equivalent grinding wheel diameter is obtained from the following relation:

$$d_e = d_s \cdot \frac{1}{1 + (d_s/d_w)} \qquad (8.7)$$

where, for a flat workpiece, $d_w = \infty$, and, consequently, $d_e = d_s$.

The contact arc length between grinding wheel and workpiece is considered to be equal to the geometric contact length and is given by the geometry of the contact areas of the grinding wheel and workpiece, as:

$$\ell_c = \sqrt{d_e a_e} \qquad (8.8)$$

The average grain diameter is evaluated as a function of grain size through a correlation with the grit number M proposed by Malkin [13]:

$$d_g = 15.2 M^{-1} \qquad (8.9)$$

When all the above-mentioned quantities are known, the number of static grains may be estimated accordingly:

$$V_g = \frac{n_s V_{grain}}{V_{tot}}$$

or finally

$$n_s = \frac{V_g V_{tot}}{V_{grain}} = 12 \cdot \frac{32 - S}{100} \cdot \frac{\ell_c b}{\pi d_g^2} \tag{8.10}$$

Consequently, the number of active grains is equal to:

$$n_a = \Phi_a n_s \tag{8.11}$$

8.2.2   Process forces

In order to calculate the thermal flux toward the workpiece, the power produced by the process should first be calculated by the determination of process forces. The two components of grinding force are the tangential F_t and the normal F_n. As the grinding wheel diameter is several orders of magnitude larger than the depth of cut, the tangential force can be considered equal to the horizontal force. The total tangential process force can be calculated as a sum of slip force, chip-forming force, and plastic deformation forces. The last two forces are also denoted as cutting forces.

$$F_t = F_{t,sl} + F_{t,c} \tag{8.12}$$

$$F_{t,c} = F_{t,ch} + F_{t,pl} \tag{8.13}$$

8.2.2.1   Slip force calculation

Slip force calculation can be conducted by determining the average contact pressure of grains on the workpiece surface, the area of contact, and the average friction coefficient.

$$F_{t,sl} = \mu p_m A_a \tag{8.14}$$

According to experimental data [13], it can be stated that friction coefficient is independent of the grinding wheel and workpiece topography and is affected only if thermal damage of the workpiece occurs due to process parameters. In the present study, no thermal damage is supposed to occur, and the friction coefficient value is proposed to be $\mu = 0.38$.

The average contact pressure is a linear function of curvature difference Δ. The curvature difference is defined as the difference between grinding wheel radius and the radius of cutting trajectory. When grinding wheel speed is significantly larger than the workpiece feed speed ($u_w \ll u_s$), the difference in curvature can be obtained as:

$$\Delta = \frac{4u_w}{d_e u_s} \tag{8.15}$$

The linear regression coefficients for the correlation of p_m and Δ are derived from the experiments conducted by Kannapan and Malkin [15]. The proposed values for the two empirical parameters are:

$$k_1 = 2.58 \cdot 10^6 \, (\text{N/mm})$$
(8.16)

$$k_2 = 35 \, (\text{N/mm})$$
(8.17)

Therefore, the average contact pressure can be calculated by the expression:

$$p_m = k_1 \Delta + k_2$$
(8.18)

The real contact area between grinding wheel and workpiece A_α depends on the number of active grains and the slip area of each grain. The average slip area per grain A_g is considered to be the area of a circle with a diameter equal to the two-thirds of the average grain diameter:

$$A_g = \frac{1}{4} \pi \ell_{\text{wf}}^2$$
(8.19)

$$\ell_{\text{wf}} = \frac{2}{3} d_g$$
(8.20)

Thus,

$$A_a = n_a A_g$$
(8.21)

The total tangential component of slip force can be obtained as follows:

$$F_{t,\text{sl}} = \mu p_m A_a = 1.3 \cdot 10^{-4} \mu \Phi_a \, b \, \ell_{\text{wf}}^2 M^2 (32 - S) \sqrt{d_e a_e} \left(k_1 \frac{4 u_w}{d_e u_s} + k_2 \right)$$
(8.22)

8.2.2.2   Cutting force

Cutting forces can be determined from the special cutting force, which is defined as the energy being consumed for the removal of a unit volume of the workpiece material. The special cutting energy is the sum of the chip-forming energy and the workpiece deformation energy, without material removal. Malkin and Guo [13] found experimentally that the special cutting energy asymptotically approaches the chip-forming energy when material removal rate increases. It has also been experimentally proven that chip-forming energy does not depend on process parameters, grinding wheel type, and workpiece material. In the majority of references, a specific cutting energy value of $u_{\text{ch}} = 13.8 \, \text{J/mm}^3$ is proposed.

According to the aforementioned, the following equation is proposed for the estimation of specific cutting energy:

$$u_c = u_{ch} + u_{pl} = u_{ch} + \frac{28.1}{u_w a_e} \tag{8.23}$$

$$MRR = u_w a_e \tag{8.24}$$

When specific cutting energy is known, the cutting force, that is, the sum of tangential forces for chip forming and workpiece deformation, can be obtained by a formula proposed by Malkin and Joseph [16]:

$$u_c = \frac{F_{t,c} u_s}{b a_e u_w}$$

or equivalently:

$$F_{t,c} = u_c \frac{u_w}{u_s} b a_e = \left(u_{ch} + \frac{28.1}{u_w a_e} \right) \frac{u_w}{u_s} b a_e \tag{8.25}$$

$$F_{t,c} = F_{t,ch} + F_{t,pl} \tag{8.26}$$

8.2.3   Production and partition of heat

In order to calculate the heat produced during the process, it is necessary to analyze the mechanism that causes the production of this amount of heat. Thus, the heat produced during grinding is generated by the following:

- Friction between grinding grains and workpiece
- Plastic deformation at slip plane during chip removal
- Plastic deformation of workpiece material without material removal

Based on experimental data, it is considered that heat amount, which is produced by plastic deformation not only at slip plane but also at the workpiece, is considered to be negligible in comparison with the heat amount produced by friction at the grain–workpiece interface. Thus, if the only heat source is attributed to friction, process power can be computed as follows:

$$P = F_t(u_s \pm u_w) \tag{8.27}$$

In the previous formula, the positive sign is used for a down-grinding process, where peripheral wheel velocity and workpiece feed velocity vector are on the same direction, and the negative sign is used in the opposite case.

However, as the two velocities differ considerably in terms of magnitude, this expression could further be simplified, and it could be supposed that velocity is equal to the peripheral wheel velocity u_s; however, it is intended not to further simplify this formula.

The heat produced is partitioned to several components of the process, as there is heat flux directed toward the workpiece, the grinding wheel, and chips. The heat amount produced can be calculated as the ratio of process power to the area of grinding zone. According to this, heat can be evaluated by the relation:

$$q_t = \frac{P}{b\ell_c} = q_w + q_s + q_{ch} \tag{8.28}$$

The heat amount that is transferred from chips is also derived from the specific energy, which is distributed in the grinding zone. Chip specific energy is defined as the heat amount required for the increase of chip temperature up to its melting point, a temperature that is often attained by chips, according to Malkin [13]. So,

$$e_{ch} = \rho_{w,T=T_{mp}} C_{w,T=T_{mp}} (T_{mp} - T_o) \tag{8.29}$$

Then, thermal flux to the chip can be calculated as:

$$q_{ch} = e_{ch} \frac{a_e u_w}{\ell_c} \tag{8.30}$$

Partition ratio of heat carried away by chips is equal to:

$$R_{ch} = \frac{q_{ch}}{q_t} \tag{8.31}$$

In this process, the largest amount of heat is produced by friction at grain–workpiece interface. Furthermore, the contact surface is considerably larger than the contact surface of other material-removal processes. In the developed model from Rowe et al. [17], grinding wheel and workpiece are approximated as two sliding bodies. By using this model, heat partition coefficient between grinding wheel and workpiece can be obtained:

$$R_{ws} = \frac{q_w}{q_w + q_s} \tag{8.32}$$

$$R_{ws} = \left(1 + \frac{\beta_s}{\beta_w} \sqrt{\frac{u_s}{u_w}} \right)^{-1} \tag{8.33}$$

where $\beta_w = \sqrt{k_w \rho_w C_w}$ and $\beta_s = \sqrt{k_s \rho_s C_s}$ are the average heat transfer coefficients for workpiece and grinding wheel, respectively.

To calculate grinding wheel properties, it is supposed that grinding wheel is a material composed of voids and grain material. Average values for its properties can be derived from a mixing rule with respect to surface porosity φ:

$$\bar{i_s} = \varphi \cdot i_g + (1 - \varphi) \cdot i_a \tag{8.34}$$

$$\varphi = \frac{A_a}{i_c \cdot b} \text{(surface porosity of a composite material)} \tag{8.35}$$

Then, heat flux and partition ratio toward the workpiece, as well as heat partition ratio toward the grinding wheel, can be estimated as follows:

$$q_w = R_{ws}(q_t - q_{ch}) \tag{8.36}$$

$$R_w = \frac{q_w}{q_t} \tag{8.37}$$

$$R_s = \frac{q_s}{q_t} \tag{8.38}$$

8.3   Analytical modeling

The analytical model used in the present work is derived from an analytical model for the computation of 3D temperature profile, proposed by Foeckerer et al. [18]. The following assumptions are made for the derivation of this model:

- Rectangular, semi-infinite workpiece. This geometry is similar to that studied in this paper, and the assumption of semi-infinite dimensions is valid, as depth of cut values are several orders of magnitude smaller than the other dimensions.
- Constant feed speed.
- Heat source, which represents the heat produced by the process, has a triangular spatial distribution. This distribution is often employed in publications relevant to grinding process modeling. Latent heat is not taken into consideration.
- Two coordinate systems are being used: one related to the workpiece (x, y, z) and the other related to the source (x', y', z'). The first one is immobile—constant, whereas the other is moving relatively to the center of workpiece—heat source contact area.

Time-dependent, 3D heat transfer in the case of a moving workpiece can be described by the following differential equation [18]:

$$k\left(\frac{\partial^2 T}{\partial x^2} + \frac{\partial^2 T}{\partial y^2} + \frac{\partial^2 T}{\partial z^2}\right) = \rho c\left(\frac{\partial T}{\partial t} - u_w \frac{\partial T}{\partial x}\right) \tag{8.39}$$

where $T = T(x, y, z, t)$ is the temporal and spatial temperature distribution. To solve this differential equation, the following data are applied:

Initial conditions: $T(x, y, z, t)|_{t=0} = T_\infty = 0°C$
Boundary conditions:

- Out of the region, affected by the heat source:

$$-k\frac{\partial T}{\partial z}\bigg|_{z=0, |x|>\frac{\ell_c}{2}, |y|>\frac{b}{2}} = -h\left(T\big|_{z=0} - T_\infty\right) \tag{8.40}$$

- Inside the region, affected by the heat source:

$$-k\frac{\partial T}{\partial z}\bigg|_{z=0, |x|\le\frac{\ell_c}{2}, |y|\le\frac{b}{2}} = \dot{q}(x') - h\left(T\big|_{z=0} - T_\infty\right) \tag{8.41}$$

Heat source is defined in terms of dimensionless coordinates and is calculated by the following relation:

$$\dot{q}(x') = \dot{q}_o\left(1 + \frac{2}{\ell_c}x'\right) = \frac{Q_o}{\ell_c b}\left(1 + \frac{2}{\ell_c}x'\right)$$

or finally:

$$\dot{q}_o = \frac{Q_o}{\ell_c b} = \frac{PR_w}{\ell_c b} \tag{8.42}$$

Furthermore, nondimensional variables are introduced according to Des Ruisseaux and Zerkle formulation [19,20]:

$$X = \frac{u_w x}{2a}, Y = \frac{u_w y}{2a}, Z = \frac{u_w z}{2a} \tag{8.43}$$

$$L = \frac{u_w \ell_c}{4a}, B = \frac{u_w b}{4a} \tag{8.44}$$

$$H = \frac{2ah}{ku_w} \tag{8.45}$$

$$\tau = \frac{u_w\sqrt{t-t'}}{2\sqrt{a}} \tag{8.46}$$

Finally, the analytical solution of Equation 8.39, for the boundary conditions imposed and using dimensionless quantities, yields:

$$
T(X,Y,Z,\tau) = \frac{2a\dot{q}_o}{\sqrt{\pi}ku_w} \int\limits_{0}^{\frac{u_w\sqrt{t}}{2\sqrt{a}}} \frac{1}{2}\left[\operatorname{erf}\left(\frac{Y+B}{2\tau}\right) - \operatorname{erf}\left(\frac{Y-B}{2\tau}\right)\right] e^{-\frac{Z^2}{4\tau^2}}
$$

$$
\cdot\left\{\left(1 + \frac{X+2\tau^2}{L}\right)\left[\operatorname{erf}\left(\frac{X+L}{2\tau} + \tau\right) - \operatorname{erf}\left(\frac{X-L}{2\tau} + \tau\right)\right]\right.
$$

$$
\left. + \frac{2\tau}{L\sqrt{\pi}}\left[e^{-\left(\frac{X+L}{2\tau} + \tau\right)^2} - e^{-\left(\frac{X-L}{2\tau} + \tau\right)^2}\right]\right\}d\tau
$$

$$\tag{8.47}$$

$$
-\frac{2a\dot{q}_o}{ku_w}H\int\limits_{0}^{\frac{u_w\sqrt{t}}{2\sqrt{a}}} \frac{\tau}{2}\left[\operatorname{erf}\left(\frac{Y+B}{2\tau}\right) - \operatorname{erf}\left(\frac{Y-B}{2\tau}\right)\right] e^{HZ+H^2\tau^2}
$$

$$
\cdot\left\{\left(1 + \frac{X+2\tau^2}{L}\right)\left[\operatorname{erf}\left(\frac{X+L}{2\tau} + \tau\right) - \operatorname{erf}\left(\frac{X-L}{2\tau} + \tau\right)\right]\right.
$$

$$
\left. + \frac{2\tau}{L\sqrt{\pi}}\left[e^{-\left(\frac{X+L}{2\tau} + \tau\right)^2} - e^{-\left(\frac{X-L}{2\tau} + \tau\right)^2}\right]\right\}
$$

$$
\cdot\operatorname{erfc}\left(\frac{Z}{2\tau} + H\tau\right)d\tau
$$

This analytical solution is computed numerically by Gauss–Kronrod algorithm, also proposed by Foeckerer et al. [18].

8.4   Numerical solution using ANSYS

8.4.1   Description of general details of the model

The heat source is considered to have a triangular spatial distribution, as in the case of analytical solution, and its velocity is equivalent to the feed speed of the workpiece (u_w). Owing to the inability of the employed ANSYS component to explicitly model moving heat sources, the moving heat source will be simulated implicitly. In the workpiece surface, special surfaces are defined, in which time-dependent heat sources equivalent to the total heat source are assigned. Each individual source, having a triangular distribution, is activated only when the real heat source, representing the heat produced by the process, passes over the surface at the point where the individual source surface starts, that is, governed by the distance between adjacent cells, feed speed, and contact length of the process.

For the system studied in the current work, the following quantities are defined:

- Contact length: ℓ_{c1}
- Heat flux: q_w
- Workpiece feed speed: u_w
- Node I position: x_i
- Distance between nodes I and $(i+1)$: $\Delta x = x_{i+1} - x_i$

An important note should be made concerning the input of heat source profile into ANSYS. As a triangular heat source is not directly supported and the value of heat source $q_{w\,\text{max}}$ is constant during each time step, a suitable correction $q'_{w\,\text{max}}$ is required to simulate a heat source with a triangular distribution. Thus, in order to not add supplementary heat during each time and calculate an erroneously increased temperature distribution, the heat flux $q_{w\,\text{max}}$ is reduced by a certain amount, with respect to the contact length and spatial discretization of workpiece, as follows:

$$q'_{w\,\text{max}} = 2q_{w\,\text{max}} - \frac{1}{2} \cdot \frac{2q_{w\,\text{max}}}{(\ell_c/0.25)} = 2q_{w\,\text{max}} - \frac{1}{4} \cdot \frac{q_{w\,\text{max}}}{\ell_c} \qquad (8.48)$$

Additional details on the computational model developed in ANSYS are given as follows:

- Time of activation of heat source at node i:

$$t_i = \frac{x_i}{u_w} \text{ or } t_{i+1} = t_i + \frac{\Delta x}{u_w} \qquad (8.49)$$

- Time duration of each heat source (constant and dependent only on process parameters):

$$t_{\text{active}} = \frac{\ell_{c1}}{u_w} \qquad (8.50)$$

- Temporal heat source distribution:

$$q_{wi}(t) = \begin{cases} 0, & \text{if } t < t_i \text{ or } t > t_i + t_{active} \\ q'_{w\,max} - \dfrac{q'_{w\,max}}{t_{active}}(t - t_i), & \text{if } t_i < t < t_i + t_{active} \end{cases} \left[\dfrac{W}{m^2} \right] \quad (8.51)$$

- Thermal and mechanical materials properties: Material proper-ties for AISI O1 and AISI D2 materials were inserted in array form, with respect to temperature [$c_p(T)$, $\rho(T)$, $k(T)$], as denoted in Table 8.1.
- Definition of workpiece geometry: An orthogonal parallelepiped workpiece with dimensions of 50 × 10 × 4 mm was created, in accor-dance with the validation experiments that were conducted. The length of 50 mm was considered sufficient, as the transient phases of the process (start and end) can be modeled accurately and a longer workpiece would only increase the computational cost. The height of 4 mm was determined from preliminary tests, which indicated that, in any case, the heat affected zone depth did not exceed 2 mm. In the grinding surface, the individual surfaces on which the heat sources would be applied had a length of 0.25 mm each, and 100 such sur-faces were defined, having a length more than two times larger than the contact length of the process.
- Computational mesh definition: A 3D computational mesh with cubic cells of 0.25 mm edge length was created. This dimension was chosen after preliminary tests with respect to the computational time.
- Definition of time step: In this work, time step is defined as the time duration required for a heat source to travel along an individual surface. Thus,

$$t_{step} = \frac{0.25}{u_w} \quad (8.52)$$

Table 8.1 Workpiece material properties

AISI O1			AISI D2		
Temperature (°C)	Thermal conductivity (W/mK)	Density (kg/m³)	Temperature (°C)	Thermal conductivity (W/mK)	Density (kg/m³)
20	32	7800	20	20	7700
200	33	7750	200	21	7650
400	34	7700	400	23	7600

8.4.2   *Verification of the model*

A rough estimation of maximum temperature in the workpiece surface can be obtained by a semi-empirical formula created by Heinzel et al. [21], relating maximum temperature to process parameters and material properties:

$$T_m = \frac{1.13 q_w a^{1/2} a_e^{1/4} d_s^{1/4}}{k_w u_w^{1/2}} \tag{8.53}$$

where $a = k\rho/c_p$ (mm^2/s) is the thermal diffusion coefficient.

This expression was also given by Malkin and Guo [22] for a triangular heat source, as follows:

$$T_m = \frac{1.06 q_w a^{1/2} a_e^{1/4} d_s^{1/4}}{k_w u_w^{1/2}} \tag{8.54}$$

8.4.3   *Thermomechanical material properties*

According to the literature data, an effort was made to model temperature-dependent properties of materials involved in the study, as it is anticipated that high temperatures will develop inside the workpiece during grind-hardening process. Material properties were derived as a function of temperature with a linear model. As can be observed in curves of Figure 8.1, a high level of fit was attained ($R^2 > 0.97$), which makes an accurate calculation of these properties possible at high temperatures. In the analytical model, an average value for temperatures observed in workpiece surface during the process was employed, whereas in ANSYS environment, the data were inserted as arrays.

Heat capacity was defined as $c_p = 650$ J/kg K, according to Figure 8.2. This figure is considered very important, as it portrays a highly nonlinear correlation of specific heat for temperatures in the range of 650°C–750°C, implicating that in this temperature range, a large amount of energy is absorbed, with no increase in the temperature.

In the calculations, the thermal properties of air and alundum are required to represent the thermal behavior of grinding wheel properly. These properties are presented in Table 8.2.

Finally, the process parameters used in the simulation are tabulated in Table 8.3.

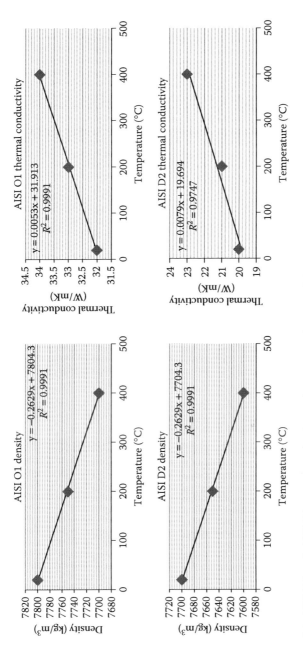

Figure 8.1 AISI D2 and AISI O1 temperature-dependent material properties.

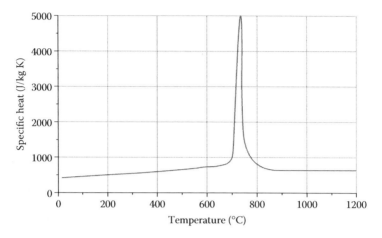

Figure 8.2 Specific heat capacity of steel.

Table 8.2 Air and grinding wheel properties

Material	Melting temperature (°C)	Density (kg/mm³)	Heat capacity (J/kg K)	Thermal conductivity (W/mmC)
Alundum	1900	0.0000033	900	0.0027
Air	—	$3.53e^{-10}$	0.001411	0.00006754

Table 8.3 Process parameters

Process parameters	
Grinding wheel	38A36-K8VG 300X50X76
Grinding wheel speed (m/s)	43.98
Workpiece feed speed (m/s)	0.195/0.2815/0.3765
Depth of cut (mm)	0.3/0.4/0.5
Grinding method	Up-/down-grinding
Cutting fluid	No
Workpiece material	AISI O1/AISI D2

8.5   Results and discussion

8.5.1   Analytical model

Table 8.4 presents the results, that is, maximum temperatures, of the analytical model and Malkin's empirical equation versus cutting parameters

Table 8.4 Results using the analytical model

Workpiece material	#	Depth of cut (mm)	Feed speed (m/s)	l_c (mm)	P (W)	q_w (W/mm²)	R_w (%)	Malkin's equation maximum temperature (°C)	Analytical model maximum temperature (°C)	Deviation (%)
AISI O1 Up-grinding	1	0.3	0.195	9.49	11400	76.30	63	1305	1310	0.36
	2	0.3	0.2815	9.49	15544	101.28	62	1442	1447	0.36
	3	0.3	0.3765	9.49	20114	128.93	61	1587	1593	0.36
	4	0.4	0.195	10.95	13165	70.06	58	1288	1292	0.36
	5	0.4	0.2815	10.95	17951	92.23	56	1411	1416	0.36
	6	0.4	0.3765	10.95	23229	116.79	55	1545	1550	0.36
	7	0.5	0.195	12.25	14721	64.56	54	1255	1259	0.36
	8	0.5	0.2815	12.25	20072	84.26	51	1363	1368	0.36
	9	0.5	0.3765	12.25	25973	106.09	50	1484	1489	0.36
Down-grinding	10	0.3	0.195	9.49	11300	75.26	63	1287	1292	0.37
	11	0.3	0.2815	9.49	15347	99.24	61	1412	1418	0.37
	12	0.3	0.3765	9.49	19773	125.39	60	1543	1549	0.37
	13	0.4	0.195	10.95	13049	69.02	58	1268	1273	0.37
	14	0.4	0.2815	10.95	17723	90.19	56	1379	1385	0.37
	15	0.4	0.3765	10.95	22834	113.25	54	1498	1503	0.37
	16	0.5	0.195	12.25	14591	63.52	53	1234	1239	0.37
	17	0.5	0.2815	12.25	19817	82.21	51	1330	1334	0.37
	18	0.5	0.3765	12.25	25532	102.54	49	1434	1439	0.37

(Continued)

Table 8.4 (Continued) Results using the analytical model

Workpiece material		#	Depth of cut (mm)	Feed speed (m/s)	l_c (mm)	P (W)	q_w (W/mm²)	R_w (%)	Malkin's equation maximum temperature (°C)	Analytical model maximum temperature (°C)	Deviation (%)
AISI D2	Up-grinding	19	0.3	0.195	9.49	11400	76.43	64	1498	1503	0.36
		20	0.3	0.2815	9.49	15544	101.59	62	1657	1663	0.36
		21	0.3	0.3765	9.49	20114	129.43	61	1825	1832	0.36
		22	0.4	0.195	10.95	13165	70.29	58	1480	1485	0.36
		23	0.4	0.2815	10.95	17951	92.68	57	1624	1630	0.36
		24	0.4	0.3765	10.95	23229	117.46	55	1780	1786	0.36
		25	0.5	0.195	12.25	14721	64.88	54	1444	1450	0.36
		26	0.5	0.2815	12.25	20072	84.82	52	1572	1577	0.36
		27	0.5	0.3765	12.25	25973	106.91	50	1713	1719	0.36
	Down-grinding	28	0.3	0.195	9.49	11300	75.41	63	1478	1483	0.36
		29	0.3	0.2815	9.49	15347	99.57	62	1624	1630	0.36
		30	0.3	0.3765	9.49	19773	125.92	60	1776	1782	0.36
		31	0.4	0.195	10.95	13049	69.27	58	1458	1464	0.36
		32	0.4	0.2815	10.95	17723	90.65	56	1589	1594	0.36
		33	0.4	0.3765	10.95	22834	113.94	55	1727	1733	0.36
		34	0.5	0.195	12.25	14591	63.85	54	1422	1427	0.36
		35	0.5	0.2815	12.25	19817	82.79	51	1534	1540	0.36
		36	0.5	0.3765	12.25	25532	103.40	50	1657	1663	0.36

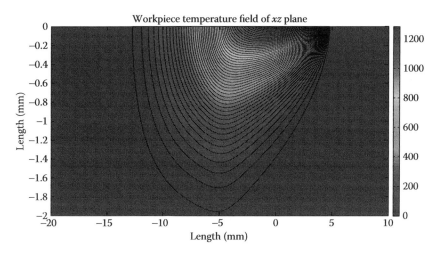

Figure 8.3 Workpiece temperature field of *xz* plane for cutting parameters $u_w = 0.195$ m/sec and $a_e = 0.3$ mm.

(depth of the cut and workpiece speed), length of the arc contact, grinding wheel spindle power, heat flux, and heat partition of the workpiece.

In Figure 8.3, the temperature contours within the workpiece at a specific time step are shown.

Comparing the earlier results, some conclusions may be drawn:

- Practically, the same maximum temperatures are estimated by using the analytical model and Malkin's equation. The deviation is between 0.36% and 0.37%.
- The heat flux depends on the depth of cut and the feed speed; higher heat flux, in most of the cases, leads to higher maximum temperatures.
- It is apparent that the differences between up- and down-grinding are insignificant (between 1% and 3%). So, it can be said that the result of the process applied was not affected by this particular parameter.
- Workpieces from AISI D2 material showed higher maximum temperatures in comparison with those of AISI O1, for the same cutting parameters. The reason is the difference in thermal conductivity. AISI O1 steel has a thermal conductivity $k = 32$ W/mK and AISI D2 steel has a thermal conductivity $k = 20$ W/mK. The 33% lower thermal conductivity coefficient indicates a *harder* heat conduct through the workpiece, causing higher topical temperature rise.

8.5.2 Results obtained from ANSYS simulations

In Figure 8.4, the workpiece geometry and the connected mesh used in ANSYS are shown.

In Figure 8.5 (a)–(c) and Figure 8.6 (a)–(c), a comparison of the ANSYS-calculated temperature field for different process parameters is portrayed.

In Table 8.5, the maximum temperature results calculated by the simulation with ANSYS are shown. In the same table, the results with empirical and analytical methods are tabulated for comparison.

The results verify the basic assumption that by increasing the feed speed, the workpiece temperature profile becomes more *narrow*, because of the decreased heat source (grinding wheel) effect time. The heat partition to the workpiece and the heat source effect time significantly affect the maximum HPD.

Figure 8.7 presents the temperature cycles on the workpiece surface, during process, as these are calculated with both the analytical model and ANSYS for $u_e = 0.195$ m/s and $a_e = 0.3$ mm. A minor deviation between the two solutions is observed, as the analytical model does not consider the generation of latent heat. No practical diversifications in HPD are expected, as martensitic transformation temperatures are at least one order of magnitude greater than the difference.

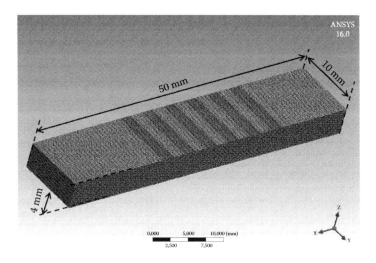

Figure 8.4 Workpiece with the adjusted mesh.

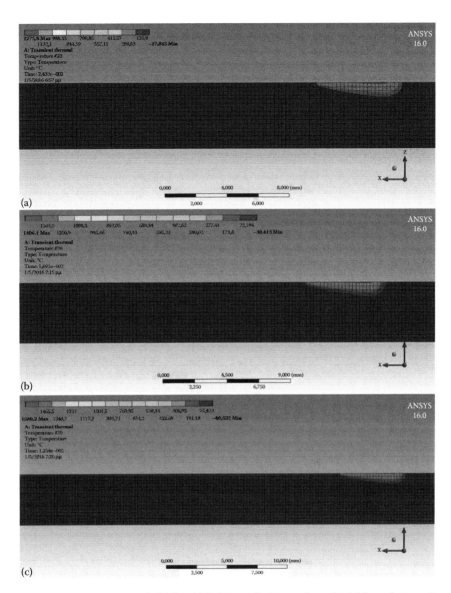

Figure 8.5 Temperature field for AISI O1 workpiece, when the 20th node is activated for depth of cut $a_e = 0.3$ mm and feed speed (a) 0.195 m/s, (b) 0.2815 m/s, and (c) 0.3765 m/s.

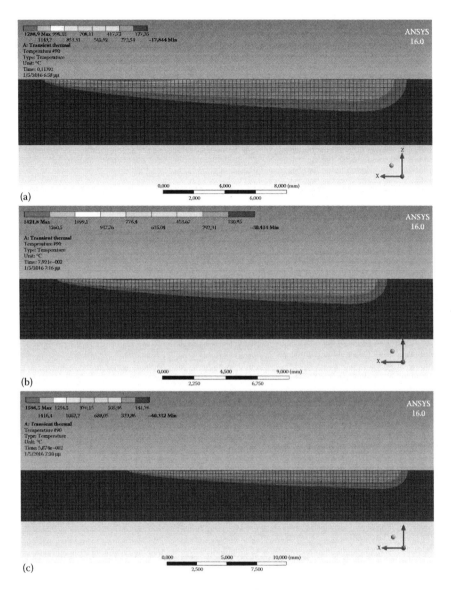

Figure 8.6 Temperature field for AISI O1 workpiece, when the 90th node is activated for depth of cut $a_e = 0.3$ mm and feed speed (a) 0.195 m/s, (b) 0.2815 m/s, and (c) 0.3765 m/s.

Table 8.5 Results using ANSYS software

Workpiece material		#	Depth of cut (mm)	Feed speed (m/sec)	l_c (mm)	q_w (w/mm²)	Malkin's equation maximum temperature (°C)	Analytical model maximum temperature (°C)	ANSYS maximum temperature (°C)	Analytical model—ANSYS deviation (%)
AISI O1	Up-grinding	1	0.3	0.195	9.49	76.30	1305	1310	1289	1.6
		2	0.3	0.2815	9.49	101.28	1442	1447	1422	1.8
		3	0.3	0.3765	9.49	128.93	1587	1593	1599	0.4
		4	0.4	0.195	10.95	70.06	1288	1292	1254	3.0
		5	0.4	0.2815	10.95	92.23	1411	1416	1389	1.9
		6	0.4	0.3765	10.95	116.79	1545	1550	1525	1.7
		7	0.5	0.195	12.25	64.56	1255	1259	1223	3.0
		8	0.5	0.2815	12.25	84.26	1363	1368	1335	2.5
		9	0.5	0.3765	12.25	106.09	1484	1489	1462	1.8
AISI D2	Up-grinding	19	0.3	0.195	9.49	76.43	1498	1503	1479	1.6
		20	0.3	0.2815	9.49	101.59	1657	1663	1630	2.0
		21	0.3	0.3765	9.49	129.43	1825	1832	1820	0.6
		22	0.4	0.195	10.95	70.29	1480	1485	1464	1.5
		23	0.4	0.2815	10.95	92.68	1624	1630	1605	1.6
		24	0.4	0.3765	10.95	117.46	1780	1786	1755	1.8
		25	0.5	0.195	12.25	64.88	1444	1450	1416	2.4
		26	0.5	0.2815	12.25	84.82	1572	1577	1536	2.7
		27	0.5	0.3765	12.25	106.91	1713	1719	1700	1.1

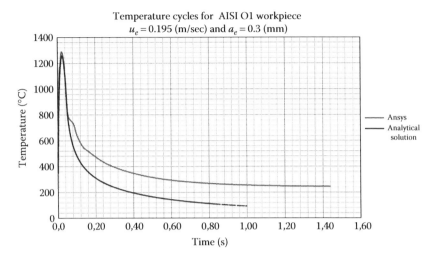

Figure 8.7 Temperature time variation.

8.5.3 Method's accuracy test

In order to ensure the accuracy and correctness of the method, experimental data of the research by Liu et al. [6] were used to compare them with the present calculation's results. In the aforementioned work, the researchers have used square-edged workpieces of 1060 steel and grinding conditions, as listed in Table 8.6, for their experiments.

Table 8.7 presents the results of the applied method, comparing them with the experimental ones from the above-mentioned publication. An acceptable deviation of less than 7% can be seen.

Table 8.6 Grinding conditions

Grinding wheel	WA46L8V P350 × 40 × 127
Wheel speed (m/s)	26.3
Table speed (m/s)	0.008
Depth of cut (mm)	0.2/0.3/0.4/0.5
Grinding method	Up-grinding
Cooling	Dry-grinding
Workpiece material	1060 steel

Table 8.7 Comparison of experimental and simulation results

Depth of a cut (mm)	Hardened layer thickness [6] (mm)	Hardened layer thickness via proposed method (mm)	Deviation (%)
0.2	0.60	0.57	5.0
0.3	0.81	0.86	6.2
0.4	1.01	1.07	5.9
0.5	1.21	1.24	2.5

8.5.4 Effect of grinding fluid use

8.5.4.1 Grinding fluid model

Grinding fluids serve a number of purposes during process, such as mechanical lubrication, chemo-physical lubrication, cooling the contact area, bulk cooling outside the contact area, wheel cleaning, and entrapment of abrasive dust and harmful vapors. In order to have an efficient and effective usage of grinding fluid, account must be taken of a set of parameters such as the fluid properties (water-based fluid or neat oils), the pumping system, the nozzle position, and the fluid speed. By using grinding fluids, great amounts of the grinding energy can be saved, achieving lower surface temperatures [23,24]. The main difficulty in the precise calculation of this amount of energy is the definition of the fluid convection factor h_f. This factor depends on the grinding wheel speed; the fluid film thickness within the contact zone, which is determined by the grinding wheel speed; porosity; grain size; fluid type; fluid rate; and nozzle size [23].

The aim of grinding-hardening process is the temperature rise above a certain level, that is, the austenitization temperature, in order to have a hardened surface layer. Taking away great amount of heat by using grinding fluid seems to have a negative effect on the process. Nevertheless, there are some special cases where grinding fluid usage is indicated. In most of the cases, the generated heat is dissipated inside the workpiece, so as to raise the surface temperature and induce metallurgical transformations. For bulky materials, this heat dissipation provides the necessary quenching of the workpiece. However, for utilizing the process with thick or small-diameter cylindrical parts, the cooling rate achieved is not so significant as to allow the martensitic transformation.

Therefore, a coolant fluid is often applied, directly after the contact area, for achieving the quenching of these parts. The application of the coolant fluid also reduces the grinding wheel temperature, thus prolonging its life [25].

In order to have a safe estimation of the fluid convection factor, the laminar flow model (LFM) is employed, by assuming that the velocity of flow within the pores of the wheel is u_s. Simplification using u_s yields [24]:

$$h_f = \frac{4}{9} k_f^{2/3} \rho_f^{1/2} C_f^{1/3} \eta_f^{1/6} \left(\frac{u_s}{\ell_c} \right)^{1/2}$$

(8.55)

and the rate of the heat flow to the fluid per unit area is calculated:

$$q_f = h_f \Delta T_{max}$$

(8.56)

As a ΔT, that is, difference between the workpiece and the fluid temperature, is needed, an iterative solution is used. To simplify the model, it is assumed that the chip and grinding wheel heat ratio do not change and only the workpiece heat ratio splits into the workpiece and the grinding fluid. As a typical grinding fluid, water at 50°C is used, the properties of which are tabulated in Table 8.8.

8.5.4.2   Results of grinding fluid use
in grinding-hardening experiments

In this subsection, results from cases with grinding fluid are presented and a comparison between these cases and cases without the use of cutting fluid is conducted. It is evident from the findings in Table 8.9, Figure 8.8, and Figure 8.9 that a much smaller HPD is predicted, as expected, and in some cases, no hardening is predicted to take place.

Table 8.8 Water properties

Water properties at 50°C			
Thermal conductivity (W/mK)	Density (kg/m³)	Specific heat capacity (J/kg K)	Dynamic viscosity (kg/ms)
0.644	988	4182	0.000532

Table 8.9 Results for using grinding fluid

Workpiece material		#	Depth of cut (mm)	Feed speed (m/s)	l_c (mm)	h_f (W/m²K)	q_f (W/mm²)	R_f (%)	T_{max} with the use of grinding fluid (°C)	T_{max} without the use of grinding fluid (°C)	T_{max} change (%)	HPD (mm)
AISI O1	Up-grinding	1	0.3	0.195	9.49	40149	29.1	24	840	1310	35.8	0
		2	0.3	0.2815	9.49	40149	34.4	21	990	1447	31.5	0.06
		3	0.3	0.3765	9.49	40149	39.8	19	1141	1593	28.4	0.11
		4	0.4	0.195	10.95	37363	27.1	23	822	1292	36.4	0
		5	0.4	0.2815	10.95	37363	31.8	19	961	1416	32.1	0.04
		6	0.4	0.3765	10.95	37363	36.6	17	1103	1550	28.9	0.1
		7	0.5	0.195	12.25	35336	25.2	21	797	1259	36.7	0
		8	0.5	0.2815	12.25	35336	29.3	18	925	1368	32.4	0.02
		9	0.5	0.3765	12.25	35336	33.6	16	1055	1489	29.1	0.09
	Down-grinding	10	0.3	0.195	9.49	40149	28.7	24	829	1292	35.8	0
		11	0.3	0.2815	9.49	40149	33.7	21	971	1418	31.5	0.05
		12	0.3	0.3765	9.49	40149	38.7	19	1110	1549	28.3	0.1
		13	0.4	0.195	10.95	37363	26.7	22	810	1273	36.4	0
		14	0.4	0.2815	10.95	37363	31.1	19	940	1385	32.1	0.03
		15	0.4	0.3765	10.95	37363	35.5	17	1070	1503	28.8	0.09
		16	0.5	0.195	12.25	35336	24.8	21	784	1239	36.7	0
		17	0.5	0.2815	12.25	35336	28.6	18	902	1334	32.4	0.01
		18	0.5	0.3765	12.25	35336	32.4	16	1020	1439	29.1	0.07

(Continued)

Table 8.9 (Continued) Results for using grinding fluid

Workpiece material	#	Depth of cut (mm)	Feed speed (m/s)	l_c (mm)	h_f (W/m²K)	q_f (W/mm²)	R_f (%)	T_{max} with the use of grinding fluid (°C)	T_{max} without the use of grinding fluid (°C)	T_{max} change (%)	HPD (mm)
AISI D2 Up-grinding	19	0.3	0.195	9.49	40149	31.7	26	944	1503	37.2	0
	20	0.3	0.2815	9.49	40149	37.7	23	1121	1663	32.6	0.07
	21	0.3	0.3765	9.49	40149	43.9	21	1298	1832	29.1	0.11
	22	0.4	0.195	10.95	37363	29.5	25	924	1485	37.8	0
	23	0.4	0.2815	10.95	37363	34.9	21	1090	1630	33.1	0.06
	24	0.4	0.3765	10.95	37363	40.4	19	1256	1786	29.7	0.11
	25	0.5	0.195	12.25	35336	27.5	23	897	1450	38.1	0
	26	0.5	0.2815	12.25	35336	32.2	20	1049	1577	33.5	0.05
	27	0.5	0.3765	12.25	35336	37.1	17	1204	1719	29.9	0.1
Down-grinding	28	0.3	0.195	9.49	40149	31.2	26	932	1483	37.2	0
	29	0.3	0.2815	9.49	40149	37.0	23	1099	1630	32.6	0.06
	30	0.3	0.3765	9.49	40149	42.7	20	1263	1782	29.1	0.1
	31	0.4	0.195	10.95	37363	29.1	24	911	1464	37.8	0
	32	0.4	0.2815	10.95	37363	34.1	21	1066	1594	33.1	0.05
	33	0.4	0.3765	10.95	37363	39.2	19	1219	1733	29.7	0.1
	34	0.5	0.195	12.25	35336	27.0	23	884	1427	38.1	0
	35	0.5	0.2815	12.25	35336	31.4	19	1025	1540	33.4	0.04
	36	0.5	0.3765	12.25	35336	35.9	17	1165	1663	30.0	0.09

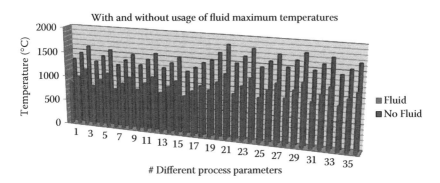

Figure 8.8 Comparison of maximum temperature by using or not using grinding fluid.

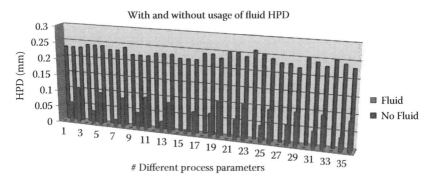

Figure 8.9 Comparison of HPD by using or not using grinding fluid.

As it is seen from Figure 8.8, process temperatures do not exceed 1300°C, and from Figure 8.9, it is clear that HPDs are less than half the depths predicted for cases without grinding fluid.

Nomenclature

A_a	Real contact surface (mm²)
A_g	Average sliding surface per grain (mm²)
a_e	Depth of cut (mm)
b	Grinding wheel width (mm)
c_p	Workpiece heat capacity (J/kg K)
c_f	Cutting fluid heat capacity (J/kg K)
d_e	Equivalent grinding wheel diameter (mm)
d_g	Average grain dimension (mm)
d_s	Grinding wheel diameter (mm)

d_w Workpiece diameter (mm)

e_{ch} Specific energy carried away by grinding chips (J/mm³)

F Force (N)

F_t Total tangential component of cutting force (N)

$F_{t,c}$ Tangential component of cutting force (N)

$F_{t,ch}$ Tangential component of chip formation force (N)

$F_{t,pl}$ Tangential component of plastic deformation force (N)

$F_{t,sl}$ Tangential component of slip force (N)

h_f Heat transfer coefficient (convection) (W/m²K)

k Thermal conductivity (W/mK)

k_f Cutting fluid thermal conductivity (W/mK)

k_1 Empirical parameter equal to 2.58×10^6 N/mm

k_2 Empirical parameter equal to 35 N/mm²

ℓ_c Length of contact arc between grinding tool and workpiece (mm)

ℓ_{wf} Equivalent diameter of slip surface (mm)

M Grit number

n_a Active grinding wheel grains

n_s Static grinding wheel grains

P Grinding wheel power (W)

p_m Average contact pressure of grains on the workpiece surface (N/mm²)

S Grinding wheel structure number

q_f Heat flux to the cutting fluid (W/m²)

q_s Heat flux to the grinding wheel (W/mm²)

q_t Produced heat flux (W/mm²)

q_w Heat flux to the workpiece for an orthogonal thermal profile (W/mm²)

$q_{w\,\text{max}}$ Maximum heat flux value to the workpiece for a triangular thermal profile (W/mm²)

q_{ch} Heat flux to the chip (W/mm²)

R_{ch} Heat partition ratio (to the chip)

R_{ws} Heat partition ratio (between workpiece and grinding wheel)

T_o Environmental temperature (°C)

T_{mp} Workpiece material melting temperature (°C)

u_c Special cutting energy (J/mm³)

u_s Grinding wheel speed (m/s²)

u_w Workpiece feed speed (m/s)

u_{ch} Specific energy for chip formation (J/mm³)

u_{pl} Specific energy for plastic deformation (J/mm³)

V_b Volumetric concentration of grinding wheel bonding material

V_g Volumetric concentration of grinding wheel grains

V_p Volumetric concentration of grinding wheel pores

β_s Grinding wheel average heat transfer coefficient (J/m²sK)

β_w Workpiece average heat transfer coefficient (J/m²sK)

Δ Curvature difference (mm⁻¹)

ΔT_{max} Maximum temperature difference between workpiece and cutting fluid (*K*)

η_f Dynamic viscosity of cutting fluid (kg/ms)

μ Friction coefficient between grinding wheel grains and workpiece

ρ Workpiece material density (kg/m³)

ρ_f Cutting fluid density (kg/m³)

φ Surface porosity of a composite material

Φ_{ref} Reference ratio equal to 3.8%

References

1. T. Brockhoff, E. Brinksmeier (1999), Grind-hardening: A comprehensive view, *Annals of the CIRP* 48(1) 255–260.
2. T. Nguyen, I. Zarudi, L.C. Zhang (2007), Grinding-hardening with liquid nitrogen: Mechanisms and technology, *International Journal of Machine Tools & Manufacture* 47(1) 97–106.
3. T. Nguyen, L.C. Zhang (2010), Grinding-hardening using dry air and liquid nitrogen: Prediction and verification of temperature fields and hardened layer thickness, *International Journal of Machine Tools & Manufacture* 50(10) 901–910.
4. Z. Han, N. Zhang, D. Gao, G. Yang (2007), Research into grinding hardening of microalloyed non-quenched and tempered steel, *Journal of China University of Mining & Technology* 17(2) 238–241.
5. J. Zhang, P. Ga, T.C. Jen, L. Zhang (2009), Experimental and numerical studies of AISI1020 steel in grind-hardening, *International Journal of Heat and Mass Transfer* 52(3/4) 787–795.
6. J. Liu, W. Yuan, S. Huang, Z. Xu (2012), Experimental study on grinding-hardening of 1060 Steel, *Energy Procedia* 16 Part A 103–108.
7. B. Kolkwitz, T. Foeckerer, C. Heinzel, M.F. Zaeh, E. Brinksmeier (2011), Experimental and numerical analysis of the surface integrity resulting from outer—diameter grind–hardening, *Procedia Engineering* 19 222–227.
8. G. Chryssolouris, K. Tsirbas, K. Salonitis (2005), An analytical, numerical, and experimental approach to grind hardening, *Journal of Manufacturing Processes* 7(1) 1–9.
9. J. Li, S. Liu, C. Du (2013), Experimental research and computer simulation of face grind–hardening technology, *Strojniški vestnik—Journal of Mechanical Engineering* 59 81–88.
10. M. Liu, T. Nguyen, L. Zhang, Q. Wu, D. Sun (2014), Effect of grinding induced cyclic heating on the hardened layer generation in the plunge grinding of a cylindrical component, *International Journal of Machine Tools & Manufacture* 89 55–63.
11. U. Alonso, N. Ortega, J.A. Sanchez, I. Pombo, B. Izquierdo, S. Plaza (2015), Hardness control of grind–hardening and finishing grinding by means of area-based specific energy, *International Journal of Machine Tools & Manufacture* 88 24–33.

12. K. Salonitis (2014), On surface grind hardening induced residual stresses, *Procedia CIRP* 13 264–269.
13. S. Malkin, C. Guo (2008), *Grinding Technology: Theory and Application of Machining with Abrasives*, Industrial Press, New York.
14. Z.B. Hou, R. Komanduri (2003), On the mechanics of the grinding process—Part I. Stochastic nature of the grinding process, *International Journal of Machine Tools & Manufacture* 43(15) 1579–1593.
15. S. Kannapan, S. Malkin (1972), Effects of grain size and operating parameters on the mechanics of grinding, *ASME Journal of Engineering for Industry* 94(3) 833–842.
16. S. Malkin, N. Joseph (1975), Minimum energy in abrasive processes, *Wear* 32(1) 15–23.
17. W.B. Rowe, M.N. Morgan, S.C.E. Black (1998), Validation of thermal properties in grinding, *Annals of the CIRP* 47(1) 275–279.
18. T. Foeckerer, M.F. Zaeh, O.B. Zhang, (2013), A three–dimensional analytical model to predict the thermo–metallurgical effects within the surface layer during grinding and grind–hardening, *International Journal of Heat and Mass Transfer* 56(1/2) 223–237.
19. N.R. DesRuisseaux, R.D. Zerkle (1970), Thermal analysis of the grinding process, *ASME Journal of Engineering for Industry* 92(2) 428–434.
20. N.R. DesRuisseaux, R.D. Zerkle (1970), Temperature in semi–infinite and cylindrical bodies subjected to moving heat sources and surface cooling, *ASME Journal of Heat Transfer* 92(3) 456–464.
21. C. Heinzel, J. Solter, S. Jermolajev, B. Kolkwitz, E. Brinksmeier (2014), A versatile method to determine thermal limits in grinding, *Procedia CIRP* 13 131–136.
22. S. Malkin, C. Guo (2007), Thermal analysis of grinding, *Annals of the CIRP* 56(2) 760–782.
23. T. Jin, D.J. Stephenson (2008), A study to the convection heat transfer coefficients of grinding fluids, *CIRP Annals—Manufacturing Technology* 57(1) 367–370.
24. W.B. Rowe (2014), *Principles of Modern Grinding Technology*, 2nd edition, Elsevier, Oxford, UK.
25. K. Salonitis (2015), *Grind Hardening Process*, SpringerBriefs in Manufacturing and Surface Engineering, Springer International Publishing, Switzerland.

chapter nine

Finite element modeling of mechanical micromachining

Samad Nadimi Bavil Oliaei and Murat Demiral

Contents

9.1 Introduction.. 245
9.2 Challenges of mechanical micromachining 247
    9.2.1 Size effect in mechanical micromachining........................... 247
    9.2.2 Minimum uncut chip thickness 248
9.3 Finite element modeling of microcutting of Ti6Al4V..................... 251
    9.3.1 Material model 251
    9.3.2 Friction modeling... 254
    9.3.3 Finite element modeling of the effect of edge radius in
        micromachining... 256
    9.3.4 Analysis of the effect of cutting speed 260
    9.3.5 Analysis of the effect of friction conditions......................... 261
    9.3.6 Finite element modeling of micromachining of Ti6Al4V
        in the presence of built-up edge 263
9.4 Finite element modeling of micromachining: Influence of
    crystallography 267
References...274

9.1 Introduction

Nowadays, there is an emerging global trend toward miniaturization of systems, equipment, and devices in almost every field of science, engineering, and technology, where the increasing demand for miniaturized products to increase design flexibility, to dwindle energy consumption, and to achieve higher degree of accuracy has resulted in the creation of new concepts such as *small equipment for small parts*, *microfactories for microscale products*, and *scaled down manufacturing systems* [1–3].

Today, this miniaturization trend is demanding the production of higher functionality structural and mechanical components with significantly decreased manufactured features in the range of a few microns to a few hundred microns in several industrial sectors such as aerospace, automotive, smart communication systems, optics, microelectronics, microsensor systems, medicine, microelectromechanical systems, biotechnology, environmental sciences, defense, and avionics, to name a few [4–8].

Generally, microfabrication techniques can be classified based on different criteria. The most comprehensive classification is based on the nature of the process. Based on this criterion, microfabrication techniques can be divided into two major groups [9]: microsystem technologies (MST), which are also known as mask-based, micro electro mechanical system (MEMS)-based, or integrated circuit (IC) fabrication techniques, and microengineering technologies (MET), in which tool-based micromachining techniques are a major part [1].

Tool-based microfabrication techniques encompass a large variety of processes, in which some of them have been developed by scaling down of conventional machining processes such as microturning, micromilling, and microdrilling. These processes are known as mechanical micromachining processes. There are other processes, which are based on advanced machining technologies, such as microelectrical discharge machining (μ-EDM) and microelectrochemical machining (μ-ECM), or a combination of these processes known as hybrid processes, such as EDM/ECM, ultrasonic-assisted EDM (US/EDM), and so on [10].

Among aforementioned microfabrication techniques, micromechanical machining has found an ever-increasing acceptance, especially in the production of discrete microcomponents, due to its unique advantages, such as its cost-effectiveness because of eliminating the need for expensive clean room facilities, the ability of producing complex three-dimensional microfeatures, the ability to process different engineering materials, and having small environmental footprint due to the elimination of chemical materials such as etching solutions [11].

Mechanical micromachining is also considered an enabling technology in bridging nanoscale to macroscale feature development. However, there are several challenges associated with mechanical micromachining processes that should be explored before they can be successfully implemented in the industrial scale for the fabrication of miniature parts [11]. According to Silva et al. [12], since mechanical micromachining processes are capable of producing high-dimensional and geometrical accuracies with desired surface quality and subsurface integrity at reasonably low costs, they should be the first choice for microcomponent manufacturers among various micromanufacturing techniques.

Although micromechanical machining is considered the scaled-down version of macro-scale machining, the mechanics of machining at microscale are quite different [13]. An in-depth understanding of various phenomena involved in the micromachining process, including mechanism of material removal, burr formation, tool wear/deflection/failure, the interaction between cutting tool microgeometry and workpiece material at microscale, the nature of cutting forces, and the evolution of micropart surface topography, are key factors in improving productivity and reducing the cost [14]. In addition, the knowledge about all these phenomena is necessary to fabricate microparts, satisfying required dimensional and geometrical tolerances and surface quality requirements. Therefore, development of predictive techniques to improve the quality of microparts has emerged as an important research area. Several approaches have been used to predict outputs of microcutting operations, including analytical modeling, numerical techniques, molecular dynamics (MD) simulation, and experimental studies. Since experimental studies are costly and time-consuming and they are only valid for the conditions and range of machining parameters used in the experiments, numerical methods are used as an alternative method to predict machining process outputs.

Among various numerical methods, finite element (FE) method is the most widely accepted numerical method for the simulation of metal cutting processes [15]. Therefore, this chapter is devoted to the application of FE method in the modeling of microcutting operations. After a general introduction of the challenges of microcutting, the FE method (FEM) of microcutting of Ti6Al4V titanium alloy will be discussed, where a material model adopted for macroscale cutting operations is used in the simulation to investigate the efficiency of the FE method for the simulation of orthogonal microcutting operations. Second, the effect of crystallographic anisotropy on the micromachining of crystalline material—fcc single-crystal copper—is studied by using crystal plasticity (CP) theory.

9.2   Challenges of mechanical micromachining

Mechanical micromachining is defined as the machining of precision parts, with features in the range of 1–999 µm, made up of a wide range of engineering materials (metallic alloys, ceramics, polymers, etc.), using complex surfaces technologies [7,16]. It may also refer to the machining that cannot be realized by conventional technologies [16]. According to Camara et al. [7], micromachining can also be defined based on the dimensions of the cutting tools used in the material removal process, which should lie within the range of 1–1000 µm. There are several distinct features that make micromachining significantly different from conventional machining. The subsequent sections briefly discuss some of these features.

9.2.1   Size effect in mechanical micromachining

According to Vollertsen [17], a deviation from intensive or proportional extrapolated extensive values of process characteristics as a result of scaling of geometrical dimensions is defined as size effect. In micromachining, the size effect can be related to two different aspects of the process. The first one is when the edge radius of the cutting tool and uncut chip thickness are in the same order of magnitude. The second one is due to the effect of material microstructure, where isotropic and homogeneous material assumptions are not valid anymore [18,19] and polycrystalline material should be considered as heterogeneous and discrete [20].

A large edge radius to uncut chip thickness ratio in micromachining results in large effective negative rake angles during cutting. This large negative rake angle promotes ploughing, rubbing, and burnishing [21,22].

Specific cutting energy, which is defined as the total energy required for removing a unit volume of material, is an important measure of machinability [23]. One important size effect in micromachining is a nonlinear increase in the specific cutting energy with decreasing uncut chip thicknesses, which becomes noticeable for uncut chip thickness values less than the cutting edge radius [24,25]. This increase in the specific cutting energy can be attributed to different phenomena involved in the cutting process. At larger edge radius to chip thickness ratios, the cutting tool acts like a blunt tool, where, according to Armarego et al. [26], in this case, the relative contribution of ploughing forces become significant, which in turn will result in the increased specific cutting energies.

In a study of Lucca et al. [27], the increase in the specific cutting energy in micromachining is attributed to the energy dissipation of ploughing by the edge radius and rubbing of the flank face of the tool due to the elastic recovery of the workpiece material.

Material strengthening effect due to the reduction in the density of movable dislocations has been reported by Backer et al. [28] as another factor influencing specific cutting energy. Filiz et al. [29] showed that material strengthening at small uncut chip thickness values occurs, because at small scales, the effect of strain hardening becomes more dominant than the effect of thermal softening of the material. The presence of short-range inhomogeneities in commercial metals has been reported by Shaw [30] as another origin of the size effect. Larsen and Oxley [31] used strain rate sensitivity of the flow stress to describe the size effect in machining. Their experimental results on plain carbon steel revealed an inverse proportionality between maximum shear strain rate and undeformed chip thickness. They came to a conclusion that as uncut chip thickness decreases, shear strain rate increases, and consequently, flow stress of the material increases, which in turn results in higher specific cutting energies.

9.2.2 Minimum uncut chip thickness

In microscale machining, the cutting tool edge radius and the amount of material being cut are in the same order of magnitude. There is a value of uncut chip thickness after which continuous chip formation ceases. This critical value is defined as minimum uncut chip thickness (MUCT), which is known to be a function of the tool material, cutting edge radius, and workpiece material [32].

Understanding MUCT for a given tool–workpiece combination is an essential part of any micromachining study, since it specifies the lower bound of mechanical micromachining, where beyond this thickness, material can be removed in a stable way under perfect performance of the machine tool [32]. In conventional machining processes, the uncut chip thickness is much larger than the cutting edge radius; therefore, the effect of cutting edge radius is negligibility small in the cutting process. However, in microcutting processes, the chip thickness is in the same order of magnitude as the cutting edge radius, which necessitates consideration of the concept of MUCT in micromachining operations. The concept of MUCT can be used to distinguish between different cutting conditions in terms of chip formation [6,33]. Figure 9.1 illustrates three different scenarios.

When undeformed chip thickness is less than MUCT (Figure 9.1a), there is no chip formation; therefore, due to the rubbing action of the cutting tool, workpiece material will deform elastically underneath the cutting tool, and elastic recovery will take place on the passing of the cutting tool. Theoretically, no chips will be formed in this case.

When undeformed chip thickness becomes same as MUCT (Figure 9.1b), chips will start to form in the presence of elastic recovery and ploughing. In this case, the actual thickness of the material being removed is less than the undeformed chip's thickness. When undeformed chip's thickness becomes greater than MUCT (Figure 9.1c), shearing becomes the dominant material removal mechanism and elastic recovery becomes negligible. It has been emphasized by Weule et al. [34] that edge sharpness and material properties have a strong contribution to MUCT.

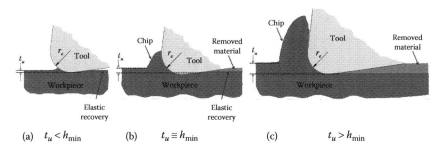

(a) $t_u < h_{min}$ (b) $t_u \cong h_{min}$ (c) $t_u > h_{min}$

Figure 9.1 Different cutting scenarios based on undeformed chip thickness value (a) $t_u < h_{min}$, (b) $t_u \cong h_{min}$, and (c) $t_u > h_{min}$.

Kim et al. [35] analyzed the periodicity of forces during micromilling and identified the transition between the noncutting and cutting regimes near the minimum chip thickness value. They concluded that the periodicity of cutting forces is affected by the minimum chip thickness, feed per tooth, and cutting position angle.

For orthogonal cutting process, a MUCT model has been developed by Son et al. [36] by dividing the workpiece into perfectly plastic and perfectly elastic regions.

Using equilibrium of forces, they calculated the stagnation angle, and assuming that at MUCT, the shear angle is equal to the stagnation angle, they calculated MUCT as:

$$\text{MUCT} = r_e \left[1 - \cos\left(\frac{\pi}{4} - \frac{\beta}{2} \right) \right] \tag{9.1}$$

where:

r_e is the edge radius
β is the friction angle

Another analytical relation that has been developed by Yuan et al. [22] to calculate MUCT as a function of cutting edge radius (r_e), horizontal (F_x) and vertical (F_y) components of the machining forces, and coefficient of friction (μ) is as follows:

$$\text{MUCT} = r_e \left(1 - \frac{F_y + \mu F_x}{\sqrt{F_x^2 + F_y^2 \left(1 + \mu^2\right)}} \right) \tag{9.2}$$

The influence of surface roughness on minimum uncut thickness has been confirmed by Oliaei and Karpat [37] in the presence of a built-up edge (BUE). They calculated a ratio of uncut chip thickness to edge radius of about 10% during the orthogonal micromachining of Ti6Al4V.

In addition to analytical models, numerical techniques are also used to determine MUCT. In a study conducted by Shi and Liu [38], an Arbitrary Lagrangian Eulerian (ALE)-based numerical modeling is utilized to determine the MUCT for copper, using cutting tools with different nominal rake angles, without employing a chip separation criterion. The FE analysis of micromachining by using the ALE method has been carried out by Woon et al. [39] to determine the critical undeformed chip thickness to cutting edge ratio (a/r) in micromachining of AISI 4340 steel. They obtained a critical a/r value of 0.2625 for AISI 4340 steel. Woon et al. [40] also studied the interaction between uncut chip thickness and edge radius by using experimental and FE-based techniques when micromachining AISI 1045.

9.3   Finite element modeling of microcutting of Ti6Al4V

Finite element modeling has been extensively used for the modeling of conventional machining processes. According to Lauro et al. [41], one important advantage of the application of FEM in micromachining studies is to make it possible to have an insight into some events that are complicated to be observed because of their smaller dimensions. However, although the application of the FEM in the conventional machining is increasing, its application in micromachining studies is still modest.

In FEM, materials are modeled as a continuum, where chemistry, atomic scale effects, lattice structure, and grain size are not included in the model. Three different frameworks are used in the FEM approach, including Lagrangian formulation, where it is assumed that mesh is attached to the workpiece and that it moves with the material [42]; Eulerian formulation, where mesh is fixed in space to define a control volume and material flow occurs through the meshes [43]; and ALE formulation, which combines the features of pure Lagrangian and Eulerian analyses [44].

One important material that shows a high potential for micropart fabrication is Ti6Al4V titanium alloy. It is among the most commonly used materials for the biomedical and aerospace industry, which accounts for about 50%–60% of the total titanium alloy production [45,46]. In 1955, Siekmann [47] mentioned that machining of titanium alloys would always be a problem, independent of chip removal process. Even with the many sophisticated developments in the cutting tool industry, this fact still remains true, and titanium alloys remain as difficult-to-machine materials [48,49]. Therefore, modeling of cutting process of Ti6Al4V is of significant importance for both research and industrial purposes.

In FEM of machining processes, the effectiveness of the model to predict field variables such as stress, strain, temperature, and velocity is highly affected by the workpiece material behavior (constitutive material model), thermomechanical properties, and contact conditions at tool–chip and tool–workpiece interfaces (friction laws) [50], which are briefly discussed in the subsequent sections.

9.3.1   Material model

As mentioned earlier, the material constitutive model employed in the FE simulation of machining processes, together with friction definition, significantly affects the process output predictions. One of the most widely used material constitutive models in the modeling of machining operations is Johnson–Cook (JC) material model [51], which is defined as:

$$\sigma = \left(A + B\varepsilon^n\right)\left[1 + C\ln\frac{\dot{\varepsilon}}{\dot{\varepsilon}_0}\right]\left[1 - \left(\frac{T - T_r}{T_m - T_r}\right)^m\right]$$ (9.3)

where:
   ε is the equivalent plastic strain
   $\dot{\varepsilon}$ is the strain rate
   $\dot{\varepsilon}_0$ is the reference plastic strain rate
   T is the instantaneous temperature
   T_r is the room temperature
   T_m is the melting temperature of workpiece material

In this model, the flow stress of the material (σ) is defined by three multiplicative, yet distinctive terms. The first term is used to define elastic–plastic behavior of the material and is used to define the strain hardening. The second term, which is called viscosity term, is used to account for the strain rate sensitivity. The last term is the temperature softening term, which shows the thermal softening behavior of the material. In Equation 9.3, A, B, C, n, and m are material constants representing the yield strength, strain sensitivity, strain rate sensitivity, strain hardening exponent, and thermal softening index, respectively.

The JC material model has been used by Afazov et al. [52] in the FE analyses of microcutting of AISI 4340 steel at different uncut chip thicknesses and cutting velocities. Jin and Altintas [53] also used JC material model to simulate microcutting of brass 260 with model parameters of $A = 90$ MPa, $B = 404$ MPa, $C = 0.009$, $n = 0.42$, $m = 1.68$, $T_m = 916°C$, and $T_r = 25°C$.

One major shortcoming of the JC material model is its inability to model strain-softening effect [54,55]. To overcome this shortcoming, a modified JC material model has been proposed by Calamaz et al. [54] in the form of:

$$\sigma = \left(A + B\varepsilon^n\left(\frac{1}{\exp(\varepsilon^a)}\right)\right)\left(1 + C\ln\frac{\dot{\varepsilon}}{\dot{\varepsilon}_0}\right)\left(1 - \left(\frac{T - T_r}{T_m - T_r}\right)^m\right)$$

$$\times\left(D + (1 - D)\tanh\left(\frac{1}{(\varepsilon + S)^c}\right)\right)$$ (9.4)

with

$$D = 1 - \left(\frac{T}{T_m}\right)^d$$

and

$$S = \left(\frac{T}{T_m}\right)^b$$

The modified JC material model has been used by Thepsonthi and Özel [56] to simulate micromilling of Ti6Al4V. The values of $A = 782.7$ MPa, $B = 498.4$ MPa, $n = 0.28$, $a = 2$, $C = 0.028$, $m = 1.0$, $d = 0.5$, $r = 2$, $b = 5$, and $s = 0.05$ have been utilized in their model for Ti6Al4V.

In this study, a material model developed by Karpat [57] is adopted to simulate microscale machining of titanium alloy Ti6Al4V. This material model has been successfully used in the simulation of macroscale machining forces under various machining conditions. This material model considers strain softening as a function of temperature, and it was developed such that the influence of strain softening decreases as temperature decreases. The model also considers the relationship between strain rate and strain. The material model is shown in Equation 9.5, and the material model parameters are given in Table 9.1.

$$\sigma(\varepsilon, \dot{\varepsilon}, T) = (a\varepsilon^{n^*} + b)(cT^{*2} + dT^* + e)\left(1 - \left[1 - \left(\frac{\ln(\dot{\varepsilon}_0)}{\ln(\dot{\varepsilon})}\right)^q\right]\left[\frac{1}{l \times \tanh(\varepsilon + p)}\right]\right) \quad (9.5)$$

with

$$T^* = \frac{T}{T_r}$$

A flow softening function, which starts after a critical strain and temperature, is integrated into Equation 9.5, as shown in Equation 9.6. After a critical

Table 9.1 The coefficients of the material model

a (MPa)	n*	b (MPa)	c	d
590	0.27	740	7.1903e-5	−0.0209
Q	1	p	$\dot{\varepsilon}_0$	e
0.035	1.1	0.08	800	1.6356

Source: Oliaei, S.N.B. and Karpat, Y., *Int. J. Adv. Manuf. Technol.*, 1–11, 2016c.

Table 9.2 The coefficients of the material model

Material property	Expression	Unit
Young's modulus	$E_Y(T) = -57.7 \times T + 111672$	MPa
Thermal expansion	$\alpha_T(T) = 3.10^{-9} \times T + 7.10^{-6}$	1/°C
Thermal conductivity	$\lambda(T) = 0.015 \times T + 7.7$	W/m.K
Heat capacity	$C_p(T) = 2.7e^{0.0002T}$	N/mm²/°C

Source: Karpat, Y., *J. Mater. Process. Technol.*, 211, 737–749, 2011.

strain value, the proposed function controls the softening behavior of the material. For strain values less than the critical strain value, Equation 9.5 is valid [57].

$$\sigma_{soft}(\varepsilon, \dot{\varepsilon}, T) = \sigma - (\sigma - \sigma_s)\left[\tanh(k\varepsilon^*)^r\right] \tag{9.6}$$

Temperature-dependent mechanical and thermophysical properties of titanium alloy Ti6Al4V are used in the simulations, as shown in Table 9.2.

9.3.2 Friction modeling

Frictional conditions at the tool–chip and the tool–workpiece interfaces are important, since they affect the heat generation at the interfaces [58], and due to the existence of very severe friction conditions in machining operations, small changes in friction modeling can cause large changes in chip formation and can highly affect the accuracy of the machining performance predictions [59,60]. The contact and frictional conditions at the tool–chip interface are influenced by several factors such as machining process parameters (cutting speed, feed rate, and depth of cut), material being machined, tool geometry parameters (rake angle, clearance angle, and edge radius), and so on [8].

In the literature, different friction models have been used in the modeling of machining operations, and their effect on the FE simulations have been investigated in detail [61,62]. Three friction models that are extensively used in the modeling of machining processes are discussed in this section.

The first low, which is based on simple Coulomb's law, relates the frictional stress (τ_f) at the tool–workpiece interface to normal stress (σ_n) via a constant coefficient of friction (μ) as [59]:

$$\tau_f = \mu \cdot \sigma_n \tag{9.7}$$

In the second model, a constant frictional stress (τ) at the tool–chip interface on the rake face; equal to a fixed percentage of the shear flow stress of the work material k, is considered [59]:

$$\tau = m \cdot k \tag{9.8}$$

The third model is based on the sticking–sliding model of Zorev [63]. Based on this model, the tool–chip interface on the rake face is divided into two regions: the sticking region, where the frictional stress is assumed to be equal to the shear flow stress of the material, and the sliding region, where Coulomb's friction low holds [59]. This law can be expressed as follows:

$$\tau = \begin{cases} \mu \cdot \sigma_n & \text{when} \quad \tau < k \\ k & \text{when} \quad \tau \geq k \end{cases} \tag{9.9}$$

Usually, in the studies of FE simulation of micromachining, a combination of these friction models is used. In the literature, different frictional conditions are used, depending on the machining conditions and tool–workpiece pair. For instance, Lai et al. [8] assumed a coefficient of friction of 0.3 for modeling of microcutting of oxygen-free high conductivity copper (OFHC). In a study conducted by Tajalli et al. [64], the mechanical interaction between the tool and the workpiece is modeled with a Coulomb's friction coefficient of 0.1 during micromachining of fcc crystalline OFHC.

A hybrid friction model that combines sticking and Coulomb's friction laws has been used by Thepsonti and Özel [65], considering a sticking contact conditions ($m = 0.9$) around the cutting edge and Coulomb's friction of $\mu = 0.7$ on the rake face for tungsten carbide–titanium pair. For cubic boron nitride (CBN) cutting tools, the Coulomb's friction is changed to $\mu = 0.4$ while micromachining Ti6Al4V. In another study, Thepsonti and Özel [56] used FE simulation to model micromilling of Ti6Al4V. In their study, the contact between the chip and the workpiece was assigned as a sliding contact, with the constant Coulomb's friction coefficient of $\mu = 0.2$.

The effect of frictional conditions on cutting and thrust force predictions has been examined by Kim et al. [66] by considering three different cases, that is, when ignoring friction effect and for two coefficient of friction of $\mu = 0.15$ and $\mu = 0.3$ during micromachining of OFHC. Their findings revealed that thrust forces are highly affected by the change in frictional conditions, where a discrepancy of 80% has been observed between the predictions and experimental results. Based on their investigations, a coefficient of friction of $\mu = 0.3$ is found as the most proper value to represent the frictional conditions at the tool–chip interface. Jin and Altintas [53] used sticking and sliding contact conditions during FEM of micromilling of brass 260. They used a Coulomb's friction $\mu = 0.15$ in their modeling. Afazov et al. [52] also assumed a Coulomb's friction coefficient of 0.4 to model micromilling of AISI 4340 steel by using a TiN-coated tungsten carbide tool. The coefficient of friction of $\mu = 0.5$ is considered by Özel and Zeren [44] in the FEM of meso-scale machining of AISI 1045 steel, using uncoated carbide cutting tool.

9.3.3   Finite element modeling of the effect of edge radius in micromachining

In this section, SFTC DEFORM-2D software is used to simulate microcutting of Ti6Al4V titanium alloy, using uncoated tungsten carbide cutting tools. This software is based on the updated Lagrangian formulation and implements implicit integration method [59]. The heat transfer coefficient at the tool–chip interface is taken as a constant, 5,000 W/m² °C. The workpiece and tool are meshed with 10,000 quadrilateral elements. The workpiece is modeled as elastic–plastic, whereas the cutting tool is modeled as rigid.

In order to analyze the effect of edge radius on micromachining forces, five edges with different edge radii in the range of 1–5 µm are considered. The simulations are conducted under the same depth of cut of 1 µm, with a Coulomb's friction factor of $\mu = 0.2$ at a cutting speed of 62 m/min [58]. Figure 9.2 illustrates the simulated cutting and thrust forces for cutting edges with different edge radii.

As it can be seen in Figure 9.2, the cutting edge radius has a significant influence on both cutting and thrust forces during micromachining. It can also be seen that at larger edge radii, thrust forces are more influenced by an increase in the edge radius compared with the cutting forces.

Another important difference that can be observed is the magnitude of the cutting and thrust forces. At an edge radius of 1 µm, the cutting forces are larger than the thrust forces, whereas for an edge radius of 2 µm, the cutting and trust forces get closer to each other and they have almost the same values. As cutting edge radius increases further, the thrust forces become larger than the cutting forces. It can be seen that at an edge radius of 5 µm, thrust forces are 1.7 times larger than the cutting forces. Therefore, it can be said that larger thrust forces and smaller cutting forces are characteristics of ploughing dominant material removal, which is mainly due to larger ratios of cutting edge radius to uncut chip thicknesses, commonly observed in microscale cutting operations. The chip morphology for each edge radius and its associated effective stress are shown in Figure 9.3. It can be seen that as the edge radius increases,

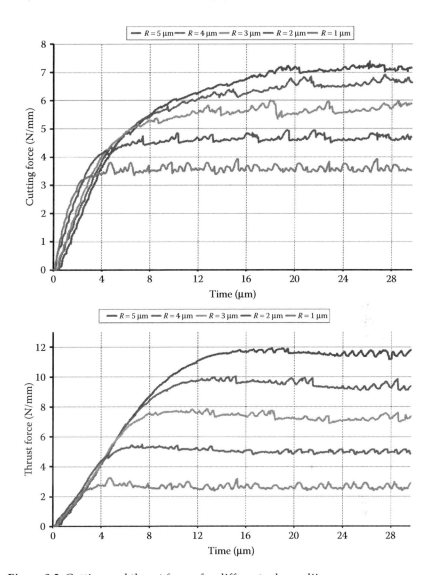

Figure 9.2 Cutting and thrust forces for different edge radii.

materials tend to pile up in front of the cutting tool, which affects the chip compression ratio, defined as the ratio between the uncut chip thickness and the deformed chip thickness.

The pile up of material in front of the cutting edge can be considered as an important characteristic of ploughing dominant cutting process. Therefore, based on the ratio of uncut chip thickness to cutting edge radius, micromachining process can be divided into three zones. In zone I,

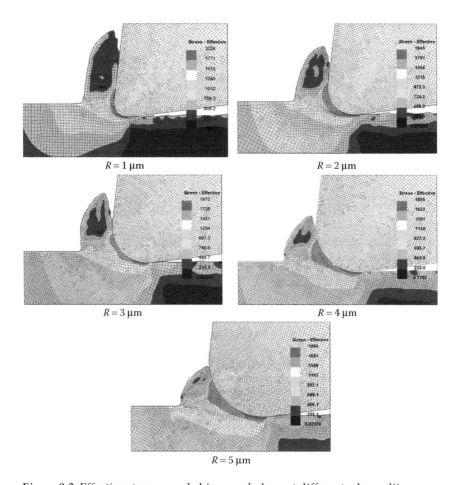

Figure 9.3 Effective stresses and chip morphology at different edge radii.

where the uncut chip thickness to cutting edge ratio is larger than one, the cutting forces are larger than the thrust forces; in this region, chips can be easily formed through a shearing dominant cutting process. In zone II, with chip thickness to edge radius ratios around 40%–50%, the cutting and thrust forces become equal, and chip formation occurs by a combination of shearing and ploughing. In zone III, thrust forces become significantly larger than the cutting forces, and material piles up in front of the cutting edge. In this zone, ploughing becomes the dominant material removal mechanism.

Another important observation that needs to be addressed is the maximum effective stress zone. As it can be seen in Figure 9.3 that as

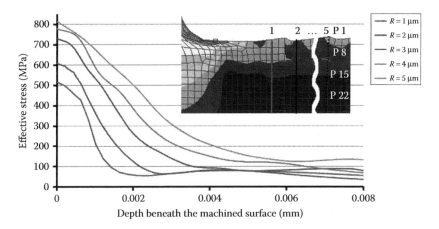

Figure 9.4 Micromachining-induced stress distributions of effective stresses with respect to depth beneath the machined layer at different edge radii.

the cutting edge radius increases, maximum effective stress zone moves from rake face toward the flank face of the tool, which means that, as the cutting edge radius increases, the cutting tool exerts more stresses to the machined surface.

The distribution of machining-induced stresses with respect to the depth beneath the machined surface are also obtained for different edge radii. For this purpose, a prescribed path is defined, starting from the machined surface and ending at a depth of about 8 μm. The average value of five measurements obtained from different locations is calculated. The profile of the effective stress distribution as a function of the depth beneath the machined surface is shown for different edge radii in Figure 9.4.

It can be seen in Figure 9.4 that the magnitude and distribution of machining-induced stresses are highly influenced by the cutting edge radius. As the cutting edge radius increases, the magnitude of machining-induced stresses significantly increases and more material is affected by the machining process beneath the machined surface.

For comparison, the temperature distributions at two different edge radii of 1 μm and 5 μm are shown in Figure 9.5 for a machining time of 30 μs. As can be seen in Figure 9.5, for an edge radius of 5 μm, the cutting temperature, which is defined as the maximum temperature at the tool–chip interface, is significantly higher than that of an edge radius of 1 μm. The higher temperatures at larger edge radii can be attributed to the larger negative rake angles and consequently sever plastic deformations. However, it can be seen that the temperatures generated during micromachining processes are substantially lower than that generated during macroscale cutting.

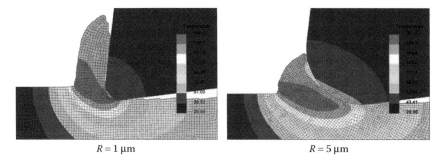

$R = 1\,\mu\text{m}$ $R = 5\,\mu\text{m}$

Figure 9.5 Temperature distributions at two different edge radii.

9.3.4 Analysis of the effect of cutting speed

In order to analyze the effect of cutting speed on micromachining forces and chip morphology, micromachining of Ti6Al4V is simulated at three different cutting speeds of 47, 62, and 124 m/min. The average cutting and thrust forces are shown for two different edge radii of 1 μm and 5 μm in Figure 9.6.

It can be seen in Figure 9.5 that both cutting and thrust forces show a decreasing trend by increasing the cutting speed; however, decrease in the cutting forces is significantly larger than that in the thrust forces. These results reveal that as far as increasing cutting speed does not affect the tool life, which can cause edge rounding, increasing cutting speed can result in smaller cutting forces. On the other hand, since the decrease in the cutting forces due to higher cutting speeds is lower than the effect of edge rounding on increasing cutting forces, smaller cutting speeds can be used if edge rounding exists.

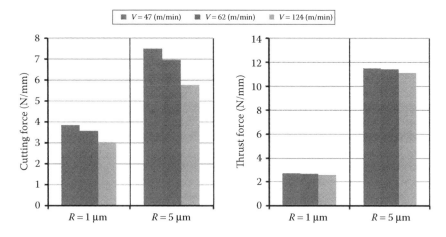

Figure 9.6 Average cutting and thrust forces at different cutting speeds.

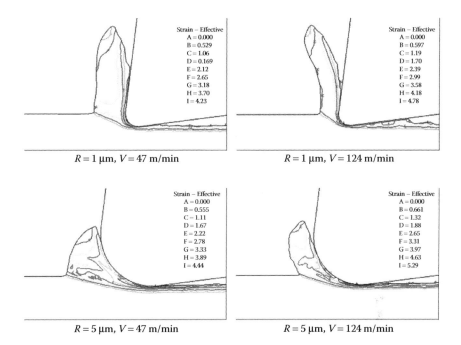

Figure 9.7 Chip morphology at different cutting speeds and edge radii.

Figure 9.7 illustrates the chip morphologies at two different cutting speeds of 47 m/min and 124 m/min for two different edge radii of 1 μm and 5 μm. It can be seen that increasing cutting speed can result in a reduced chip compression ratio. In addition, it can be seen that at lower cutting speeds, the chip tends to stay straight, which results in a larger chip–tool contact length; however, at higher cutting speeds, the contact length decreases, since the chip tends to curl away from the tool.

9.3.5   Analysis of the effect of friction conditions

As mentioned earlier, the friction at the tool–chip and tool–workpiece interfaces can be defined by using different friction laws. To further analyze the effect of friction conditions, simulations are conducted at three different frictional conditions: two coulomb frictions of $\mu = 0.1$ and $\mu = 0.3$ and a shear factor of $m = 0.95$. The results of FE simulations (Figure 9.8) are compared with the experimental data obtained by Oliaei and Karpat [37] at a cutting speed of 62 m/min.

As can be seen in Figure 9.8, with increasing Coulomb's friction from 0.1 to 0.3, a significant increase in the cutting forces can be observed. Despite cutting forces, the thrust forces decrease by increasing Coulomb's friction factor. It can also be seen that when using a shear friction of

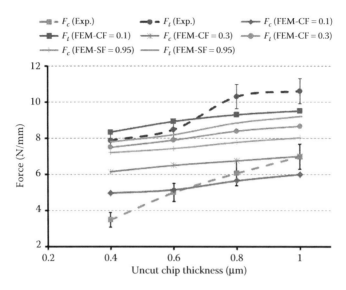

Figure 9.8 Comparison of measured and predicted micromachining forces for different frictional conditions.

$m = 0.95$, thrust forces lie between predictions using two Coulomb's friction factors of $\mu = 0.1$ and $\mu = 0.3$; however, the use of shear friction of $m = 0.95$ results in a very large overprediction of the cutting forces. It should be mentioned that in this simulation, a cutting edge radius of 4 μm is used, as reported by Oliaei and Karpat [37].

It can also be seen in Figure 9.8 that, although in some uncut chip thickness values, the FE model is capable of producing good predictions, in some uncut chip thicknesses, it ceases to give accurate predictions. This discrepancies between predictions and experiments are attributed to the BUE formation in front of the cutting tool, which results in a change in cutting tool geometry, that is, cutting edge radius, rake and clearance angles, and contact and frictional conditions [58].

In FE simulation of machining processes, the velocity field obtained at high shear factors (m) can be used to create sticking conditions at the tool–chip interface to give some idea about the formation of dead-metal zone in front of the cutting tool. This low-velocity region, which resembles a dead-metal zone, is assumed to behave like a BUE [67,68]. In this study, a high shear factor value of $m = 0.95$ is used in the FE simulations, in order to examine the low-velocity region in front of the cutting edge. The velocity fields for two different edge radii of 1 and 5 μm with $m = 0.95$ are shown in Figure 9.9.

Using velocity field information, there is also a possibility of determining a stagnation point on the cutting edge, where flow velocity of the materials becomes zero. An important distinction that can be observed

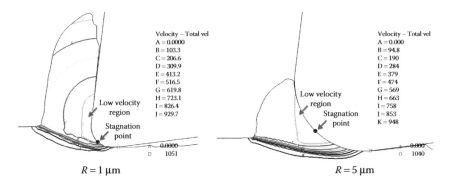

Figure 9.9 Velocity field in front of the cutting edge at two different edge radii.

from velocity field predictions is the extension of the low-velocity region. As can be seen in Figure 9.9, at a large cutting edge radius, the low-velocity region extends well below the stagnation point and toward the flank face of the cutting tool, which further promotes BUE formation.

It should be mentioned that the lengths of the dead-metal zone obtained by simulations are quite smaller than the experimental values, according to Oliaei and Karpat [58]. Therefore, in the next section, an attempt has been made to include BUE effects in the FE simulations of micromachining of Ti6Al4V titanium alloy by modifying the cutting edge geometry based on the geometry of the BUE.

9.3.6   Finite element modeling of micromachining of Ti6Al4V in the presence of built-up edge

Built-up edge is an important phenomenon that has been observed in micromilling experiments on Ti6Al4V [37]. A BUE consists of material layers that are deposited onto the tool surface, changing the tool geometry and hence the mechanics of the process. A stable and thin BUE is known to protect the cutting edge [69].

In conventional machining processes, BUE is known to be a commonly observed phenomenon, which appears during continuous chip formation, when machining ductile materials such as aluminum, steel, and titanium. It is known to affect surface roughness and tool wear. However, the effect of BUE on the process outputs of microscale cutting requires special attention, and its effects in microscale cutting needs to be carefully explored. Therefore, a solid understanding of the mechanics of cutting at microscale with the presence of BUE seems to be a crucial step in building predictive models and controlling the quality of microparts made of Ti6Al4V.

Childs [70] developed an FE model to predict BUE formation during machining of steel by integrating a damage model. He also reported some preliminary results on simulating BUE during microscale machining. In this section, the influence of BUE on microscale machining of Ti6Al4V is investigated. As mentioned in the previous section, the use of dead-metal zone to represent BUE formation significantly underestimates the BUE size, which highly affects the accuracy of the predictions. The edge modification based on the geometry of the BUE has been studied by Oliaei and Karpat [58].

The geometry of the BUE obtained from laser scanning microscopy (Figure 9.10) is mapped into the FE model by defining BUE geometric parameters depicted in Figure 9.11. The FE model of the modified cutting edge is also shown in Figure 9.11. The dimensions of BUE are shown in Table 9.3.

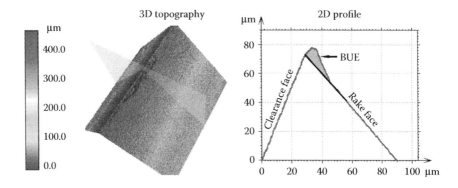

Figure 9.10 Laser scanning microscope image of BUE. (Data from Oliaei, S.N.B. and Karpat, Y., *Int. J. Adv. Manuf. Technol.*, 1–11, 2016c. With permission.)

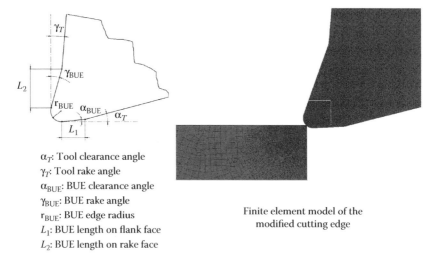

α_T: Tool clearance angle
γ_T: Tool rake angle
α_{BUE}: BUE clearance angle
γ_{BUE}: BUE rake angle
r_{BUE}: BUE edge radius
L_1: BUE length on flank face
L_2: BUE length on rake face

Finite element model of the
modified cutting edge

Figure 9.11 Modified geometry of the cutting edge including BUE. (Data from Oliaei, S.N.B. and Karpat, Y., *Int. J. Adv. Manuf. Technol.*, 1–11, 2016c. With permission.)

Table 9.3 BUE parameters obtained at a cutting speed of 62 m/min

Uncut chip thickness (μm)	α_{BUE} (°)	γ_{BUE} (°)	r_{BUE} (μm)	L_1 (mm)	L_2 (mm)
0.4	2	22	3	11	18
0.6	3	18	3	13	20
0.8	3	18	4	16	26
1	3	20	4	19	35

Source: Oliaei, S.N.B. and Karpat, Y., *Int. J. Adv. Manuf. Technol.*, 1–11, 2016c.

In this section, a hybrid friction model is used, where different combinations of Coulomb's friction (μ) and shear friction (*m*) values are considered. Shear friction is varied between 0.7 and 0.95, whereas Coulomb's friction is varied between 0.1 and 0.3. Figure 9.12 illustrates the influence of hybrid friction on the chip formation at uncut chip thickness of 1 μm and cutting speed of 62 m/min. It can be seen that using high shear factors (sticking conditions), chips tend to slide over the rake face of the tool.

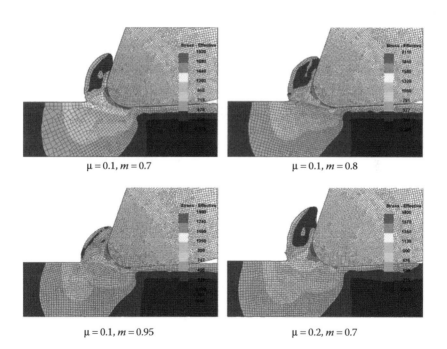

μ = 0.1, *m* = 0.7   μ = 0.1, *m* = 0.8

μ = 0.1, *m* = 0.95   μ = 0.2, *m* = 0.7

Figure 9.12 Finite element simulations of chip morphology and effective stresses for different frictional conditions.   (*Continued*)

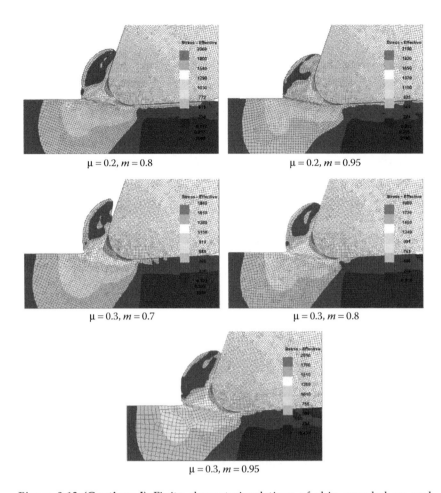

$\mu = 0.2, m = 0.8$

$\mu = 0.2, m = 0.95$

$\mu = 0.3, m = 0.7$

$\mu = 0.3, m = 0.8$

$\mu = 0.3, m = 0.95$

Figure 9.12 (**Continued**) Finite element simulations of chip morphology and effective stresses for different frictional conditions.

Figure 9.13 illustrates the FE force predictions at different combinations of Coulomb's friction and shear factor for uncut chip thickness of 1 μm at a cutting speed of 62 m/min. A Coulomb's friction of 0.2 and a shear factor of 0.8 resulted in the closest predictions compared with the experimental results of Oliaei and Karpat [58]. It can be seen that by modifying the cutting edge radius based on the actual shape of the BUE and using hybrid friction models, it is possible to come up with acceptable predictions.

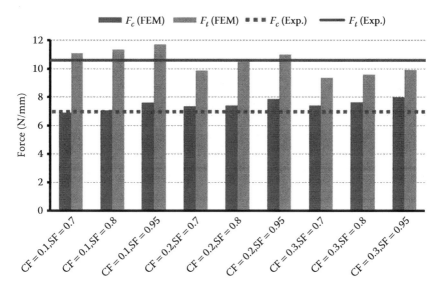

Figure 9.13 Micromachining force predictions at different frictional conditions (uncut chip thickness of 1 μm and cutting speed of 62 m/min). (Data from Oliaei, S.N.B. and Karpat, Y., *Int. J. Adv. Manuf. Technol.*, 1–11, 2016c. With permission.)

9.4   Finite element modeling of micromachining: Influence of crystallography

In the manufacturing and forming of single-crystal metals, which are increasingly used in the aerospace, biomedical, automotive, and optics industry, process zone is limited to a single or a few grains of machined material. Since they are known to be highly anisotropic in their physical properties, it is not astonishing that machining response relies on the crystallographic orientation.

Microscale material removal processes were investigated experimentally from different points of view. Earlier studies suggested a strong dependence of process parameters, such as chip shape, cutting force, surface finish, shear angle, and so on, on the orientation of the crystal being cut [71–74]. In some turning experiments, coordination of the cutting plane crystal and cutting direction varied with workpiece rotation, but the zone axis remained unchanged [75,76]. The behavior of the machining forces was observed to be stable, reducing, monotonically increasing, and periodically varying for different cutting conditions [71–73,77–79].

Crystallographic symmetries for the machining forces and shear angles of fcc metals were observed with changing cutting orientation [75,76]. The variation of dynamic shear stress (DSS)—the ratio of shear force to shear plane area—with grain orientation was also investigated [79]. In the in situ (inside a scanning electron microscope [SEM]) turning experiments of aluminum and copper samples, the DSS was found to vary with crystal orientation [72,75], whereas in a study on copper, not only the cutting plane but also the cutting direction was observed to affect DSS [71].

In micromachining process, there occurs a subsurface damage, altering the mechanical properties of the material at the surface. It was discovered that the surface and subsurface damage caused in the finished material can be reduced as the undeformed chip thickness is decreased [80,81], and process parameters such as depth of cut, feed rate [81,82], rake and clearance angle [83], and tilting of workpiece [84] were reported as parameters to achieve transition between ductile and brittle modes of material removal. Feng et al. [85] investigated femtosecond laser micromachining on a single-crystal superalloy, where material damage and laser-induced microstructural change were characterized and compared with nanosecond laser machining. Kota et al. [86] demonstrated that the deformation depth was approximately equal to the cutting depth for aluminum single crystal, using electron backscatter diffraction (EBSD). Nahata et al. [87] machined a single-crystal sample at the microscale along multiple crystallographic orientations and compared the extent of damage caused by it to understand/predict the behavior of coarse-grained crystals.

In micromachining of various crystal materials under different cutting conditions, the size effect, a nonlinear increase in the specific cutting energy as the uncut chip thickness is decreased, was observed [28,88–91]. This was linked mostly to material strengthening due to an increase in the strain rate in the primary shear zone at smaller uncut chip thickness [31], an increase in the shear strength of the material due to a decrease in the tool–chip interface temperature as the uncut chip thickness is decreased [88], dependence of flow stress of the metal on the strain gradient in the deformation zone, subsurface deformation of the workpiece material, tool edge radius effects, and energy required to create new surfaces through ductile fracture [31,91,92].

Compared with experimental studies, a limited number of modeling studies for single-grain micromachining is reported in the literature, owing to an inherent difficulty in modeling large deformation processes at high strain rates numerically. Among the analytical models, Tsutiya [93] used a Schmid factor to characterize active slip systems during machining. Studies in [94,95] used a modified Taylor factor with a texture softening factor to predict the shear angle uniquely in single-crystal cutting. Kota and Ozdoganlar [96] considered minimization of the total power, including rake-face friction power and the shearing in the process zone, to determine the shear angle and specific cutting energy. Venkatachalam et al. [97] proposed a closed-form analytical model for ductile-regime machining process of single-crystal brittle materials to determine the transition undeformed chip thickness in reducing the production time and enhancing the productivity. Liu and Melkote [98] developed a strain-gradient plasticity-based FE model of orthogonal microcutting to examine the influence of tool edge radius on the size effect, without considering crystallographic effects. Pen et al. [99] and Komanduri et al. [100] used quasi-continuum and molecular-dynamics simulation methods, respectively, in modelling of nano-cutting process to observe the influence of crystal orientation and cutting direction on the deformation mechanism. Zahedi et al. [101] used hybrid FE and mesh-free methods to model single-grain cutting process. Tajalli et al. [64] studied the orthogonal microcutting of fcc materials based on CP. Demiral et al. [92,102] used a numerical implementation of an enhanced model of strain-gradient CP [103] to demonstrate the influence of strain gradients and their evolution in the micromachining process for different cutting directions of fcc single-crystal copper and bcc single-crystal β-brass, respectively. Liu et al. [104] presented an FE modeling approach for the microcutting process of single-crystal metals, incorporating a new shear–strain-based criterion accounting for the partial and full activations of slip systems.

In this part, am FEM of orthogonal micromachining of fcc copper single crystal was developed. The CP theory was used in the simulations. The essential equations of CP are summarized as follows.

In the CP model, the stress rate ($\dot{\sigma}_{ij}$) is related to the elastic strain rate ($\dot{\varepsilon}_{ij}^e$) via $\dot{\sigma}_{ij} = C_{ijkl}\dot{\varepsilon}_{kl}^e = C_{ijkl}(\dot{\varepsilon}_{ij} - \dot{\varepsilon}_{kl}^p)$, where plastic strain rate $\dot{\varepsilon}_{ij}^p$ equals to $\sum_{\alpha=1}^{N} \mu_{ij}^{\alpha}\dot{\gamma}^{\alpha}$; N is the total number of available slip systems, μ_{ij}^{α} is the Schmid tensor and is equal to a dyadic product of the slip direction s^{α} and the slip plane normal m^{α} in the initial unloaded configuration, and $\dot{\gamma}^{\alpha}$ is the shearing rate on the slip system α. A power-law representation was chosen for $\dot{\gamma}^{\alpha}$:

$$\dot{\gamma}^{\alpha} = \dot{\gamma}^{\alpha} \, \mathrm{sgn}(\tau^{\alpha}) \left|\frac{\tau^{\alpha}}{g^{\alpha}}\right|^n \tag{9.10}$$

where:

$\dot{\gamma}_0^{\alpha}$ is the reference strain rate

n is the macroscopic rate-sensitivity parameter

τ^{α} is the resolved shear stress, and sgn (Ψ) is the signum function of Ψ

The strength of the slip system α at the current time (g^α), equal to a sum of the *critical resolved shear stress* (CRSS) ($g^\alpha|_{t=0}$) and the evolved slip resistance due to strain hardening (Δg^α), is as follows:

$$g^\alpha = g^\alpha|_{t=0} + \Delta g^\alpha, \text{ where } \Delta g^\alpha = \sum_{\beta=1}^{N} h_{\alpha\beta}\Delta\gamma^\beta \qquad (9.11)$$

Here, $h_{\alpha\beta}$ corresponding to the slip-hardening modulus is represented by the model proposed by Peirce et al. [105], as follows:

$$h_{\alpha\alpha} = h_0 \operatorname{sech}^2\left|\frac{h_0\tilde{\gamma}}{g^\alpha|_{\text{sat}} - g^\alpha|_{t=0}}\right|, h_{\alpha\beta} = qh_{\alpha\alpha}(\alpha \neq \beta), \tilde{\gamma} = \sum_\alpha \int_0^t |\dot{\gamma}^\alpha| dt \qquad (9.12)$$

where:

  h_0 is the initial hardening parameter
  $g^\alpha|_{\text{sat}}$ is the saturation stress of the slip system α
  q is the latent hardening ratio
  $\tilde{\gamma}$ is the Taylor cumulative shear strain on all slip systems

The CP theory was implemented in the FE code Abaqus/Explicit, using the user-defined material subroutine (VUMAT) [106].

An FE model of orthogonal micromachining cutting was developed. Dimensions of the workpiece sample, discretized with 29,600 eight-node linear brick elements, used in the FE model were $20 \times 20 \times 0.48\ \mu m$ (Figure 9.15). The cutting tool, modeled as a rigid body with rake and clearance angles of $0°$, was displaced in the negative x-direction with velocity of 1300 mm/s (Figure 9.14). The maximum cutting length of $0.5\ \mu m$ with $0.8\ \mu m$ depth of cut (a_p) was considered in the simulations. The modeling technique of element deletion was employed to simulate chip separation from the workpiece material. Material parameters used in the simulations are listed in Table 9.4. More details about the model can be found in Demiral et al. [92,102].

Here, machining in a single-crystal of copper, which has an fcc crystalline structure, is studied. In such materials, slip may occur on 12 individual slip systems, represented by the family {1 1 1} <1 1 0>. Five cutting directions, viz. $0°$, $30°$, $45°$, $60°$, and $90°$, on (1 1 0) crystal plane were investigated. The corresponding values are listed in Table 9.5.

Evolution of the calculated cutting forces with an increasing cutting length for various cutting directions is shown in Figure 9.15. Fluctuations in the cutting force were observed due to, on the one hand, dynamic response associated with stress waves moving through the material and reflecting at boundaries, and on the other hand, reorientation of the

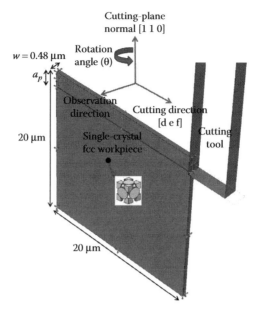

Figure 9.14 Dimensions and orientations for orthogonal machining of single-crystal workpiece material.

Table 9.4 Material parameters of single-crystal copper

Elastic parameters		Plastic parameters	
C_{11} (GPa)	168.0	$\dot{\gamma}_0^\alpha (s^{-1})$	0.001
C_{12} (GPa)	121.4	N	20
C_{44} (GPa)	75.4	h_0 (MPa)	180
		$g^\alpha|_{t=0}$ (MPa)	60.84
		$g^\alpha|_{sat}$ (MPa)	109.51

Source: Tajalli, S.A. et al., *Comput. Mater. Sci.*, 86, 79–87, 2014; Demiral, M. et al., *Mater. Sci. Eng. A*, 608, 73–81, 2014; Huang, Y. et al., *J Mater. Res.*, 15, 1786–1796, 2000.

Table 9.5 Cutting direction setup ([d e f]) for (1 1 0) crystal plane (Figure 9.15)

Rotation angle (θ)	Cutting directions [d e f]
0°	[1 −1 0]
30°	[1 −1 −0.816]
45°	[1 −1 −1.414]
60°	[1 −1 2.449]
90°	[0 0 −1]

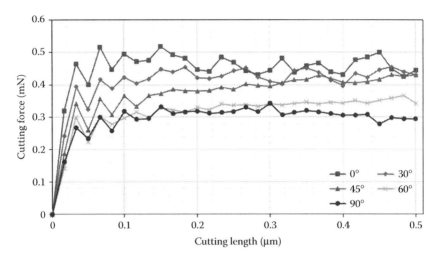

Figure 9.15 Evolution of cutting forces for different cutting directions of (1 1 0) plane.

local mesh during the process, leading to a variation in the shear angle [64,101]. The cutting force was found to vary with different crystallographic cutting directions. The measured value of the cutting force at the [1 –1 0] direction is the largest, whereas it reached a minimum value in the [0 0 –1] direction. Table 9.6 presents the averages of force magnitudes for the cutting lengths of 0.10 μm and 0.50 μm. A decreasing trend in the average cutting force value was observed when the rotation angle was changed from 0° to 90°. We compared our results with the experiments performed by Zhou and Ngoi [109] for cutting of single-crystal copper qualitatively, due to differences in the cutting conditions such as tool geometry and contact conditions between the tool and the workpiece. A similar tendency in the averages of force magnitudes with varying

Table 9.6 Average cutting energies obtained with CP theory for different rotation angles of (1 1 0) plane

Rotation angle (θ)	Average cutting force (mN)
0°	463.25
30°	430.05
45°	397.59
60°	335.07
90°	310.44

cutting direction was observed experimentally. For cutting in the (1 1 0) crystal orientation, the [0 0 –1] cutting direction is preferable to [1 –1 0], owing to a lower cutting force imposed on the cutting tool, which, in turn, leads to a longer tool life. Results here demonstrate that variation in the microcutting force can be predicted if the crystallographic texture of the material is known.

Figure 9.16 demonstrates the obtained chip morphologies for $\theta = 0°$, 30°, 45°, 60°, and 90° of (1 1 0) single-crystal copper. The chip shape was observed to be heavily influenced by crystallographic cutting direction. A larger shear angle with a smaller post-turning chip thickness was obtained when $\theta = 0°$, and this became smaller for larger θ values. In ultra-precision machining, continuous chip formation and good surface finish demand a large shear angle [110]. Therefore, among the cutting directions investigated here, [1 –1 0] is preferable for microcutting of (1 1 0) copper single crystals for a better surface finish.

It was thus concluded that an accurate prediction of micromachining in the presence of crystallographic anisotropy requires the development of models that incorporate crystal plasticity in their constitutive description.

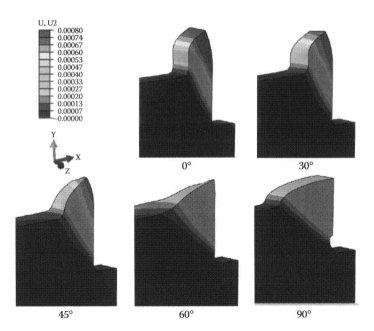

Figure 9.16 Chip morphologies at cutting length of 0.5 μm for different rotation angles of (1 1 0) plane.

References

1. Oliaei, S. N. B. (2016). Design and fabrication of micro end mills for the machining of difficult-to-cut materials, Doctoral dissertation, Bilkent University, Turkey.
2. Ehmann, K. F., Bourell, D., Culpepper, M. L., Hodgson, T. J., Kurfess, T. R., Madou, M., & DeVor, R. E. (2005). *International Assessment of Research and Development in Micromanufacturing.* World Technology Evaluation Center Inc, Baltimore, MD.
3. Okazaki, Y., Mishima, N., & Ashida, K. (2004), Microfactory-concept, history, and developments, Technical Briefs.
4. Lauro, C. H., Ribeiro Filho, S. L., Brandão, L. C., & Davim, J. P. (2016). Analysis of behaviour biocompatible titanium alloy (Ti-6Al-7Nb) in the micro-cutting. *Measurement*, 93, 529–540.
5. Liu, X., DeVor, R. E., Kapoor, S., & Ehmann, K. (2004). The mechanics of machining at the microscale: Assessment of the current state of the science. *Journal of Manufacturing Science and Engineering*, 126(4), 666–678.
6. Chae, J., Park, S. S., & Freiheit, T. (2006). Investigation of micro-cutting operations. *International Journal of Machine Tools and Manufacture*, 46(3), 313–332.
7. Camara, M. A., Rubio, J. C., Abrão, A. M., & Davim, J. P. (2012). State of the art on micromilling of materials, a review. *Journal of Materials Science & Technology*, 28(8), 673–685.
8. Lai, X., Li, H., Li, C., Lin, Z., & Ni, J. (2008). Modelling and analysis of micro scale milling considering size effect, micro cutter edge radius and minimum chip thickness. *International Journal of Machine Tools and Manufacture*, 48(1), 1–14.
9. Jackson, M. J. (2007). *Micro-and Nanomanufacturing.* Springer, Boston, MA. 635–685.
10. Asad, A. B. M. A., Masaki, T., Rahman, M., Lim, H. S., & Wong, Y. S. (2007). Tool-based micro-machining. *Journal of Materials Processing Technology*, 192, 204–211.
11. Weck, M., Fischer, S., & Vos, M. (1997). Fabrication of microcomponents using ultraprecision machine tools. *Nanotechnology*, 8(3), 145.
12. Silva, L. R., Abrão, A. M., Faria, P., & Davim, J. P. (2012). Machinability study of steels in precision orthogonal cutting. *Materials Research*, 15(4), 589–595.
13. Oliaei, S. N. B., & Karpat, Y. (2016a). Influence of tool wear on machining forces and tool deflections during micro milling. *The International Journal of Advanced Manufacturing Technology*, 84(9–12), 1963–1980.
14. Oliaei, S. N. B., & Karpat, Y. (2014). Experimental investigations on micro milling of Stavax stainless steel. *Procedia CIRP*, 14, 377–382.
15. Astakhov, V. P., & Outeiro, J. C. (2008). *Metal Cutting Mechanics, Finite Element Modelling. In Machining.* Springer, London, 1–27.
16. Masuzawa, T., & Tönshoff, H. K. (1997). Three-dimensional micromachining by machine tools. *CIRP Annals-Manufacturing Technology*, 46(2), 621–628.
17. Vollertsen, F. (2008). Categories of size effects. *Production Engineering*, 2(4), 377.
18. Dornfeld, D., Min, S., & Takeuchi, Y. (2006). Recent advances in mechanical micromachining. *CIRP Annals-Manufacturing Technology*, 55(2), 745–768.
19. Abouridouane, M., Klocke, F., & Lung, D. (2014). Microstructure-based 3D FE modeling for micro cutting ferritic-pearlitic carbon steels. In *ASME 2014 International Manufacturing Science and Engineering Conference collocated with*

the *JSME 2014 International Conference on Materials and Processing and the 42nd North American Manufacturing Research Conference.* American Society of Mechanical Engineers, V001T01A008–V001T01A008.

20. Moronuki, N., Liang, Y., & Furukawa, Y. (1994). Experiments on the effect of material properties on microcutting processes. *Precision Engineering,* 16(2), 124–131.

21. Lucca, D. A., Seo, Y. W., & Komanduri, R. (1993). Effect of tool edge geometry on energy dissipation in ultraprecision machining. *CIRP Annals-Manufacturing Technology,* 42(1), 83–86.

22. Yuan, Z. J., Zhou, M., & Dong, S. (1996). Effect of diamond tool sharpness on minimum cutting thickness and cutting surface integrity in ultraprecision machining. *Journal of Materials Processing Technology,* 62(4), 327–330.

23. Bayoumi, A. E., Yücesan, G., & Hutton, D. V. (1994). On the closed form mechanistic modeling of milling: specific cutting energy, torque, and power. *Journal of Materials Engineering and Performance,* 3(1), 151–158.

24. Wu, X., Li, L., He, N., Hao, X., Yao, C., & Zhong, L. (2016). Investigation on the ploughing force in microcutting considering the cutting edge radius. *The International Journal of Advanced Manufacturing Technology,* 86(9–12), 2441–2447.

25. Vollertsen, F., Biermann, D., Hansen, H. N., Jawahir, I. S., & Kuzman, K. (2009). Size effects in manufacturing of metallic components. *CIRP Annals-Manufacturing Technology,* 58(2), 566–587.

26. Armarego, E. J. A., & Brown, R. H. (1961). On the size effect in metal cutting. *The International Journal of Production Research,* 1(3), 75–99.

27. Lucca, D. A., Rhorer, R. L., & Komanduri, R. (1991). Energy dissipation in the ultraprecision machining of copper. *CIRP Annals-Manufacturing Technology,* 40(1), 69–72.

28. Backer, W. R., Marshall, E. R., & Shaw, M. C. (1952). The size effect in metal cutting. *Transactions of the American Society of Mechanical Engineers,* 74(1), 61.

29. Filiz, S., Conley, C. M., Wasserman, M. B., & Ozdoganlar, O. B. (2007). An experimental investigation of micro-machinability of copper 101 using tungsten carbide micro-endmills. *International Journal of Machine Tools and Manufacture,* 47(7), 1088–1100.

30. Shaw, M. C. (2003). The size effect in metal cutting. *Sadhana,* 28(5), 875–896.

31. Larsen-Basse, J., & Oxley, P. L. B. (1973). Effect of strain-rate sensitivity on scale phenomena in chip formation. In *Proceedings of the Thirteenth International Machine Tool Design and Research Conference.* Macmillan Education, UK, 209–216.

32. Ikawa, N., Shimada, S., & Tanaka, H. (1992). Minimum thickness of cut in micromachining. *Nanotechnology,* 3(1), 6.

33. Saedon, J. B., Halim, A. H. A., Husain, H., Meon, M. S., & Othman, M. F. (2013). Influence of cutting edge radius in micromachining AISI D2. *Applied Mechanics and Materials Trans Tech Publications,* 393, 253–258.

34. Weule, H., Hüntrup, V., & Tritschler, H. (2001). Micro-cutting of steel to meet new requirements in miniaturization. *CIRP Annals-Manufacturing Technology,* 50(1), 61–64.

35. Kim, C. J., Mayor, J. R., & Ni, J. (2004). A static model of chip formation in microscale milling. *Transactions of the ASME-B-Journal of Manufacturing Science and Engineering,* 126(4), 710–718.

36. Son, S. M., Lim, H. S., & Ahn, J. H. (2005). Effects of the friction coefficient on the minimum cutting thickness in micro cutting. *International Journal of Machine Tools and Manufacture,* 45(4), 529–535.

37. Oliaei, S. N. B., & Karpat, Y. (2016b). Investigating the influence of built-up edge on forces and surface roughness in micro scale orthogonal machining of titanium alloy Ti6Al4V. *Journal of Materials Processing Technology*, 235, 28–40.

38. Shi, Z. Y., & Liu, Z. Q. (2011). Numerical Modeling of Minimum Uncut Chip Thickness for Micromachining With Different Rake Angle. In *ASME 2011 International Manufacturing Science and Engineering Conference*. American Society of Mechanical Engineers, 403–407.

39. Woon, K. S., Rahman, M., Fang, F. Z., Neo, K. S., & Liu, K. (2008a). Investigations of tool edge radius effect in micromachining: A FEM simulation approach. *Journal of Materials Processing Technology*, 195(1), 204–211.

40. Woon, K. S., Rahman, M., Neo, K. S., & Liu, K. (2008b). The effect of tool edge radius on the contact phenomenon of tool-based micromachining. *International Journal of Machine Tools and Manufacture*, 48(12), 1395–1407.

41. Lauro, C. H., Brandão, L. C., Ribeiro Filho, S. L., Valente, R. A., & Davim, J. P. (2015). Finite element method in machining processes: A review. In *Modern Manufacturing Engineering*. Springer International Publishing, New York, 65–97.

42. Shih, A. J. (1995). Finite element simulation of orthogonal metal cutting. *Transactions-American Society of Mechanical Engineers Journal of Engineering for Industry*, 117, 84–84.

43. Carroll, J. T., & Strenkowski, J. S. (1988). Finite element models of orthogonal cutting with application to single point diamond turning. *International Journal of Mechanical Sciences*, 30(12), 899–920.

44. Özel, T., and Zeren, E. (2007). Numerical modelling of meso-scale finish machining with finite edge radius tools. *International Journal of Machining and Machinability of Materials*, 2(3/4), 451–468.

45. Baltzer, N., & Copponnex, T. (Eds.). (2014). *Precious Metals for Biomedical Applications*. Elsevier.

46. Boyer, R. R. (1995). Titanium for aerospace: rationale and applications. *Advanced Performance Materials*, 2(4), 349–368.

47. Siekmann, H. J. (1955) How to machine titanium. *The Tool Engineer*, 34(1), 78–82.

48. Zhao, X., Ke, W., Zhang, S., & Zheng, W. (2016). Potential failure cause analysis of tungsten carbide end mills for titanium alloy machining. *Engineering Failure Analysis*, 66, 321–327.

49. Shokrani, A., Dhokia, V., & Newman, S. T. (2016). Investigation of the effects of cryogenic machining on surface integrity in CNC end milling of Ti–6Al–4V titanium alloy. *Journal of Manufacturing Processes*, 21, 172–179.

50. Arrazola, P. J., & Özel, T. (2010). Investigations on the effects of friction modeling in finite element simulation of machining. *International Journal of Mechanical Sciences*, 52(1), 31–42.

51. Johnson, G. R., & Cook, W. H. (1983). A constitutive model and data for metals subjected to large strains, high strain rates and high temperatures. In *Proceedings of the 7th International Symposium on Ballistics*, 21(1983): 541–547.

52. Afazov, S. M., Ratchev, S. M., & Segal, J. (2010). Modelling and simulation of micro-milling cutting forces. *Journal of Materials Processing Technology*, 210(15), 2154–2162.

53. Jin, X., & Altintas, Y. (2012). Prediction of micro-milling forces with finite element method. *Journal of Materials Processing Technology*, 212(3), 542–552.

54. Calamaz, M., Coupard, D., & Girot, F. (2008). A new material model for 2D numerical simulation of serrated chip formation when machining titanium alloy Ti–6Al–4V. *International Journal of Machine Tools and Manufacture*, 48(3), 275–288.

55. Sima, M., & Özel, T. (2010). Modified material constitutive models for serrated chip formation simulations and experimental validation in machining of titanium alloy Ti–6Al–4V. *International Journal of Machine Tools and Manufacture*, 50(11), 943–960.

56. Thepsonthi, T., & Özel, T. (2015). 3-D finite element process simulation of micro-end milling Ti-6Al-4V titanium alloy: experimental validations on chip flow and tool wear. *Journal of Materials Processing Technology*, 221, 128–145.

57. Karpat, Y. (2011). Temperature dependent flow softening of titanium alloy Ti6Al4V: An investigation using finite element simulation of machining. *Journal of Materials Processing Technology*, 211(4), 737–749.

58. Oliaei, S. N. B., & Karpat, Y. (2016c). Investigating the influence of friction conditions on finite element simulation of microscale machining with the presence of built-up edge. *The International Journal of Advanced Manufacturing Technology*, 1–11.

59. Filice, L., Micari, F., Rizzuti, S., & Umbrello, D. (2008). Dependence of machining simulation effectiveness on material and friction modelling. *Machining Science and Technology*, 12(3), 370–389.

60. Outeiro, J. C., Umbrello, D., M'Saoubi, R., & Jawahir, I. S. (2015). Evaluation of present numerical models for predicting metal cutting performance and residual stresses. *Machining Science and Technology*, 19(2), 183–216.

61. Masuzawa, T. (2000). State of the art of micromachining. *CIRP Annals-Manufacturing Technology*, 49(2), 473–488.

62. Zeng, Z., Wang, Y., Wang, Z., Shan, D., & He, X. (2012). A study of micro-EDM and micro-ECM combined milling for 3D metallic micro-structures. *Precision Engineering*, 36(3), 500–509.

63. Zorev, N. N. (1963). Inter-relationship between shear processes occurring along tool face and shear plane in metal cutting. *International Research in Production Engineering*, 49, 143–152.

64. Tajalli, S. A., Movahhedy, M. R., & Akbari, J. (2014). Simulation of orthogonal micro-cutting of FCC materials based on rate-dependent crystal plasticity finite element model. *Computational Materials Science*, 86, 79–87.

65. Thepsonthi, T., & Özel, T. (2013). Experimental and finite element simulation based investigations on micro-milling Ti-6Al-4V titanium alloy: Effects of cBN coating on tool wear. *Journal of Materials Processing Technology*, 213(4), 532–542.

66. Kim, K. W., Lee, W. Y., & Sin, H. C. (1999). A finite-element analysis of machining with the tool edge considered. *Journal of Materials Processing Technology*, 86(1), 45–55.

67. Kim, J. D., Marinov, V. R., & Kim, D. S. (1997). Built-up edge analysis of orthogonal cutting by the visco-plastic finite-element method. *Journal of Materials Processing Technology*, 71(3), 367–372.

68. Atlati, S., Haddag, B., Nouari, M., & Moufki, A. (2015). Effect of the local friction and contact nature on the Built-Up Edge formation process in machining ductile metals. *Tribology International*, 90, 217–227.

69. Kalpakjian, S., & Schmid, S. R. (2014). *Manufacturing Engineering and Technology*. K. V. Sekar (Ed.). Pearson, Upper Saddle River, NJ, 913.

70. Childs, T. H. C. (2013). Ductile shear failure damage modeling and predicting built-up edge in steel machining. *Journal of Materials Processing Technology,* 213(11), 1954–1969.

71. Williams, J. A., & Horne, J. G. (1982). Crystallographic effects in metal cutting. *Journal of Materials Science,* 17(9), 2618–2624.

72. Ueda, K., Iwata, K., & Nakayama, K. (1980). Chip formation mechanism in single crystal cutting of β-brass. *CIRP Annals-Manufacturing Technology,* 29(1), 41–46.

73. Sato, M., Kato, Y., Aoki, S., & Ikoma, A. (1983). Effects of crystal orientation on the cutting mechanism of the aluminum single crystal: 2nd report: on the (111) plane and the (112) end cutting. *Bulletin of JSME,* 26(215), 890–896.

74. Sato, M., Kato, Y., & Tsutiya, K. (1979). Effects of crystal orientation on the flow mechanism in cutting aluminum single crystal. *Transactions of the Japan Institute of Metals,* 20(8), 414–422.

75. Cohen, P. H. (1982). The orthogonal in-situ machining of single and polycrystalline aluminum and copper. Doctoral dissertation, The Ohio State University.

76. To, S., Lee, W. B., & Chan, C. Y. (1997). Ultraprecision diamond turning of aluminium single crystals. *Journal of Materials Processing Technology,* 63(1–3), 157–162.

77. Lawson, B. L., Kota, N., & Ozdoganlar, O. B. (2008). Effects of crystallographic anistropy on orthogonal micromachining of single-crystal aluminum. *Journal of Manufacturing Science and Engineering,* 130(3), 031116.

78. Moriwaki, T., Okuda, K., & Shen, G. J. (1993). Study of ultraprecision orthogonal microdiamond cutting of single-crystal copper. *JSME International Journal. Ser. C, Dynamics, Control, Robotics, Design and Manufacturing,* 36(3), 400–406.

79. Hazra, J. (1973). Dynamic shear stress—Analysis of single crystal machining studies. *Journal of Engineering for Industry,* 95(4), 939–944.

80. Bifano, T. G., Dow, T. A., & Scattergood, R. O. (1991). Ductile-regime grinding: a new technology for machining brittle materials. *Journal of Engineering for Industry,* 113(2), 184–189.

81. Blackley, W. S., & Scattergood, R. O. (1991). Ductile-regime machining model for diamond turning of brittle materials. *Precision Engineering,* 13(2), 95–103.

82. Blackley, W. S., & Scattergood, R. O. (1990). Crystal orientation dependence of machning damage–a stress model. *Journal of the American Ceramic Society,* 73(10), 3113–3115.

83. Patten, J. A., & Gao, W. (2001). Extreme negative rake angle technique for single point diamond nano-cutting of silicon. *Precision Engineering,* 25(2), 165–167.

84. Chao, C. L., Ma, K. J., Liu, D. S., Bai, C. Y., & Shy, T. L. (2002). Ductile behaviour in single-point diamond-turning of single-crystal silicon. *Journal of Materials Processing Technology,* 127(2), 187–190.

85. Feng, Q., Picard, Y. N., Liu, H., Yalisove, S. M., Mourou, G., & Pollock, T. M. (2005). Femtosecond laser micromachining of a single-crystal superalloy. *Scripta Materialia,* 53(5), 511–516.

86. Kota, N., & Ozdoganlar, O. B. (2012). Orthogonal machining of single-crystal and coarse-grained aluminum. *Journal of Manufacturing Processes,* 14(2), 126–134.

87. Nahata, S., Picard, Y. N., Kota, N., & Ozdoganlar, O. B. (2014). Experimental investigation of sub-surface deformation using EBSD in single crystal aluminum during orthogonal micromachining. *Microscopy and Microanalysis*, 20(S3), 1472–1473.

88. Kopalinsky, E. M., & Oxley, P. L. B. (1984). Size effects in metal removal processes. *Mechanical Properties at High Rates of Strain*, 1984, 389–396.

89. Furukawa, Y., & Moronuki, N. (1988). Effect of material properties on ultra precise cutting processes. *CIRP Annals-Manufacturing Technology*, 37(1), 113–116.

90. Nakayama, K., & Tamura, K. (1968). Size effect in metal-cutting force. *Journal of Engineering for Industry*, 90(1), 119–126.

91. Liu, K. (2005). Process modeling of micro-cutting including strain gradient effects. Doctoral dissertation, Georgia institute of technology.

92. Demiral, M., Roy, A., El Sayed, T., & Silberschmidt, V. V. (2014). Numerical modelling of micro-machining of fcc single crystal: Influence of strain gradients. *Computational Materials Science*, 94, 273–278.

93. Tsutıya, K. (1981). Effects of crystal orientation on the cutting mechanism of aluminum single crystal. *Bulletin of JSME*, 24(196), 1864–1870.

94. Lee, W.B., Cheung, C. F., & To, S. (2002). A microplasticity analysis of micro-cutting force variation in ultra-precision diamond turning. *Journal of Manufacturing Science and Engineering—Transactions ASME*, 124, 170–177.

95. Lee, W. B., & Zhou, M. (1993). A theoretical analysis of the effect of crystallographic orientation on chip formation in micromachining. *International Journal of Machine Tools and Manufacture*, 33(3), 439–447.

96. Kota, N., & Ozdoganlar, B. (2010). A model-based analysis of orthogonal cutting for single-crystal fcc metals including crystallographic anisotropy. *Machining Science and Technology*, 14(1), 102–127.

97. Venkatachalam, S., Li, X., & Liang, S. Y. (2009). Predictive modeling of transition undeformed chip thickness in ductile-regime micro-machining of single crystal brittle materials. *Journal of Materials Processing Technology*, 209(7), 3306–3319.

98. Liu, K., & Melkote, S. N. (2007). Finite element analysis of the influence of tool edge radius on size effect in orthogonal micro-cutting process. *International Journal of Mechanical Sciences*, 49(5), 650–660.

99. Pen, H. M., Liang, Y. C., Luo, X. C., Bai, Q. S., Goel, S., & Ritchie, J. M. (2011). Multiscale simulation of nanometric cutting of single crystal copper and its experimental validation. *Computational Materials Science*, 50(12), 3431–3441.

100. Komanduri, R., Chandrasekaran, N., & Raff, L. M. (2000). MD Simulation of nanometric cutting of single crystal aluminum–effect of crystal orientation and direction of cutting. *Wear*, 242(1), 60–88.

101. Zahedi, S. A., Demiral, M., Roy, A., & Silberschmidt, V. V. (2013). FE/SPH modelling of orthogonal micro-machining of fcc single crystal. *Computational Materials Science*, 78, 104–109.

102. Demiral, M., Roy, A., & Silberschmidt, V. V. (2016). Strain-gradient crystal-plasticity modelling of micro-cutting of bcc single crystal. *Meccanica*, 51(2), 371–381.

103. Demiral, M., Roy, A., & Silberschmidt, V. V. (2013). Indentation studies in bcc crystals with enhanced model of strain-gradient crystal plasticity. *Computational Materials Science*, 79, 896–902.

104. Liu, Q., Roy, A., Tamura, S., Matsumura, T., & Silberschmidt, V. V. (2016). Micro-cutting of single-crystal metal: Finite-element analysis of deformation and material removal. *International Journal of Mechanical Sciences*, 118, 135–143.

105. Peirce, D., Asaro, R. J., & Needleman, A. (1982). An analysis of nonuniform and localized deformation in ductile single crystals. *Acta Metallurgica*, 30(6), 1087–1119.

106. Simulia, D. (2011). ABAQUS 6.11 analysis user's manual. *Abaqus* 6, 22-2.

107. Demiral, M., Roy, A., El Sayed, T., & Silberschmidt, V. V. (2014). Influence of strain gradients on lattice rotation in nano-indentation experiments: A numerical study. *Materials Science and Engineering: A*, 608, 73–81.

108. Huang, Y., Xue, Z., Gao, H., Nix, W. D., & Xia, Z. C. (2000). A study of micro-indentation hardness tests by mechanism-based strain gradient plasticity. *Journal of Materials Research*, 15(08), 1786–1796.

109. Zhou, M., & Ngoi, B. K. A. (2001). Effect of tool and workpiece anisotropy on microcutting processes. *Proceedings of the Institution of Mechanical Engineers, Part B: Journal of Engineering Manufacture*, 215(1), 13–19.

110. Lee, W. B. (1990). Prediction of microcutting force variation in ultra-precision machining. *Precision Engineering*, 12(1), 25–28.

chapter ten

Modeling of materials behavior in finite element analysis and simulation of machining processes

Identification techniques and challenges

Walid Jomaa, Augustin Gakwaya, and Philippe Bocher

Contents

10.1 Introduction: Background and motivations..................................... 282
10.2 Material constitutive equations used in machining modeling 283
    10.2.1 Plasticity constitutive equations 284
        10.2.1.1 Phenomenological plasticity models 284
        10.2.1.2 Physical-based models 289
        10.2.1.3 Microstructure-based models 291
    10.2.2 Damage modeling 296
10.3 Identification techniques 297
    10.3.1 Dynamic tests 297
    10.3.2 Machining-based inverse methods 298
    10.3.3 Hybrid/advanced methods 299
10.4 Case study: Effect of material modeling on finite element
    analysis of near-micromachining of Inconel 718 300
    10.4.1 Machining tests 300
    10.4.2 Finite element modeling of near-micromachining 301
        10.4.2.1 Finite element formulation and boundary
                conditions 301
        10.4.2.2 Contact modeling 301
        10.4.2.3 Material modeling 302

10.4.3   Results and discussion 304
          10.4.3.1   Chip morphology 304
          10.4.3.2   Machining forces...306
          10.4.3.3   Cutting temperature distribution 307
          10.4.3.4   Machined surface alteration: State variables
                     evolution 308
10.5   Conclusions and recommendations...311
References..312

In metal cutting processes, the work materials undergo intense deformation process, involving large strain (up to 10 and more), high strain rate (up to 10^6 s^{-1}), high stress (ultimate stress), and elevated temperature (up to 90% the melting temperature). However, there is no devoted experimental setup for describing the flow stress relevant to practical machining conditions. Hence, adequately selecting the constitutive equation and its calibration method constitute a challenge. This chapter presents a critical review of the different approaches used for describing the material behavior modeling in the finite element simulation of machining processes. Specifically, the authors emphasize the critical issues encountered in the selection of the constitutive equations and their calibration methods. Finally, a case study is presented to highlight the effects of material models and their calibration techniques on the reliability of the finite element method (FEM) of near-micromachining process.

10.1   Introduction: Background and motivations

Over the last decades, analytical and FEM of machining processes have received increasing attention to model and simulate machining processes. So, adequate and reliable constitutive equations, describing the material behavior under extreme deformation process, are needed. To realize a successful process modeling, two critical issues need to be solved: first, the selection of the adequate constitutive equations for the material in use, and second, the fine tuning of material constants.

Earlier in the 1960s, researchers [1–4] attested that the flow stress data obtained on static tests were not adequate for machining modeling due to the low strain rates achieved as compared with those reached in machining processes. In fact, material constants have to be determined at high strain rates (up to 10^6 S^{-1}), elevated temperatures (up to $0.9 \times T_m$), and large strains (up to 10), commonly achieved during machining [5]. Hence, many techniques have been developed to calibrate constitutive equations for machining modeling and simulations.

Dynamic mechanical tests such as the Split-Hopkinson pressure bar (SHPB) test [6] and Taylor test [7] are the most used techniques for the identification of the strain rate-dependent material constants. However, the achieved strain and strain rate are still lower than those encountered in machining. So, the constitutive equations based on dynamic tests results were often extrapolated well beyond their test range to cover the metal cutting range [8].

More recently, inverse methods based on machining tests [9,10] have been developed to determine material constants representing machining processes. These methods are considered one of the most reliable techniques, since these methods can provide material constants at high strain rates and temperatures, representing the actual material behavior during machining [11]. Although promising results have been obtained, all these techniques suffer from several shortcomings.

The challenge is not only in finding the adequate experimental technique used for the calibration of the material constitutive equations but also in the selection of the constitutive equation itself. In most of the cases, the metal cutting simulations were performed using the well-known Johnson–Cook (JC) material constitutive equation, thanks to its simplicity. In addition, it is implemented in most of the commercial software. However, the JC equation may not always accurately describe the behavior of all existing metallic materials [12,13], and other models can be more accurate [14].

In this context, different approaches and techniques used in material behavior modeling for finite element simulation of machining processes will be discussed in this chapter. The authors will focus on the critical issues encountered in the selection of the constitutive equations and their calibrations. The chapter will end with a case study, emphasizing the effects of the identification techniques on the reliability of FEM in the case of near-micromachining of a superalloy Inconel 718.

10.2 Material constitutive equations used in machining modeling

Modeling work materials behavior during the cutting process depends on several parameters, mainly related to the work material characteristics, machining conditions to be simulated, and their interactions (material/machining conditions). In particular, the chip formation process is very sensitive to materials characteristics, including the chemical composition, grain size, and hardness, among others. On the other hand, for a given material, varying machining conditions can significantly affect the chip formation process, resulting in chip variation from continuous to serrated/elemental chip. To simulate serrated/elemental chips, a damage model should be applied in conjunction with the plasticity model. In the following sections, a brief description of materials plasticity and damage models used in machining modeling and simulation will be presented.

10.2.1 Plasticity constitutive equations

In machining, since the cutting tool action generates a complex thermomechanical loading, most of the time, the materials behavior is described by a thermoviscoplastic constitutive equation. As discussed earlier, several constitutive equations have been applied to the machining process modeling; here, the authors will focus on the most accepted and used ones, as they are implemented in most of the commercial codes such as DEFORM™, Abaqus™, and AdvantEdge™.

10.2.1.1 Phenomenological plasticity models

The phenomenological constitutive models are considered as empirical models, describing the flow stress evolution under specific experimental conditions.

10.2.1.1.1 Power-law models Earlier, in developing a predictive machining theory, Oxley and coworkers [15] have used a modified power-law material model, as:

$$\sigma = \sigma_y \left(T_{mod} \right) \varepsilon^{n \left(T_{mod} \right)} \qquad (10.1)$$

where the material constants σ_y and n are assumed to be dependent on the velocity-modified temperature parameter T_{mod}, proposed by MacGregor and Fisher [16]:

$$T_{mod} = T \left[1 - \vartheta \log \left(\frac{\dot{\varepsilon}_p}{\dot{\varepsilon}_0} \right) \right] \qquad (10.2)$$

where:
  ϑ is a constant
  σ_y and n are polynomial functions of T_{mod}

High-order polynomials are used for better accuracy, and they can vary for different range of T_{mod}.

Marusich and Ortiz [17] proposed a modified power-law model for high-speed machining of materials experiencing a transition from low to high strain rate sensitivity, such as aluminum alloys [18] and structural steel [17]. This material model is implemented in Third Wave Systems' AdvantEdge™ software. However, the users cannot access the material constants database, making the software less flexible for exploring new material models. This model [17] involves a stepwise variation of the rate sensitivity exponents (m_1 and m_2), while maintaining the continuity of the flow stress as follows:

$$\begin{cases} \left(1+\dfrac{\dot{\varepsilon}_p}{\dot{\varepsilon}_0}\right) = \left[\dfrac{\sigma}{g(\varepsilon_p)}\right]^{m_1} & \text{if } \dot{\varepsilon}_p < \dot{\varepsilon}_t \\[4mm] \left(1+\dfrac{\dot{\varepsilon}_p}{\dot{\varepsilon}_0}\right)\left(1+\dfrac{\dot{\varepsilon}_t}{\dot{\varepsilon}_0}\right)^{(m_2/m_1)-1} = \left[\dfrac{\sigma}{g(\varepsilon_p)}\right]^{m_2} & \text{if } \dot{\varepsilon}_p \geq \dot{\varepsilon}_t \end{cases} \tag{10.3}$$

They also have used a function capable of capturing both hardening and softening effects, following Lemonds and Needleman [19]:

$$g(\varepsilon_p, T) = \sigma_y \Theta(T)\left(1+\dfrac{\varepsilon_p}{\varepsilon_0}\right)^{1/n_{NL}} \tag{10.4}$$

More recently, a piecewise strain-hardening function was also used by [20,21]:

$$\begin{cases} g(\varepsilon_p) = \sigma_y \Theta(T)\left(1+\dfrac{\varepsilon_p}{\varepsilon_0}\right)^{1/n_{NL}} & \text{if } \varepsilon_p < \varepsilon_c \\[4mm] g(\varepsilon_p) = \sigma_y \Theta(T)\left(1+\dfrac{\varepsilon_c}{\varepsilon_0}\right)^{1/n_{NL}} & \text{if } \varepsilon_p \geq \varepsilon_c \end{cases} \tag{10.5}$$

where:

n_{NL} denotes the strain hardening exponent

m_1 and m_2 are the low and high strain rate sensitivity exponents

α_{NL} is the thermal softening coefficient

$\Theta(T)$ is the thermal softening parameter

$\dot{\varepsilon}_t$ is the threshold strain rate, separating the low- and high-strain regimes

ε_c is the cut-off strain rate regime

Different forms of the thermal softening parameter $\Theta(T)$ have been utilized in the literature. Marusich and Ortiz [17] have adopted a linear thermal softening parameter $\Theta(T)$, that is:

$$\Theta(T) = \left[1 - \alpha_{NL}(T - T_0)\right] \tag{10.6}$$

While other research studies [21–23] proposed a five-order polynomial function for the thermal softening parameter $\Theta(T)$:

$$\begin{cases} \Theta(T) = \displaystyle\sum_{i=1}^{5} C_i T^i & \text{if } T_0 \leq T \leq T_{\text{cut}} \\[4mm] \Theta(T) = \left(1 - \dfrac{T - T_{\text{cut}}}{T_m - T_{\text{cut}}}\right) \displaystyle\sum_{i=1}^{5} C_i T_{\text{cut}}^i & \text{if } T_{\text{cut}} < T \leq T_m \end{cases} \qquad (10.7)$$

where:

C_0 to C_5 are material constants

T_{cut} is the critical temperature, which is often taken as the recrystallization temperature of the work material

10.2.1.1.2 *Johnson–Cook's constitutive equation and modified versions* The JC's material model [24] is given by:

$$\sigma = \left[A + B\left(\varepsilon_p\right)^n\right]\left[1 + C \ln\left(\frac{\dot{\varepsilon}_p}{\dot{\varepsilon}_0}\right)\right]\left[1 + \left(\frac{T - T_0}{T_m - T_0}\right)^m\right] \qquad (10.8)$$

where:

A denotes the yield strength coefficient

B is the hardening modulus

C is the strain rate sensitivity coefficient

n is the hardening coefficient

m is the thermal softening coefficient

The main drawback of the JC's model is that no interaction between work hardening, strain sensitivity, and thermal softening effects can be considered. This allow the various coefficients to be calibrated separately. Nevertheless, this is physically not true, especially when it comes to machining conditions. Thus, the JC's constitutive equation has been modified/improved to capture these effects.

Wang et al. [25] proposed a modified JC's constitutive equation for high strain rates and temperatures representative of those encountered during machining of Inconel 718. They found that the strain rate softening parameter C is dependent on strain rate and temperature, that is:

$$C(\dot{\varepsilon}_p, T) = a_1 - \left[a_2 + a_3 \sin\left(\frac{\dot{\varepsilon}_p - 5000}{3000}\pi\right)\right]\sin\left(\frac{T - 500}{150}\pi\right) \qquad (10.9)$$

Where a_1 to a_3 are additional material constants, which are calibrated using dynamic mechanical tests.

Similarly, Chen et al. [26] proposed a modified JC's constitutive equation, where both the work hardening and thermal softening terms are dependent on strain rate and temperature, as follows:

$$\sigma = \left[A + B\left(\varepsilon_p\right)^n \left(\frac{1 - \left(\dot{\varepsilon}_p/\dot{\varepsilon}_{\max}\right)^{p_2} \tanh \varepsilon_p}{\exp\left(\varepsilon_p^{p_1}\right)} \right) \left(\frac{T_m}{T} \right)^{p_3} \right]$$

$$\left[1 + C\ln\left(\frac{\dot{\varepsilon}_p}{\dot{\varepsilon}_0} \right) \right] \left[1 - \left(1 - \frac{\ln \dot{\varepsilon}_{\max} - \ln \dot{\varepsilon}_p}{\ln \dot{\varepsilon}_{\max} - \ln \dot{\varepsilon}_{\min}} \right)^q \left(\frac{T - T_0}{T_m - T_0} \right)^m \right]$$

(10.10)

where p_1 to p_3 and q are additional material constants.

In another context, Calamaz et al. [27] developed a material model based on a modified JC's constitutive equation, in order to simulate segmented chip formation, without the need for a material damage model. The proposed model is given as:

$$\sigma = \left[A + B\left(\varepsilon_p\right)^n \left(\frac{1 -}{\exp\left(\varepsilon_p^a\right)} \right) \right] \left[1 + C\ln\left(\frac{\dot{\varepsilon}_p}{\dot{\varepsilon}_0} \right) \right]$$

$$\left[1 - \left(\frac{T - T_0}{T_m - T_0} \right)^m \right] \left[D + (1-D)\tanh\left(\frac{1}{(\varepsilon + S)^c} \right) \right]$$

(10.11)

with

$$D = 1 - \left(\frac{T}{T_m} \right)^d$$

(10.12)

and

$$S = \left(\frac{T}{T_m} \right)^b$$

(10.13)

where a, b, c, and d are additional material constants.

This model was used in several research works dealing with FEM simulations of machining titanium and nickel-based alloys [28–32].

10.2.1.1.3 *Zerilli–Armstrong model* and modified versions

Zerilli and Armstrong (ZA) [33] developed a constitutive model with more physical meaning compared with the power-law and JC's models, by introducing the microstructure characteristics based on dislocation-mechanics theory (fcc and bcc). Further, their model accounts for the coupling effects of strain rate and temperature, as follows:

$$\sigma = \begin{cases} C_0 + C_1 \exp\left(-C_3 T + C_4 \ln \dot{\varepsilon}_p\right) + C_5 \varepsilon_p^n & \text{For bcc materials} \\ C_0 + C_2 \varepsilon_p^n \exp\left(-C_3 T + C_4 T \ln \dot{\varepsilon}_p\right) & \text{For fcc materials} \end{cases} \quad (10.14)$$

where C_0 to C_5 and n are material constants.

The original ZA's model (Equation 10.15) suffers from some serious issues, especially when it comes to the machining modeling. The original ZA's model is valid at temperatures below $0.6T_m$. However, the temperatures greater than $0.6T_m$ are often achieved during machining, especially at the tool–chip interface. In addition, the model assumes that C_0, which represents the yield stress, only depends on the grain size of the work material, while it is known to vary with temperature for many engineering materials [34]. In addition, the determination of the coefficients C_0 and n in Equation 10.15 needs flow stress data at 0 K, which are not straightforward available for most used materials.

Few attempts have been made to modify the ZA's model, in order to overcome the above-mentioned issues and to predict the material behavior at elevated temperatures, exceeding $0.6T_m$.

In particular, Samantaray et al. [34] proposed the following modified ZA's model for an fcc material:

$$\sigma = \left(C_1 + C_2 \varepsilon_p^n\right) \exp\left[-\left(C_3 + C_4 \varepsilon_p\right) T^* + \left(C_5 + C_6 T^*\right) \ln \dot{\varepsilon}_p\right] \quad (10.15)$$

The constants C_4 and C_6 define the coupled effect of temperature/strain and temperature/strain rate, respectively. Here, the yield stress is no longer constant, but it is sensitive to the work hardening through the constants C_2 and n. In this way, there is no longer a need for extrapolating the flow stress to 0 K for determining the constant C_2, as proposed in the original ZA's model's Equation 10.15.

10.2.1.1.4 Maekawa's model   It is worth highlighting that all the earlier models do not take into account the load history. Maekawa et al. [35] developed an empirical material model that includes the coupling effect of strain rate and temperature as well as the history effects (strain path), as follows:

$$\sigma = A\left(\frac{\dot{\varepsilon}_p}{1000}\right)^M e^{aT} \left(\frac{\dot{\varepsilon}_p}{1000}\right)^m \times \left[\int_{\text{strain path}} e^{-\frac{at}{N}} \left(\frac{\dot{\varepsilon}_p}{1000}\right)^{-m/n} d\varepsilon_p\right]^N \quad (10.16)$$

where $A, M, N, a,$ and m are five temperature-dependent material constants.

10.2.1.2 Physical-based models

In contrast to phenomenological models, physical-based models are developed based on physical processes causing the deformation rather than a curve fitting of empirical data. However, it is worth noticing that some relations of the physical phenomena may end up being phenomenological due to the need for averaging and limited knowledge about the phenomena to be described [36].

The physical-based models can be classified into two groups: the first explicitly includes the physical model as an evolution equation in the constitutive model, and the second is based on describing the constitutive equation by using knowledge about the physical process causing the deformation. The validity domain of both modeling approaches is significantly larger than that of the phenomenological models, as they are based on processes description and not only on a curve fitting of the domain.

10.2.1.2.1 Mechanical threshold stress model   The mechanical threshold stress (MTS) model, originally proposed by Kocks [37] and later modified for various materials by many other researchers, is basically developed using dislocation concepts, assuming that material behavior is dictated by its microstructure evolution.

For simplicity, the MTS model developed by Follansbee and Kocks [38] for copper alloys will be presented here, as it was later used in machining modeling [39]. This model is expressed as follows:

$$\sigma = \hat{\sigma}_a + \left(\hat{\sigma} + \hat{\sigma}_a\right)\left\{1 - \left[\frac{KT\ln\left(\dot{\varepsilon}_0/\dot{\varepsilon}_p\right)}{g_0\mu b^3}\right]^{\frac{1}{q}}\right\}^{\frac{1}{p}}$$

(10.17)

$$\frac{d\sigma}{d\varepsilon} = \theta_0\left(1 - \frac{\tanh\left\{2\left[\left(\hat{\sigma} - \hat{\sigma}_a\right)/\left(\hat{\sigma}_s + \hat{\sigma}\right)\right]\right\}}{\tanh(2)}\right)$$

where:

$\hat{\sigma}$ denotes the threshold stress

K is the Boltzmann constant

g_0 is the activation energy

μ is the shear modulus

b is the magnitude of the Burgers vector

p and q are material constants

θ_0 and $\hat{\sigma}_s$ are strain rate-dependent material constants

10.2.1.2.2 BCJ material model   Bammann and coworkers [40] developed a dislocation mechanics-based internal state variable (ISV) model, named Bammann-Chiesa-Johnson's (BCJ) model, which is able to describe the strain rate and temperature histories. The BCJ model involves a yield surface, which follow a hardening-minus-recovery format, where deformation gradients are associated with both thermal expansion and damage. The relevant constitutive relationships of the BCJ model are given as follows [40]:

$$\dot{\sigma} = \lambda tr(D^e)I + 2\mu D^e \tag{10.18}$$

$$D^e = D - D^{in} \tag{10.19}$$

$$D^{in} = f(T)\sinh\left[\frac{\|\sigma - \alpha\| - (R + Y(T))}{V(T)}\right]\frac{\sigma - \alpha}{\|\sigma - \alpha\|} \tag{10.20}$$

$$\alpha = \left\{ h(T)D^{in} - \left[\sqrt{\frac{2}{3}}r_d(T)\|D^{in}\| + r_s(T)\right]\|\alpha\|\alpha \right\} \tag{10.21}$$

$$\dot{R} = \left\{ H(T)D^{in} - \left[\sqrt{\frac{2}{3}}R_d(T)\|D^{in}\| + R_s(T)\right]R^2 \right\} \tag{10.22}$$

where:
  D^e and D^{in} denote the elastic and inelastic rates of deformation, respectively
  α is the kinematic hardening ISV
  R is the isotropic hardening ISV
  λ and μ are the elastic Lame constants

And

$$V(T) = C_1 \exp\left(-\frac{C_1}{T}\right) \tag{10.23}$$

$$Y(T) = C_3 \exp\left(-\frac{C_4}{T}\right)\left\{\frac{\left[\tanh\left(C_{19}\left(C_{20} - T\right)\right)\right]}{2}\right\} \tag{10.24}$$

$$f(T) = C_5 \exp\left(-\frac{C_6}{T}\right) \tag{10.25}$$

$$r_d(T) = C_7 \exp\left(-\frac{C_8}{T}\right) \tag{10.26}$$

$$h(T) = C_9 - C_{10}T \tag{10.27}$$

$$r_s(T) = C_{11} \exp\left(-\frac{C_{12}}{T}\right) \tag{10.28}$$

$$R_d(T) = C_{13} \exp\left(-\frac{C_{14}}{T}\right) \tag{10.29}$$

$$H(T) = C_{15} - C_{16}T \tag{10.30}$$

$$R_s(T) = C_{17} \exp\left(-\frac{C_{18}}{T}\right) \tag{10.31}$$

where:
   The parameters $V(T)$, $Y(T)$, and $f(T)$, define the yield strength
   $h(T)$ and $H(T)$ are the hardening moduli
   $r_d(T)$ and $R_d(T)$ are the dynamic recovery functions
   $r_s(T)$ and $R_s(T)$ are the static recovery functions

For simplicity, in some research works [41], these temperature-dependent model parameters were considered as constants. The parameters C_1–C_{20} are material constants.

   10.2.1.2.3   Lindgren and coworkers' model   Lindgren et al. [36] developed a flow stress model based on a coupled set of evolution equations for dislocation density (DD) and vacancy concentration. This model also takes account for dynamic strain ageing through diffusing solutes parameter.
   Kalhori et al. [42] improved the Lindgren and coworkers' model [36] by extending its applicability to high strain rates (up to 10^4 s⁻¹), in order to simulate orthogonal machining of SANMAC 316L stainless steel. Similarly, Wedberg et al. [43] further improved this model by introducing the contribution of viscous drag (phonon contribution) on flow stress to account for high strain rate. For simplicity, the reader can refer to [36,42,43] for more details about these models.

10.2.1.3   Microstructure-based models

Earlier in the 1960s, experimental research studies on surface integrity carried out by the pioneers Field and Kahles [44,45] revealed that machining processes may induce microstructural changes/alterations at the surface layers of the machined parts. The most studied feature of microstructure alteration is the well-known *white layer*, commonly observed when machining hard-to-cut material such as hardened steels, superalloys, and titanium alloys. It was only at the end of the 1990s that researchers started to model and predict the white layer formation by using empirical and analytical modeling [46]. Later, FEM of microstructural alterations during machining were developed, thanks to the development of robust computational techniques and advanced computers.
   Various microstructure-based material models have been implemented into FEM software, in order to predict phase transformations

(changes) during machining. These material models can be classified into three groups: empirical, semi-empirical, and physical-based models.

10.2.1.3.1 Empirical models   Umbrello and Filice [47] were the first to introduce an empirical material model to predict the white layer formation during machining, using a FEM. They used a hardness-based constitutive equation previously developed in [48], as follows:

$$\sigma\left(\varepsilon_p, \dot{\varepsilon}_p, T, HRC\right) = B(T)\left[C\left(\varepsilon_p\right)^n + J + K\varepsilon_p\right]\left[1 + \ln\left(\dot{\varepsilon}_p\right)^m - A\right] \quad (10.32)$$

where:
  A, C, n, and m are material constants
  $B(T)$ is a polynomial function of temperature
  J and K are hardness-dependent coefficients

To predict the white and dark layer formations commonly observed in hard machining of steels, Umbrello and Filice [47] have used the following empirical equations to describe the hardness variations associated with the quenching ($\Delta HRC_{quenching}$) and tempering ($\Delta HRC_{tempering}$) processes:

$$HRC_{quenching} = F\left(\frac{67 - HRC_{INITIAL}}{1030 - T_{WLSTART}}\right)\left(T - T_{WLSTART}\right)$$

$$\Delta HRC_{tempeing} = G\left(\frac{HRC_{INITIAL} - HRC_{TWL}}{T_{WLSTART} - T_{DLSTART}}\right)\left(T_{DLSTART} - T\right)$$

$$(10.33)$$

where:
  $HRC_{INITIAL}$ and HRC_{TWL} are the initial and fully tempered hardness, respectively.
  $T_{WLSTART}$ and $T_{DLSTART}$ are the austenite-start and tempering-start temperatures.
  F and G are constants, which are numerically calibrated using finite element simulation.

10.2.1.3.2 Semi-empirical models   Rotella et al. [49] proposed a semi-empirical material model to predict the microstructural changes during the machining of an aluminum alloy. The authors developed a routine based on Zener–Hollomon and Hall–Petch equations to describe the dynamic recrystallization process and the hardness modification, respectively.

The constitutive Zener–Hollomon parameter is given as follows [50]:

$$Z = A\left(\sinh \alpha\sigma\right)^n \quad (10.34)$$

where α and n are material constants.

The Zener–Hollomon parameter Z is related to the recrystallized grain size d by the following equation [51]:

$$d = d_0 b Z^m \tag{10.35}$$

where:
  d_0 is the initial grain size
  b and m are material constants

The Hall–Petch equation has been used to relate the recrystallized grain size and the hardness, as follows [52]:

$$HV = H_0 + k_u d^{-0.5} \tag{10.36}$$

where H_0 and k_u are material constants.

In addition, the authors have used the critical strain ε_{crit} as a criterion for the onset of the dynamic recrystallization process, following Quan et al. [53], as:

$$\varepsilon_{crit} = k_1 \left(\frac{Z}{K_2} \right)^{k_3} \tag{10.37}$$

where k_1 to k_3 are material constants.

Later, the previous approach has been adopted by several researchers [54,55]. However, some of these equations have been modified to adapt the model to the work materials studied.

More recently, Arizoy and Özel [31] have used the Avrami equation to predict the dynamic recrystallization kinetics induced by machining of a titanium alloy. The Avrami model allows to calculate the recrystallized volume fraction X_{DRX} as a function of temperature, strain, and strain rate, using an Arrhenius-type equation, as:

$$X_{DRX} = 1 - \exp\left[-\beta_d \left(\frac{\varepsilon_p - a_1 \varepsilon_{peak}}{\varepsilon_{0.5}} \right)^{k_d} \right] \tag{10.38}$$

where $\varepsilon_{0.5}$ denotes the strain level for $X_{DRX} = 0.5$ and is defined as follows:

$$\varepsilon_{0.5} = a_5 d_0^{h_5} \varepsilon_p^{n_5} \dot{\varepsilon}_p^{m_5} \exp\left(\frac{Qm_5}{RT} \right) + c_5 \tag{10.39}$$

and ε_{peak} is the peak strain given by:

$$\varepsilon_{peak} = a_1 d_0^{h_5} \dot{\varepsilon}_p^{m_1} \exp\left(\frac{Qm_1}{RT} \right) + c_1 \tag{10.40}$$

The recrystallized grain size d_{DRX} is calculated using the following equation:

$$d_{DRX} = a_8 d_0^{h_8} \varepsilon_p^{n_8} \dot{\varepsilon}_p^{m_8} \exp\left(\frac{Qm_8}{RT}\right) + c_8 \tag{10.41}$$

Then, the average grain size can be estimated as follows:

$$d_{avg} = d_0 \left(1 - X_{DRX}\right) + d_{DRX} X_{DRX} \tag{10.42}$$

where:

β_d, k_d, a_1, a_2, a_5, a_8, a_{10}, c_1, c_5, c_8, h_1, h_5, h_8, m_1, m_5, m_8, n_5, and n_8 are material constants, which can be determined experimentally and/or calibrated using FEM

Q denotes the apparent activation energy

10.2.1.3.3 Physical-based models   Ramesh and Melkote [56] developed a physical-based model, taking account of the effect of stress, strain, transformation plasticity, and volume expansion accompanying phase transformation on the transformation temperature. The model is capable of predicting white layer depth in hard turning. The authors assumed that the white layers in hard machining are formed under thermally dominant machining conditions, favoring the occurrence of phase transformation. Hence, the model basically describes the quenching process.

To estimate the effect of pressure on the phase transformation temperature on heating stage (austenitization), the authors have used the Clausius–Clapeyron equation as:

$$\frac{dp}{dT} = \frac{\Delta H_{tr}}{T \Delta V_{tr}} \tag{10.43}$$

where:

ΔH_{tr} is the heat of transformation involved in the ferrite/martensite–austenite transformation

ΔV_{tr} is the volume change per mole due to transformation

The variation of the nominal temperature ΔM_s for martensite formation starts M_s was modeled as follows:

$$\Delta M_s = A\sigma_{kk} + B\bar{\sigma} \tag{10.44}$$

where:

σ_{kk} and $\bar{\sigma}$ are the hydrostatic and effective stresses, respectively

A and B are material constants

Once the material temperature exceeds the austenitization temperature A_s, the fraction of martensite formed on cooling F_m is calculated using the following equation [57]:

$$F_m = \left[1 - \exp\left(-\lambda\left(M_s - T\right)\right)\right] \tag{10.45}$$

where λ is a material constant.

The model also considers some transformation plasticity as an additional source of strain, and its evolution is given by [58]:

$$d\varepsilon^{TP} = 2K_{TP}\bar{\sigma}\left(1 - F_m\right)dF_m \tag{10.46}$$

where K_{TP} is a material constant.

The component-wise transformation plasticity strain is then determined as follows [58]:

$$d\varepsilon_{ij}^{TP} = \frac{3}{2}\left(\frac{d\varepsilon^{TP}}{\bar{\sigma}}\right)S_{ij} \tag{10.47}$$

where S_{ij} denotes the components of the stress tensor.

The volume change effect on the strain due to martensite formation is taken into account as:

$$d\varepsilon_{ij}^{TF} = \frac{3}{2}\left(0.044dF_m\right)\delta_{ij} \tag{10.48}$$

where:
$d\varepsilon_{ij}^{TF}$ is the volumetric strain
δ_{ij} is the Kronecker delta

Finally, the strain summation equation, involving the strains induced by transformation plasticity and phase changes, is given as:

$$\varepsilon_{ij}^{el} = \varepsilon_{ij}^{el_{old}} + d\varepsilon_{ij} - d\varepsilon_{ij}^{pl} - d\varepsilon_{ij}^{TP} - d\varepsilon_{ij}^{TF} \tag{10.49}$$

The above-mentioned equations were implemented as a user-defined subroutine (VUMAT) linked to Abaqus™, where a full coupling between phase transformation effects and thermoelastoplastic material behavior is considered [56].

More recently, other microstructure-based models have been developed using transformation plasticity with DD-based grain refinement model [59–61], visco-plastic asymmetry with phase transformation model [62], and ISV model with microstructure evolution equation [63].

10.2.2   Damage modeling

The machining of metallic materials often generates serrated chips, also known as segmented/saw-toothed chips, especially when machining hard-to-cut materials such as hardened steels, superalloys, and titanium alloys [1], and when machining ductile materials such as aluminum alloys under cutting speeds higher than a critical value [2]. To model serrate chip formation, in addition to the plasticity model, a damage model should be considered to describe the material behavior when damage occurs.

There exist several ways to model the damage in machining simulation. In the present chapter, we will focus on the most used ones, named the JC damage model and Cockcroft–Latham model.

The JC fracture model [64] is one of the most used models in machining simulation. It uses the equivalent plastic strain at failure ε_f as a criterion for damage initiation. ε_f is defined as follows:

$$\varepsilon_f = \left[D_1 + D_2 \exp\left(D_3 \frac{\sigma_p}{\sigma} \right) \right] \left[1 + D_4 \ln\left(\frac{\dot{\varepsilon}_p}{\dot{\varepsilon}_0} \right) \right]$$

$$\left[1 - D_5 \left(\frac{T - T_0}{T_{\text{melt}} - T_0} \right) \right]$$

(10.50)

where:

   σ_p denotes the pressure stress
   σ is the von Mises stress
   D_1 to D_5 are the damage constants

The damage initiation threshold is modeled in Abaqus/Explicit v6.13 [21] software according to a cumulative damage law:

$$D = \sum \frac{\Delta\varepsilon}{\varepsilon_f}$$

(10.51)

where $\Delta\varepsilon$ is the increment of the equivalent plastic strain.

In FEM involving damage phenomenon, once the damage initiates ($D = 1$), the flow stress will no longer be solely governed by the constitutive equation but will be affected by the damage evolution, and a strong mesh dependency of the state variables occurs due to severe strain localization [3]. In practice, there are different ways to describe the evolution of the flow stress after the damage takes place. The damage evolution criterion is defined by a scalar variable denoted here as d (varying between 0 and 1), which can be described by either a linear or exponential function of the equivalent plastic strain.

Another damage model also widely used in manufacturing process modeling is the Cockroft–Latham model [65]. This model is developed based on the plastic energy dissipated during the process, as:

$$D = \int_0^{\overline{\varepsilon}} \sigma_1 d\varepsilon \qquad (10.52)$$

where:
  $\overline{\varepsilon}$ is the effective strain
  σ_1 is the maximum principal stress

When a critical value is reached, that is, $D = D_{crit}$, the flow stress is reduced to a lower value σ_l, which is expressed as a percentage of the original flow stress.

This criterion is easy to use because only one material constant has to be determined, while two or more material constants have to be identified for the other damage criteria, such as the JC damage model. It is should be noticed that D_{crit} and σ_1 were most of the time calibrated numerically by using finite element simulation [66,67].

10.3   Identification techniques

In this section, the authors focused on the experimental, analytical, and computational techniques used in the determination of material constants utilized in machining modeling.

10.3.1   Dynamic tests

In developing a machining theory, Oxley and Hastings [68] were the first to introduce a constitutive equation involving strain, strain rate, and temperature effects. They used high-speed compression tests on plain carbon steels, covering a strain rate of up to 450 s^{-1} and a temperature range of 0°C–1100°C. Although the extrapolation of the material constants to machining resulted in acceptable predictions, this was not considered accurate enough to model the machining processes.

Later, advanced methods such as the SHPB [69] have been used in studying the material behavior at high strain rate (up to 10^4 s^{-1}) and temperature (up to $0.9 \times T_m$) [6]. However, the strain reached was still too low (up to 0.5) to be compared with cutting conditions. Furthermore, the SHPB technique experiences some technical difficulties, which may affect the accuracy of the final results; thus, the testing data need to be carefully interpreted [70].

Taylor tests were also used [7], allowing the strain rates raised up to 10^5 s^{-1} for the identification of the JC constitutive equation [7]. As for the

SHPB technique, the strains recorded were less than 1. Moreover, this technique is costly, complex, and difficult to run and interpret [71].

Although the high strain rates and elevated temperature are achieved during dynamic tests, they are still far from representing the real thermomechanical loading encountered in machining. Thus, in the last two decades, researchers have focused on developing advanced techniques to overcome the aforementioned issues.

10.3.2 Machining-based inverse methods

The absence of any devoted mechanical test method for describing the flow stress relevant to machining and the lack of high strain rate data have prompted many researchers to use machining tests themselves to determine the material behavior. This method, known as the inverse method, is simple and fast, and it provides better prediction than dynamic tests methodology. It is well known that the interaction of the cutting tool with the work materials results in three shearing zones, namely primary, secondary, and tertiary shear zones. As actually there is no reliable analytical model that can predict state variables accurately at the secondary and tertiary shear zones, due to the complexity in describing friction at these zones, the original formulation of the inverse method is only based on the primary shear zone. It consists of extracting machining data such as cutting forces and chip thickness from simple orthogonal machining tests and using them as an input of an analytical machining theory [9,72] to determine the flow stress, strain, strain rate, and temperature in the primary shear zone. The material constants are then identified using least-squares approximation [9,10].

However, previous studies pointed out some critical issues in the application of the inverse method, impairing the accuracy of the results. These critics can be summarized as follows:

- When assessing all constitutive parameters at the same time, optimization algorithms may converge to local solutions, resulting in the non-uniqueness of material parameter sets [73]. This is primarily due to the high nonlinearity of the constitutive equations.
- The low variation of state variables calculated in the primary shear zone during conventional orthogonal cutting also provides critical issues during the optimization.
- The initial values used to determine a large number of constants at the same time [74] influence the solution found by the optimization algorithms based on least-squares approximation.
- The accuracy of the solution depends on the analytical machining model used in the calculation of state variables from machining tests.

- The applicability of the determined constitutive equation may be restricted to the range of strains, strain rates, and temperatures of the primary shear zone, limiting the extrapolation of such equations to the secondary and tertiary shear zones.

To overcome the foregoing issues, the inverse technique has undergone several enhancements. In fact, robust algorithms such as nonlinear regression solution based on the Gauss–Newton algorithm with Levenberg–Marquardt modifications [10], Levenberg–Marquardt search algorithm [75], genetic algorithm [76], evolutionary computational algorithms [77], and neural networks algorithm [78] have been used to get better global convergence. However, the enhancements were found to be limited by the weak variability of the state variables calculated within the primary shear zone, reducing the experimental domain on which the material constants can be optimized.

On the other hand, researchers [74] have further revised the inverse method by considering the secondary shear zone to enlarge the range of strain, strain rate, and temperature. Besides, they have used a regularly distributed function, combined with an FEM simulations of machining, to reach the absolute minimum of the objective function and improve the ability of searching unique solutions of material constants. More recently, Daoud et al. [79] have used the response surface methodology (RSM) as a technique, reducing the error induced by the optimization procedure. In addition, they have found that material constants obtained by inverse method are very sensitive to the tool rake angle used in orthogonal machining tests and that the 0 rake angle gives the better results. The RSM approach has been recently adopted by Malakizadi et al. [80].

The authors believe that the major revision that has revolutionized the inverse method is that developed by Shi and coworkers [81]. Their main contribution is the use of the distribution of the state variables (strain, strain rate, temperature, and stress) at the primary shear zone [82], rather than their average values (original formulation of the inverse method), to determine the flow stress data. By doing this, the authors generated larger range of state variables across the primary shear zone, allowing better estimation of the material constants.

10.3.3   *Hybrid/advanced methods*

In order to further improve the determination of material constants representing machining conditions, some nonconventional method have been developed, involving hybrid and advanced computation methods. The former is a combination of experimental methods and/or computational methods.

Surprisingly, the first versions of the hybrid method involved quasi-static mechanical tests and orthogonal machining tests. In fact, Stevenson and Stephenson [83] and Stevenson [4] have shown that under soft machining conditions (low cutting speed), the strain rate is slow and the cutting temperature is approximated as room temperature. In these conditions, the flow stresses obtained in quasi-static tests were found in close agreement with those obtained in machining tests. This approach cannot be longer used when practical machining conditions are used, where the strain rate hardening and softening effects are not negligible. For this reason, in later studies [67,70,76,81,82,84], quasi-static tests have been utilized only in determining the strain hardening parameters, while the strain rate and softening parameters were extracted from machining tests. In addition, in some studies [81], machining tests were performed under hot conditions to extend the working temperature, leading to better estimation of the material softening parameter. Other researchers [8,18,77,85] have used dynamic tests such as SHPB technique rather than quasi-static tests, together with the inverse method.

Advanced computational techniques based on FEM in conjunction with orthogonal cutting tests or mechanical tests have been also proposed in the literature [67,73,75,84,86,87]. In these methods, the material constants were partly or totally identified by fitting the numerical prediction of machining data (cutting forces, chip thickness, contact length, and so on) obtained by FEM with those measured during cutting tests. Moreover, these techniques have been criticized and considered time-consuming. Some researchers [87] highlighted critical issues related to the uniqueness of the solution when FEM-based inverse method is applied for material constants' identification.

10.4   Case study: Effect of material modeling on finite element analysis of near-micromachining of Inconel 718

In this section, the authors propose a sensitivity analysis based on an FEM simulation of near-micromachining of age-hardened Inconel 718. This case study aims to emphasize the effect of material constitutive equations and calibration methods on the performance of finite element analysis (FEA) of cutting process.

10.4.1   Machining tests

The experimental data used in this chapter are adopted from machining tests obtained in [67]. In these experiments, the authors have used an uncoated cemented carbide insert (WC-6CO) with 0° rake angle,

7° clearance angle, and flat rake face (no chip breakers). The cutting tool edge radius was around 30 μm. For the cutting conditions, the present work will be limited to one cutting condition used in [67], which is defined as follows: cutting speed of 20 m/min and uncut chip thickness of 50 μm. This results in uncut chip thickness/edge radius ratio, known as the relative tool sharpness (RTS), equal to 1.66, satisfying near-micromachining conditions [88].

The work material is an age-hardened Ni-based superalloy Inconel 718 widely used in the manufacturing of heat-resistant components of aircraft engines. The work piece geometry is a disc machined with 3-mm thickness.

10.4.2   Finite element modeling of near-micromachining

10.4.2.1   Finite element formulation and boundary conditions

Finite element simulations of orthogonal near-micromachining were performed using an implicit updated Lagrangian formulation under a plane–strain coupled thermomechanical analysis. This model is developed under the commercial FE software DEFORM-2D™ version 10. The workpiece deformation behavior is modeled as viscoplastic, and the tool is considered to be rigid.

The workpiece and cutting tool were meshed with about 7,500 and 1,500 isoparametric quadrilateral elements, respectively. Using the mesh window option in DEFORM-2D™, fine mesh with average element size of 5–7 μm has been applied in the uncut chip region in the vicinity of the cutting tool (Figure 10.1). The typical workpiece dimensions used in the simulations are 6 mm × 1 mm. As seen in Figure 10.1, the bottom and left sides of the workpiece are fully constrained, whereas the top and right sides of the tool are fixed in the Y direction. A velocity load is applied to the tool in the X direction to simulate the cutting speed. For the thermal boundary conditions, the bottom and right sides of the workpiece and the top and left sides of the cutting tool are constrained by room temperature (20°C), whereas heat transfer with environment is allowed for the remaining part of the tool and the workpiece (h_{conv} = 0.0512 W/m²K). The coefficient of heat transfer at the chip–tool interface h_{int} was set to 55,000 kW/m²K [49].

10.4.2.2   Contact modeling

The friction at the chip–tool interface is modeled using a hybrid model, already implemented in DEFORM-2D™, combining sliding and shearing friction phenomena, as:

$$\tau = \begin{cases} \mu p, & \text{if } \mu p < mk \\ mk, & \text{if } \mu p \geq mk \end{cases} \tag{10.53}$$

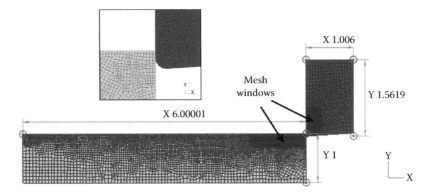

Figure 10.1 Finite element model: geometry and mesh.

where:

τ is the shear frictional stress
p is the normal stress at the tool–chip interface
k denotes the shear flow stress of the work material
The frictional shear factor m is set equal to 1
The apparent friction coefficient μ is set equal to 0.32 based on [67]

10.4.2.3 Material modeling

As mentioned earlier, the work material studied is the aged superalloy Inconel 718. The main goal of this study is to investigate the effect of flow stress modeling and identification techniques on the chip formation process and the performance of the FEM. To this end, the authors propose the original and the modified versions of the Johnson–Cook plasticity (JCP) models as material constitutive equations for Inconel 718 obtained from previous works [25,67,81]. The material constants of these models have been identified using different techniques, involving quasi-static tests, dynamic tests, FEM-based inverse method, and analytical-based inverse method. As seen in Table 10.1, the domains of validity vary with the applied identification techniques. The obtained material constants of the JCP equations are displayed in Table 10.2. The JCP1 model is a modified JCP equation (Equation 10.10), where the strain rate sensitivity is considered dependent on strain rate and temperature. In addition, the JCP4 model is another modified JCP equation (Equation 10.12), originally developed by Calamaz [27] and later adopted by Özel and coworkers [29]. The latter combined the JCP constants from Lorentzon [89] with their own flow softening parameters for Inconel 718 alloy, which are depicted in Table 10.3.

As experimental analysis of the chips has led to the formation of serrated chips [67], a damage model is therefore needed in the FEM. In the present study, the Cockroft–Latham's criterion (Equation 10.53) is

Table 10.1 Identification techniques and validity domains of selected
JCP models for aged Inconel 718 alloy

Model	Material condition	Identification technique	Applicability domain
JCP1 [25]	Aged	quasi-static (QS) and SHPB tests	$0.001 \leq \dot{\varepsilon}_p \leq 11 \times 10^3 s^{-1}$ $20 \leq T \leq 800°C$ $0 \leq \varepsilon_p \leq 0.35$
JCP2 [81]	Aged (35 HRC)	QS and analytical-based inverse method	$QS \leq \dot{\varepsilon}_p \leq 6 \times 10^3 s^{-1}$ $RT \leq T \leq 800°C$ $0 \leq \varepsilon_p \leq 1.5$
JCP3 [67]	Aged (46.1 HRC)	QS and FEM-based inverse method	Machining conditions in [67]

Table 10.2 JCP parameters for aged Inconel 718 alloy

Model	A (MPa)	B (MPa)	n	C	m	$\dot{\varepsilon}_0(s^{-1})$	T_0 (°C)
JCP1 [25]	963	937	0.333	$C = f(\dot{\varepsilon}_p, T)^a$	1.3	0.001	20
JCP2 [81]	789	700	0.22	0.0074	2.31	0.001	20
JCP3 [67]	1485	904	0.777	0.015	1.689	1	20

$$^a \quad C = 0:0232 - \left[0.00372 + 0.0021 \sin\left(\frac{\dot{\varepsilon}_p - 5000}{3000} \pi \right) \right] \sin\left(\frac{T - 500}{150} \pi \right)$$

Table 10.3 JCP4 parameters for aged Inconel 718 alloy

	A (MPa)	B (MPa)	n	C	m	$\dot{\varepsilon}_0(s^{-1})$	T_0 (°C)
JCP [89]	1241	622	0.6522	0.0134	1.3	1	20
Flow softening parameters [29]	D	S	s	r			
	0.6	0	5	1			

Source: Ozel, T. et al., *Mach. Sci. Tech.*, 15, 2011, 21–46.

employed to predict the effect of stress on the chip segmentation. For the JCP1, JCP2, and JCP3 models, the values of critical damage coefficients D_{crit} and P were set equal to 97.8% and 43.13%, respectively, following Klocke and coworkers [67]. In order to assess the effect of damage model on the chip serration phenomenon, three configurations of damage modeling were tested with the modified JCP4 model, as follows: For JCP4-1: $D_{crit} = 97.8$, $P = 43.13\%$, as in [67]; for JCP4-2: $D_{crit} = 510$, $P = 43.13\%$, as in [89], and for JCP4-3: $D_{crit} = 0$, no damage, as in [29].

Table 10.4 Physical properties of the superalloy Inconel 718[a] and cutting insert[b]

Parameter	Work material (Inconel 718)	Uncoated carbide insert (WC)
Density, ρ (kgm^{-3})	8470	11900
Elastic modulus, E (GPa)	206	612
Poisson's coefficient, v	0.3	0.22
Specific heat, C_p (JKg^{-1} K^{-1})	0.2 T + 421.7	0.12 T + 334.01
Thermal conductivity, λ (Wm^{-1} K^{-1})	0.015 T + 11.002	0.042 T + 35.95
Thermal expansion, α (μmm^{-1} K^{-1})	11.5	4.9
Melting temperature, T_m (K)	1550	–

[a] Malakizadi, A. et al., *Simul. Model. Pract. Theor.*, 60, 40–53, 2016; Kitagawa, T. et al., *Wear*, 202, 142–148, 1997; Anderson, M. et al., *Int. J. Mach. Tool Manufac.*, 46, 1879–1891, 2006.
[b] Jomaa, M. et al., *J. Manufac. Proc.*, 26, 446–458, 2017.

All input parameters relative to the physical properties of the work material and cutting tool are presented in Tables 10.4. It is worth noting that critical properties such as the thermal conductivity and heat capacity are considered temperature-dependent.

10.4.3   Results and discussion

10.4.3.1   Chip morphology

The chip morphology defined by the type and shape of the chip is an important parameter that should be considered in any validation of FEM simulation of machining. However, this is not the case for many research works where this criterion is often neglected. In the present work, the authors consider both macroscopic and microscopic geometries of the chips. Figure 10.2 displays the macroscopic feature of the chip, known as the chip up-curling. As seen in Figure 10.2, the results can be classified into two groups. The first group includes chips predicted by JCP1–JCP3 models (Figures 10.2a–c), where the chips experience similar chip up-curling radii, with some degree of segmentation. The second group includes chips predicted with JCP4-1–JCP4-3 models, where the chips are continuous, with high degree of curling, particularly for JCP4-2 and JCP4-3. The predicted chip curling of the first group is roughly similar to the experimental chip curling (Figure 10.2g). As the machining time at which the experimental photo was captured is unknown, the foregoing statement should be taken as preliminary observation.

Another important parameter to consider is the microgeometry of the chip. Figure 10.3 compares the predicted and experimental results in terms of the maximum chip height (h_{max}) and the minimum chip height (h_{min}) with the maximum chip height ratio (h_{min}/h_{max}). The latter represents the

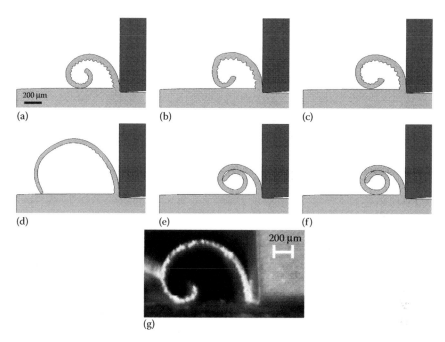

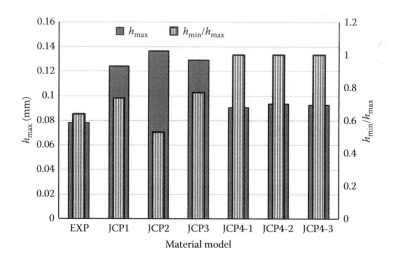

Figure 10.2 Predicted chip curling for material models: (a) JCP1, (b) JCP2, (c) JCP3, (d) JCP4-1, (e) JCP4-2, and (f) JCP4-3 after 0.012 s machining time, and (g) experimental result. (From Klocke, F. et al., *Procedia CIRP*, 8, 212–217, 20,13. With permission.)

Figure 10.3 Comparison of chip characteristics.

chip segmentation intensity and varies between 0 and 1. Continuous chips will result in h_{min}/h_{max} equal to 1. It is clear that the chips predicted with the first group of material models are thicker than those predicted with the second group of material models. The latter results in chip thickness closer to the experimental ones; however, they are of continuous chip type. Conversely, reasonable agreement is obtained between experimental and chip segmentation intensities predicted with the first group of material models.

10.4.3.2   *Machining forces*

Figures 10.4 and 10.5 display predicted cutting and thrust force signals versus machining time. As expected, the cutting force signals (Figure 10.4) generated by the first group of material models experience significant fluctuations (oscillations). The amplitude of oscillations is a bit higher in the case of JCP1 and JCP3 models than in JCP3 model. However, smooth cutting force signals are predicted with the second group of material models. The same observations can be drawn for the thrust force (Figure 10.5). These trends are expected regarding the difference in the chip formation process observed, which varies from serrated to continuous process, depending on the material model applied. Figure 10.6 quantitatively compares the predicted and experimental average cutting and thrust forces. Good agreement is obtained between the experimental and cutting forces predicted with JCP1 and JCP3 models, whereas the remaining models underestimate it. However, the predicted thrust force is underestimated for all tested material models. This trend was reported in several research studies [84,93], where it was stated that this discrepancy is independent of the material modeling used, and error related to finite element discretization is the main cause.

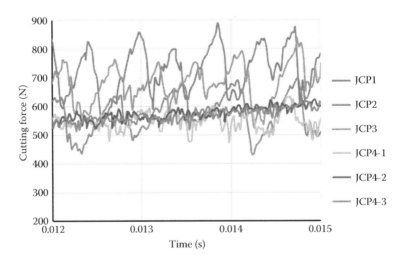

Figure 10.4 Comparison of cutting force signals.

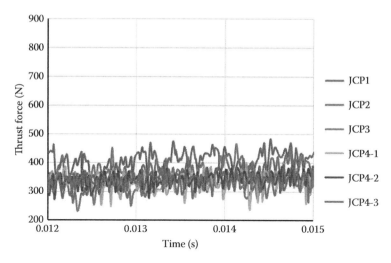

Figure 10.5 Comparison of thrust force signals.

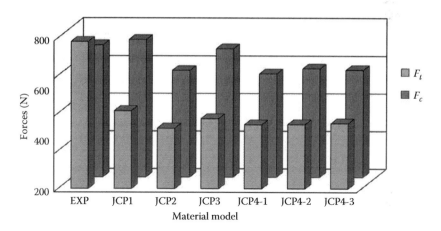

Figure 10.6 Comparison of experimental and predicted average force values.

10.4.3.3 Cutting temperature distribution

In machining process, the prediction of the cutting temperature is of great importance, as the cutting tool life significantly depends on its maximum value. Figure 10.7 displays the predicted temperature distributions at the cutting zone. Again, the highest temperatures at the tool–chip interface (501°C–570°C) are recorded with JCP1 and JCP3 models, whereas the maximum temperature values obtained with the remaining models are between 364°C and 433°C. The same trend is observed about the chip temperature.

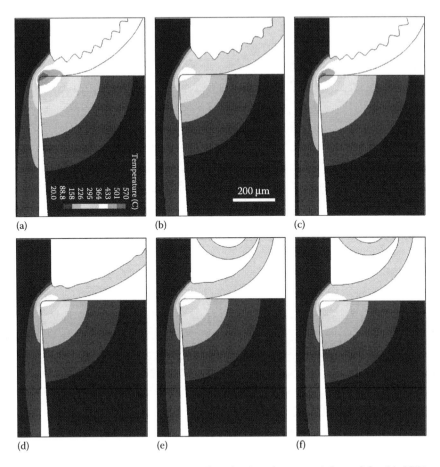

Figure 10.7 Predicted temperature distribution for material models: (a) JCP1, (b) JCP2, (c) JCP3, (d) JCP4-1, (e) JCP4-2, and (f) JCP4-3 after 0.012 s machining time.

10.4.3.4   Machined surface alteration: State variables evolution

In this section, the authors discuss the effect of material models on the thermomechanical state of the machined surface. Figure 10.8 depicts the effect of material models on the evolution of temperature at 10 μm below the machined surface. Obviously, the highest temperature values reached were 413°C and 383°C when using JCP1 and JCP3, whereas they were around 320°C for the remaining models. This can be explained by the mechanical loading described by the stress–strain curves in Figure 10.9. In fact, JCP1 and JCP3 models produce the highest stress levels compared with the other models, resulting in higher amount of plastic energy dissipated, and consequently, high amount of heat is generated.

Based on the earlier results, it is clear that constitutive equations and their calibration techniques affect the FEM performance significantly.

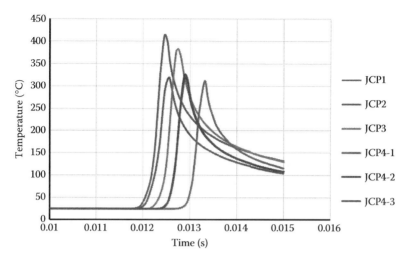

Figure 10.8 Comparison of predicted cutting temperature history at 10 μm beneath the machined surface.

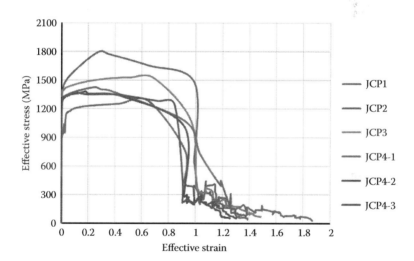

Figure 10.9 Comparison of predicted stress–strain curves at 10 μm beneath the machined surface.

In order to understand the phenomena observed, first, it is worth highlighting how the material models tested behave under similar thermomechanical tests. Figure 10.10 compares the material models under quasi-static and dynamic testing conditions. Under quasi-static conditions, flow stresses predicted with JCP1 and JCP3 models are in good agreement with experiments, whereas under dynamic conditions, JCP2 model

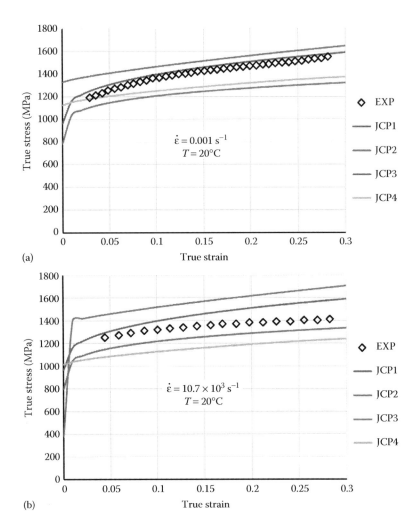

Figure 10.10 Comparison of predicted and experimental stress–strain curves under (a) quasi-static and (b) dynamic conditions. (Adopted from Wang, X. et al., *Mat. Sci. & Eng. A*, 580, 385–390, 2013. With permission.)

performs better than the other models. Specifically, JCP4 underestimate the flow stress under both quasi-static and dynamic testing conditions.

So, as a first answer to the trends observed in the predicted machining data, the authors can argue that the material strength described by JCP1, JCP2, and JCP3 is higher than that predicted by JCP4 model, favoring the formation of serrated chips and higher cutting forces. The lower cutting force and temperature generated by JCP2 model as compared with JCP1 and JCP3 model can be explained by the fact that this model is developed for an aged Inconel 718 but with hardness of 35 HRC (see Table 10.1), which is lower than the studied Inconel 718 hardness (46.1 HRC).

Although JCP1 model was calibrated using quasi-static and dynamic tests, it performed better than JCP2 and was comparable with the JCP3, which were calibrated under machining conditions. The authors rely on these results to perform successful modification of the JC model by implementing a strain rate- and temperature-dependent strain rate constant C in JCP1 model (see Table 10.2). These results confirm previous findings that the JC model can further be enhanced by taking account of the interaction between the thermal, strain, and strain rate hardening effects during calibration, particularly when it comes to the machining modeling.

From machining modeling point of view, JCP4 model failed to accurately predict the machining data, and none of the damage conditions applied was reliable. The authors suggest that the hyperbolic function representing the material flow softening factor is not adequate, as no experimental date available in the open literature have clearly shown such behavior of the Inconel 718 alloy under dynamic tests. In addition, this modification (flow softening parameter) of the JC model was developed to predict the chip serration phenomenon, without the need for damage model. However, this was not the case, as all machining simulations with JCP4 model have shown continuous chip formation, even when using additional damage criterion (Cockcroft–Latham). Furthermore, similar predictions of machining data have been obtained when using critical value $D_{crit} = 510$ and 0. This can be explained by the fact that there is a critical damage value, beyond which the material will no longer be damaged, as a very high density of energy is needed (Cockcroft–Latham, Equation 10.53). Hence, this case is equivalent to D_{crit} set equal to 0 (no damage criterion is used). It is worth recalling that the value $D_{crit} = 510$ was adopted based on experimental results from a previous study [89].

10.5   Conclusions and recommendations

In the present case study, FEM of near-micromachining of the superalloy Inconel 718 was carried out. The main goal was to study the effect of material constitutive equation and corresponding calibration methods on the performance of finite element predictions of machining outputs, such as chip morphology, cutting forces, cutting temperature, and machined surface layer alterations. The material models were selected based on the identification techniques used in their calibration. The material models used included the original and modified JC models. The identification techniques involved quasi-static tests, dynamic test (SHPB), analytical-based inverse method, and FEM-based inverse method.

Considering the forgoing results and discussions, it is evident that the material modeling not only affects the chip formation process but also alters the machined surface layer differently. The latter is a critical parameter, particularly when surface integrity characteristics such as residual stresses and microstructural changes need to be predicted.

The results showed that to achieve a successful machining modeling, it is imperatively suggested to carefully select the constitutive equation and the method used to identify its coefficients. The constitutive equations should be mathematically and physically coherent and should faithfully describe as much as possible the material behavior under machining conditions. Although, actually, there is no devoted experimental setup for describing the flow stress relevant to machining, there are some relevant precautions that can be taken to reduce material model-induced errors. Some of them can be summarized as follows:

- Work material heat treatment: When material models are adopted from the literature, it is imperatively requested to check the hardness of the work material modeled.
- The material model should be validated using experimental data (quasi-static and dynamic tests) before it is implemented in the FEM.
- The strain hardening parameters can be calibrated using quasi-static tests.
- Thermal softening and strain rate parameters can be identified by either dynamic tests or inverse method. If the former is the choice, it is suggested to define the strain rate and thermal softening coefficients as dependent on strain rate and/or temperature.

References

1. Shaw MC, *Metal Cutting Principles*. Oxford University Press, New York, 2005.
2. Kececioglu D, Shear-zone size, compressive stress, and shear strain in metal-cutting and their effects on mean shear-flow stress, *Journal of Engineering for Industry*, 1960, 82:79.
3. Okushima K, Hitomi K, An analysis of the mechanism of orthogonal cutting and its application to discontinuous chip formation, *Journal of Engineering for Industry*, 1961, 83:545.
4. Stevenson R, Study on the correlation of workpiece mechanical properties from compression and cutting tests, *Machining Science and Technology*, 1997, 1:67–79.
5. Sartkulvanich P, Altan T, Soehner J, Flow stress data for finite element simulation in metal cutting: A progress report on madams, *Machining Science and Technology*, 2005, 9:271–288.
6. Lesuer DR, Kay G, LeBlanc M, Modeling large-strain, high-rate deformation in metals. *Third Biennial Tri-Laboratory Engineering Conference Modeling and Simulation*, Pleasanton, CA, November 3–5, 1999, 2001.
7. Rule WK, A numerical scheme for extracting strength model coefficients from Taylor test data, *International Journal of Impact Engineering*, 1997, 19:797–810.

8. Chandrasekaran H, M"saoubi R, Chazal H, Modelling of material flow stress in chip formation process from orthogonal milling and split Hopkinson bar tests, *Machine Science and Technology*, 2005, 9:131–145.

9. Tounsi N, Vincenti J, Otho A, Elbestawi M, From the basic mechanics of orthogonal metal cutting toward the identification of the constitutive equation, *International Journal of Machine Tools and Manufacture*, 2002, 42:1373–1383.

10. Özel T, Zeren E, A methodology to determine work material flow stress and tool-chip interfacial friction properties by using analysis of machining, *Journal of Manufacturing Science and Engineering*, 2006, 128:119–129.

11. Li B, Wang X, Hu Y, Li C, Analytical prediction of cutting forces in orthogonal cutting using unequal division shear-zone model, *The International Journal of Advanced Manufacturing Technology*, 2011, 54:431–443.

12. Jaspers S, Dautzenberg J, Material behaviour in conditions similar to metal cutting: Flow stress in the primary shear zone, *Journal of Materials Processing Technology*, 2002, 122:322–330.

13. Abed F, Makarem F, Comparisons of constitutive models for steel over a wide range of temperatures and strain rates, *Journal of Engineering Materials and Technology*, 2012, 134:021001–021001.

14. Fang N, A new Quantitative sensitivity analysis of the flow stress of 18 engineering materials in machining, *Journal of Engineering Materials and Technology*, 2005, 127:192–196.

15. Oxley PLB, Young H, *The Mechanics of Machining: An Analytical Approach to Assessing Machinability*, Ellis Horwood Publisher, UK, 1989, 136–182.

16. MacGregor C, Fisher J, A velocity-modified temperature for the plastic flow of metals, *Journal of Applied Mechanics-Transactions of the ASME*, 1946, 13:A11–A16.

17. Marusich T, Ortiz M, Modelling and simulation of high speed machining, *International Journal for Numerical Methods in Engineering*, 1995, 38:3675–3694.

18. Jomaa W, Songmene V, Bocher P, An hybrid approach based on machining and dynamic tests data for the identification of material constitutive equations, *Journal of Materials Engineering and Performance*, 2016, 25:1010–1027.

19. Lemonds J, Needleman A, Finite element analyses of shear localization in rate and temperature dependent solids, *Mechanics of Materials*, 1986, 5:339–361.

20. Childs T, Friction modelling in metal cutting, *Wear*, 2006, 260:310–318.

21. Jiang F, Li J, Sun J, Zhang S, Wang Z, Yan L, Al7050-T7451 turning simulation based on the modified power-law material model, *The International Journal of Advanced Manufacturing Technology*, 2010, 48:871–880.

22. Yu J, Jiang F, Rong Y, Xie H, Suo T, Numerical study the flow stress in the machining process, *The International Journal of Advanced Manufacturing Technology*, 2014, 74:509–517.

23. Nieslony P, Grzesik W, Laskowski P, Habrat W, FEM-based modelling of the influence of thermophysical properties of work and cutting tool materials on the process performance, *Procedia CIRP*, 2013, 8:3–8.

24. Johnson GR, Cook WH, A constitutive model and data for metals subjected to large strains, high strain rates and high temperatures, in: *Proceedings of the 7th International Symposium on Ballistics*, The Hague, the Netherlands: International Ballistics Committee, 1983, 541–547.

25. Wang X, Huang C, Zou B, Liu H, Zhu H, Wang J, Dynamic behavior and a modified Johnson–Cook constitutive model of Inconel 718 at high strain rate and elevated temperature, *Materials Science and Engineering: A*, 2013, 580:385–390.

26. Chen G, Ren C, Ke Z, Li J, Yang X, Modeling of flow behavior for 7050-T7451 aluminum alloy considering microstructural evolution over a wide range of strain rates, *Mechanics of Materials*, 2016, 95:146–157.

27. Calamaz M, Coupard D, Girot F, A new material model for 2D numerical simulation of serrated chip formation when machining titanium alloy Ti–6Al–4V, *International Journal of Machine Tools and Manufacture*, 2008, 48:275–288.

28. Özel Tr, Experimental and finite element investigations on the influence of tool edge radius in machining nickel-based alloy, in: *ASME 2009 International Manufacturing Science and Engineering Conference*, American Society of Mechanical Engineers, 2009, 493–498.

29. Ozel T, Llanos I, Soriano J, Arrazola P-J, 3D finite element modelling of chip formation process for machining Inconel 718: Comparison of FE software predictions, *Machining Science and Technology*, 2011, 15:21–46.

30. Ulutan D, Özel T, Determination of constitutive material model parameters in FE-based machining simulations of Ti-6Al-4V and IN-100 alloys: An inverse methodology, *Proceedings of NAMRI/SME*, 41, 2013.

31. Arısoy YM, Özel T, Prediction of machining induced microstructure in Ti–6Al–4V alloy using 3-D FE-based simulations: Effects of tool microgeometry, coating and cutting conditions, *Journal of Materials Processing Technology*, 2015, 220:1–26.

32. Sima M, Özel T, Modified material constitutive models for serrated chip formation simulations and experimental validation in machining of titanium alloy Ti–6Al–4V, *International Journal of Machine Tools and Manufacture*, 2010, 50:943–960.

33. Zerilli FJ, Armstrong RW, Dislocation mechanics based constitutive relations for material dynamics calculations, *Journal of Applied Physics*, 1987, 61:1816–1825.

34. Samantaray D, Mandal S, Borah U, Bhaduri A, Sivaprasad P, A thermo-viscoplastic constitutive model to predict elevated-temperature flow behaviour in a titanium-modified austenitic stainless steel, *Materials Science and Engineering: A*, 2009, 526:1–6.

35. Maekawa K, Shirakashi T, Usui E, Flow stress of low carbon steel at high temperature and strain rate. ii: Flow stress under variable temperature and variable strain rate, *Bulletin of the Japan Society of Precision Engineering*, 1983, 17:167–172.

36. Lindgren L-E, Domkin K, Hansson S, Dislocations, vacancies and solute diffusion in physical based plasticity model for AISI 316L, *Mechanics of Materials*, 2008, 40:907–919.

37. Kocks U, Laws for work-hardening and low-temperature creep, *Journal of Engineering Materials and Technology*, 1976, 98:76–85.

38. Follansbee P, Kocks U, A constitutive description of the deformation of copper based on the use of the mechanical threshold stress as an internal state variable, *Acta Metallurgica*, 1988, 36:81–93.

39. Adibi-Sedeh AH, Madhavan V, Bahr B, Extension of Oxley's analysis of machining to use different material models, *Transactions-American Society*

of Mechanical Engineers Journal of Manufacturing Science and Engineering, 2003, 125:656–666.

40. Bammann D, Chiesa M, Johnson G, Modeling large deformation and failure in manufacturing processes, *Theoretical and Applied Mechanics*, 1996, 359–376.

41. Chuzhoy L, DeVor RE, Kapoor SG, Beaudoin AJ, Bammann DJ, Machining Simulation of ductile iron and its constituents, Part 1: Estimation of material model parameters and their validation, *Journal of Manufacturing Science and Engineering*, 2003, 125:181–191.

42. Kalhori V, Wedberg D, Lindgren L-E, Simulation of mechanical cutting using a physical based material model, *International Journal of Material Forming*, 2010, 3: 511–514.

43. Wedberg D, Svoboda A, Lindgren L-E, Modelling high strain rate phenomena in metal cutting simulation, *Modelling and Simulation in Materials Science and Engineering*, 2012, 20: 085006.

44. Field M, Kahles JF, The surface integrity of machined and ground high strength steels, *DMIC Report*, 2010, 1964, 54.

45. Field M, Kahles JF, Review of surface integrity of machined components, *Annals of the CIRP*, 1972, 20:153–162.

46. Chou YK, Evans CJ, White layers and thermal modeling of hard turned surfaces, *International Journal of Machine Tools and Manufacture*, 1999, 39:1863–1881.

47. Umbrello D, Filice L, Improving surface integrity in orthogonal machining of hardened AISI 52100 steel by modeling white and dark layers formation, *CIRP Annals—Manufacturing Technology*, 2009, 58: 73–76.

48. Umbrello D, Hua J, Shivpuri R, Hardness-based flow stress and fracture models for numerical simulation of hard machining AISI 52100 bearing steel, *Materials Science and Engineering: A*, 2004, 374:90–100.

49. Rotella G, Dillon O, Umbrello D, Settineri L, Jawahir I, Finite element modeling of microstructural changes in turning of AA7075-T651 alloy, *Journal of Manufacturing Processes*, 2013, 15:87–95.

50. Sheppard T, Tunnicliffe P, Patterson S, Direct and indirect extrusion of a high strength aerospace alloy (AA 7075), *Journal of Mechanical Working Technology*, 1982, 6:313–331.

51. Yanagimoto J, Karhausen K, Brand A, Kopp R, Incremental formulation for the prediction of flow stress and microstructural change in hot forming, *Journal of Manufacturing Science and Engineering*, 1998, 120:316–322.

52. Furukawa M, Horita Z, Nemoto M, Valiev RZ, Langdon TG, Microhardness measurements and the Hall-Petch relationship in an Al+Mg alloy with submicrometer grain size, *Acta Materialia*, 1996, 44:4619–4629.

53. Quan G-z, Mao Y-p, Li G-s, Lv W-q, Wang Y, Zhou J, A characterization for the dynamic recrystallization kinetics of as-extruded 7075 aluminum alloy based on true stress–strain curves, *Computational Materials Science*, 2012, 55:65–72.

54. Pu Z, Umbrello D, Dillon O, Lu T, Puleo D, Jawahir I, Finite element modeling of microstructural changes in dry and cryogenic machining of AZ31B magnesium alloy, *Journal of Manufacturing Processes*, 2014, 16:335–343.

55. Jafarian F, Ciaran MI, Umbrello D, Arrazola P, Filice L, Amirabadi H, Finite element simulation of machining Inconel 718 alloy including microstructure changes, *International Journal of Mechanical Sciences*, 2014, 88:110–121.

56. Ramesh A, Melkote SN, Modeling of white layer formation under thermally dominant conditions in orthogonal machining of hardened AISI 52100 steel, *International Journal of Machine Tools and Manufacture*, 2008, 48:402–414.

57. Koistinen D, Marburger R, A general equation prescribing the extent of the austenite-martensite transformation in pure iron-carbon alloys and plain carbon steels, *Acta Metallurgica*, 1959, 7:59–60.

58. Chandrasekar S, Yang H, Experimental and computational study of the quenching of carbon steel, *Journal of Manufacturing Science and Engineering*, 1997, 119:257.

59. Ding H, Shin YC, Dislocation density-based grain refinement modeling of orthogonal cutting of titanium, *Journal of Manufacturing Science and Engineering*, 2014, 136:041003.

60. Ding H, Shin YC, A metallo-thermomechanically coupled analysis of orthogonal cutting of AISI 1045 steel, *Journal of Manufacturing Science and Engineering*, 2012, 134:051014.

61. Ding H, Shin YC, Multi-physics modeling and simulations of surface microstructure alteration in hard turning, *Journal of Materials Processing Technology*, 2013, 213:877–886.

62. Mahnken R, Wolff M, Cheng C, A multi-mechanism model for cutting simulations combining visco-plastic asymmetry and phase transformation, *International Journal of Solids and Structures*, 2013, 50:3045–3066.

63. Liu R, Salahshoor M, Melkote S, Marusich T, A unified internal state variable material model for inelastic deformation and microstructure evolution in SS304, *Materials Science and Engineering: A*, 2014, 594:352–363.

64. Johnson GR, Cook WH, Fracture characteristics of three metals subjected to various strains, strain rates, temperatures and pressures, *Engineering Fracture Mechanics*, 1985, 21:31–48.

65. Cockcroft M, Latham D, Ductility and the workability of metals, *Journal of the Institute of Metals*, 1968, 96:33–39.

66. Umbrello D, Rizzuti S, Outeiro J, Shivpuri R, 'M'Saoubi R, Hardness-based flow stress for numerical simulation of hard machining AISI H13 tool steel, *Journal of Materials Processing Technology*, 2008, 199:64–73.

67. Klocke F, Lung D, Buchkremer S, Inverse identification of the constitutive equation of Inconel 718 and AISI 1045 from FE machining simulations, *Procedia CIRP*, 2013, 8:212–217.

68. Oxley P, Hastings W, Predicting the strain rate in the zone of intense shear in which the chip is formed in machining from the dynamic flow stress properties of the work material and the cutting conditions, *Proceedings of the Royal Society of London A Mathematical and Physical Sciences*, 1977, 356:395–410.

69. Hopkinson B, A method of measuring the pressure produced in the detonation of high explosives or by the impact of bullets, *Philosophical Transactions of the Royal Society of London Series A, Containing Papers of a Mathematical or Physical Character*, 1914, 213:437–456.

70. Guo Y, An integral method to determine the mechanical behavior of materials in metal cutting, *Journal of Materials Processing Technology*, 2003, 142:72–81.

71. Panov V, *Modelling of Behaviour of Metals at High Strain Rates*, Cranfield University, 2006.

72. Incropera F, Material constitutive modeling under high strain rates and temperatures through orthogonal machining tests, *Journal of Manufacturing Science and Engineering*, 1999, 121:577.

73. Shrot A, Bäker M, A study of non-uniqueness during the inverse identification of material parameters, *Procedia CIRP*, 2012, 1:72–77.
74. Pujana J, Arrazola P, M'saoubi R, Chandrasekaran H, Analysis of the inverse identification of constitutive equations applied in orthogonal cutting process, *International Journal of Machine Tools and Manufacture*, 2007, 47:2153–2161.
75. Shrot A, Bäker M, Determination of Johnson–Cook parameters from machining simulations, *Computational Materials Science*, 2012, 52:298–304.
76. Ma Y, Yu D, Feng P, Wu Z, Zhang J, Finite element method study on the influence of initial stress on machining process, *Advances in Mechanical Engineering*, 2015, 7:1687814015572457.
77. Özel T, Karpat Y, Identification of constitutive material model parameters for high-strain rate metal cutting conditions using evolutionary computational algorithms, *Materials and Manufacturing Processes*, 2007, 22:659–667.
78. Sheikh-Ahmad J, Twomey J, ANN constitutive model for high strain-rate deformation of Al 7075-T6, *Journal of Materials Processing Technology*, 2007, 186:339–345.
79. Daoud M, Jomaa W, Chatelain J, Bouzid A, A machining-based methodology to identify material constitutive law for finite element simulation, *The International Journal of Advanced Manufacturing Technology*, 2015, 77:2019–2033.
80. Malakizadi A, Cedergren S, Sadik I, Nyborg L, Inverse identification of flow stress in metal cutting process using response surface methodology, *Simulation Modelling Practice and Theory*, 2016, 60:40–53.
81. Shi B, Attia H, Tounsi N, Identification of material constitutive laws for machining—part II: generation of the constitutive data and validation of the constitutive law, *Journal of Manufacturing Science and Engineering*, 2010, 132:051009.
82. Shi B, Attia H, Tounsi N, Identification of material constitutive laws for machining—part I: an analytical model describing the stress, strain, strain rate, and temperature fields in the primary shear zone in orthogonal metal cutting, *Journal of Manufacturing Science and Engineering*, 2010, 132:051008.
83. Stevenson R, Stephenson D, The mechanical behavior of zinc during machining, *Journal of Engineering Materials and Technology*, 117, 1995.
84. Klocke F, Lung D, Buchkremer S, Jawahir I, From orthogonal cutting experiments towards easy-to-implement and accurate flow stress data, *Materials and Manufacturing Processes*, 2013, 28:1222–1227.
85. Xu D, Feng P, Li W, Ma Y, An improved material constitutive model for simulation of high-speed cutting of 6061-T6 aluminum alloy with high accuracy, *The International Journal of Advanced Manufacturing Technology*, 2015, 5:1043–1053.
86. Özel T, Altan T, Determination of workpiece flow stress and friction at the chip–tool contact for high-speed cutting, *International Journal of Machine Tools and Manufacture*, 2000, 40:133–152.
87. Shrot A, Bäker M, Is it possible to identify Johnson-Cook law parameters from machining simulations? *International Journal of Material Forming*, 2010, 3:443–446.
88. Liu K, Melkote SN, Finite element analysis of the influence of tool edge radius on size effect in orthogonal micro-cutting process, *International Journal of Mechanical Sciences*, 2007, 49:650–660.
89. Lorentzon J, Järvstråt N, Josefson B, Modelling chip formation of alloy 718, *Journal of Materials Processing Technology*, 2009, 209:4645–4653.

90. Kitagawa T, Kubo A, Maekawa K, Temperature and wear of cutting tools in high-speed machining of Inconel 718 and Ti+ 6Al+ 6V+ 2Sn, *Wear*, 1997, 202, 142–148.

91. Anderson M, Patwa R, Shin YC, Laser-assisted machining of Inconel 718 with an economic analysis, *International Journal of Machine Tools and Manufacture*, 2006, 46:1879–1891.

92. Jomaa W, Mechri O, Lévesque J, Songmene V, Bocher P, Gakwaya A, Finite element simulation and analysis of serrated chip formation during high–speed machining of AA7075–T651 alloy, *Journal of Manufacturing Processes*, 2017, 26:446–458.

93. Jomaa W, Songmene V, Bocher P, An investigation of machining-induced residual stresses and microstructure of induction-hardened AISI 4340 steel, *Materials and Manufacturing Processes*, 2016, 31:838–844.

Index

Note: Page numbers followed by f and t refer to figures and tables respectively.

2D and 3D modeling, 102–103, 103f
3D-integrated modeling, 132

A

Abrasive particles, 149, 150f, 151–152
Adaptive TF, 189
AdvantEdge, 37–38
Aerostatic bearing, dynamic modeling
 approach, 124–125, 126f
AISI 1045 steel, 213
AISI 1060 steel experiments, 214
AISI 4340 serrated chip, 38, 38t
AISI O1 and AISI D2 steel, 215
 temperature-dependent material
  properties, 227f
 temperature field, 233f, 234f
 thermal conductivity difference in, 231
Al 6061-T6 chip, 37, 38t
ALE (Arbitrary Lagrangian–Eulerian),
 30–31, 250
α-shape method, 3
Analytical model, 26–27
 grind-hardening, 221–223, 228–231
  results using, 229t–230t
  workpiece temperature field, 231f
Annealing temperature, 149
ANSYS software, numerical solution
 general detail, 223–225
 results, 232–236, 232f, 235t, 236f
 thermomechanical material properties,
  226–228
 verification, 226
Arbitrary Lagrangian–Eulerian (ALE),
 30–31, 250

ASR. *See* Average surface roughness (ASR)
Atom categories, 154, 154f
Atomic displacement, 160–161, 161f
Atomistic simulation of nanomachining
 contact area determination, 156–157
 dynamically identifying removed
  matter, 153–156, 156f
 external constraints, boundary
  conditions, and simulation
  procedure, 151–152
 preparing model, 146–151, 151f
 removing heat, 152–153
 topography, evaluating workpiece,
  157–158
Average surface roughness (ASR),
 181–182
 centerline, 184
 current and pulse-off time on, 205f
 current and pulse-on time on, 205f
 estimation
  concurrent, 202f
  MATE in, 201f
  testing, 203t
 normalized, 196, 198–199
 pulse-on and pulse-off time, 205f
Avrami equation, 293

B

Bammann-Chiesa-Johnson's (BCJ) model,
 289–291
Bearing, FE distribution, 130, 130f
Benzeggagh–Kenane criterion, 62
Block Lanczos method, 110
Bouncing-back phenomenon, 68, 69f

Boundary conditions, EDM, 96–97
  discharge cycles and cooling cycles,
      99, 99f
  maximum discharge gap, 100
  mesh end points, 97–98, 97f
  simultaneous sparks, number, 98–99
Built-up edge (BUE), 250, 262
  effect of, 263
  geometry, 264
    cutting edge, 264f
  laser scanning microscopy, 264f
  micromachining of Ti6Al4V, FEM,
      263–267
  parameters, 265t

C

Carbon fiber reinforced polymer (CFRP),
      failure analysis, 51–78
  drilling operation, simulation, 76–78,
      76f, 77f, 78f
  elastoplastic-damage model, 59–61
  numerical modeling
    interface delamination modeling,
        61–62
    machining parameters and
        boundary conditions, 54–57, 56t
  orthogonal cutting, simulation
    chip formation process, 62–64
    clearance angle effect, 68–70, 69f
    cutting forces, prediction,
        64–66, 65f
    cutting speed effect, 73–74, 73f
    depth of cut a_p effect, 71–73, 72f
    fiber orientations effect,
        interlaminar delamination,
        74–76, 74f, 75f
    subsurface damage induced,
        prediction, 65f, 66–67, 66f
    tool edge radius effect, 70, 70f
    tool rake angle effect, 67–68,
        67f, 68f
  overview, 52–53
  T300/914 composite, mechanical
      properties, 55, 55t
Centrosymmetry (CS) parameter, 157
Chip formation process
  orientation case of θ = −45°, 63–64
  orientation case of θ = 45°, 62–63
  orientation case of θ = 90°, 63
  progressive failure analysis, 62–63, 63f,
      64f, 65f

Chip morphology, 256, 273, 273f,
      304–306, 304f
  at edge radius, 258f
    cutting speeds and, 261f
  FE simulations, 265f–266f
Chip specific energy, 220
Clearance angle effect, 68–70, 69f
Clustering algorithm, 155
Cockcroft–Latham model, 296–297
Cohesive-zone elements (CZE), 52
Computational mesh, 225
Computer tomographs, 158
Conduction heat transfer, 90–91, 90f
Continuum numerical method, 33
Convection heat transfer, 91–92, 92f
Conventional machining processes,
      249, 263
Coolant fluid, 238
Coulomb's friction model, 18, 254–255, 265
Crystallography influence, 267–273,
      271f, 271t
Crystal plasticity (CP) theory, 247,
      269–270, 272t
CS (centrosymmetry) parameter, 157
Curvature difference, 217
Cutting and feed forces, 20–21, 20f, 20t
Cutting and thrust forces, 255–256
  at cutting speeds, 260f
  different edge radii, 257f
  evolution, 272f
Cutting depth effect
  chip size, 71, 72f
  damage depth, 72, 73f
  machining forces, 71, 71f
Cutting force, 218–219
  modeling, 119–120, 120f, 121f
  prediction, 64–66, 65f
  signals, 306, 306f
Cutting path generation, 120–123, 123f
Cutting speed effect, analysis, 260–261,
      260f, 261f
Cutting temperature distribution,
      307–308, 308f
CZE (cohesive-zone elements), 52

D

DEFORM-2D™ software, 301
DEM (discrete element method), 53
Dielectric fluid turbulence, 91, 91f
Discharge cycles and cooling
      cycles, 99, 99f

Discrete element method (DEM), 53
Drilling operation, simulation, 76–78, 76f, 77f, 78f
Dynamic shear stress (DSS), 268

E

EAM (embedded atom method), 145
EBSD (electron backscatter diffraction), 158
Edge radius effect, FEM, 256–260
   cutting and thrust forces, 257f
   effective stresses and chip morphology, 258f
   micromachining-induced stresses, 259, 259f
   temperature distributions, 260f
Effective plastic strain rate, 21, 21f
Effective stress
   at edge radius, 258f
   FE simulations, 265f–266f
   micromachining-induced stress, 259f
EFG (element-free Galerkin) method, 33
Elastoplastic damage behavior law and interface delamination
   plastic model, 59–61
   progressive damage analysis, 57–59
Elastoplastic-damage model, 59–61
Electric discharge machining (EDM), 180
   scheme of, 183f
   TLBO, 180–206
      experiment, 182–184, 183t
      multiobjective, 189–199
      overview, 180–182
      unified learning system development, 184–206
Electrodischarge machining (EDM), numerical modeling, 81–105
   2D and 3D modeling, 102–103, 103f
   elements, 84
   formulation, 86–92, 87f
      heat transfer, 87–92
   material properties, 100, 100t
   modeling of large parts, 103–104
   objectives, 83–86
   overview, 82–83
   plasma channel, 84–85, 85f
   precision of meshing, limits, 104–105
   sinker process, 84, 84f
   structure, 92–102
      boundary conditions, 96–100
      process parameters, 100–102, 101t

simulation mesh, 93–94, 93f
temperature transfer equation and equivalent temperature concept, 94–96, 95f, 96f
Electron backscatter diffraction (EBSD), 158
Electron–phonon coupling approach, 153
Element-free Galerkin (EFG) method, 33
Embedded atom method (EAM), 145
Empirical model, 292
ε-insensitive loss function, 186, 186f
Equation of state (EOS), 36
Euler angles advantages, 159

F

FEA. *See* Finite element analysis (FEA)
Fiber orientation, 52, 64, 66
Fine-grained structure, 171
Finite element analysis (FEA), 300
   Inconel 718, material modeling on near-micromachining
      chip morphology, 304–306
      contact modeling, 301–302
      cutting temperature distribution, 307–308
      finite element formulation and boundary conditions, 301
      machined surface alteration, 308–311, 309f, 310f
      machining forces, 306–307
      machining tests, 300–301
      material modeling, 302–304
   and simulation, material behavior modeling, 282–311
      identification techniques, 297–300
      machining modeling, material constitutive equations, 283–297
      overview, 282–283
Finite element discretization, 10–12
Finite element formulation and boundary conditions, 301
Finite element method (FEM), 28–32, 53, 108–109, 247, 282, 302f
   3D, 213
   application, 251
   concept, 28, 28f
   fly-cutting machine tool, 132f
   frameworks, 251
   machine tool, 131–134, 133f
   metal cutting, 30–32

Finite element method (FEM) (*Continued*)
    microcutting of Ti6Al4V
        in BUE, 263–267, 267f
        cutting speed effect, 260–261, 260f, 261f
        edge radius effect, 256–260
        friction conditions, effect, 261–263,
            262f, 263f
        friction modeling, 254–256
        material model, 251–254, 253t, 254t
    micromachining, 267–273, 271f, 271t
    MPMT, 131–134, 133f
    overview, 28–30
    SPH *versus*, 32, 33f
    state space model on, 123–131
        air spindle, 125f, 130–131
        fly-cutting machine tool
            configuration, 123–124, 124f
        stiffness equivalence principle,
            127–130, 130f
    tool tip response comparison, 133, 133f
The Finnis–Sinclair interaction
    potential, 153
Flow stress model, 291
Fourier's law, 153
Fractal dimension, 148
Fracture mechanics model, 38
Friction and deformation processes, 152
Friction energy effect, 168
Friction modeling, 254–256

G

Gaussian radial basis function, 188
Gaussian size-distribution, 149, 150f
Gimbal lock, 159
Grind-hardening/grinding-hardening
    process, 211–241
    analytical model, 221–223, 228–231
    ANSYS software, numerical solution
        general detail, 223–225
        results, 232–236, 232f, 235t, 236f
        thermomechanical material
            properties, 226–228
        verification, 226
    grinding fluid model, 237–241
    grinding wheel topology, 215–217
    heat production and partition, 219–221
    method's accuracy test, 236–237,
        236t, 237t
    overview, 211–215
    process forces
        cutting force, 218–219
        slip force calculation, 217–218

Grinding fluid model, 237–238
    effect, 237–241
    in grinding-hardening experiments,
        238–241, 238t
    HPD comparison, 241f
    maximum temperature comparison, 241f
    results, 239t–240t
Grinding polycrystalline ferrite, 162–172,
    163f, 165f, 167f
Grinding wheel topology model, 215–217

H

Hardness penetration depths (HPDs),
    213–214
    grinding fluid model, 241, 241f
    zones in, 214
Heat transfer
    electrical discharges produced, 87–89
    in workpiece, 89–92
        conduction, 90–91, 90f
        convection, 91–92, 92f
        heat input distribution, 88, 88f
Homogeneous equivalent material (HEM),
    53–54
HPDs. *See* Hardness penetration depths
    (HPDs)
Huber loss function, 186
Hue–saturation–value (HSV) scheme, 160
Hybrid/advanced methods, 299–300
Hydrodynamics, 32

I

ICF (inertial confinement fusion)
    program, 116
Identification techniques, machining
    modeling
    dynamic tests, 297–298
    hybrid/advanced methods, 299–300
    machining-based inverse methods,
        298–299
IMPMTS. *See* Interaction between the
    machining process and the
    machine tool structures
    (IMPMTS)
Inconel 718 alloy, FEA of
    near-micromachining
    contact modeling, 301–302
    finite element formulation and
        boundary conditions, 301
    JCP models and parameters, 303t
    machining tests, 300–301

material modeling, 302–304
physical properties, 304t
results
    chip morphology, 304–306, 304f
    cutting temperature distribution,
        307–308, 308f
    machined surface alteration, 308–311
    machining forces, 306–307, 307f
Inertial confinement fusion (ICF)
    program, 116
Integrated circuit (IC) fabrication
    techniques, 246
Integrated method
    dynamic performance analysis, MPMT,
        109–115, 110f
    modeling process
        cutting force modeling, 119–120,
            120f, 121f
        cutting path generation,
            120–123, 123f
        KDP crystal modeling, 116–119
Interaction between the machining process
    and the machine tool structures
    (IMPMTS), 108
    fly-cutting machining, 115–131, 122f
    FE model, 132, 132f
    FEM, state space model on, 123–131
    integrated method, modeling
        process, 116–123
    simulation, 134–137, 134f, 135f, 136f, 137f
Internal variables, data transfer, 5
Interpolation/shape/smoothing functions,
    33–34
Inverse methods, machining tests, 283
Inverse pole figure (IPF) coloring
    scheme, 160
Isotropic constitutive model, 13

J

Johnson–Cook (JC)
    constitutive model, 13–14, 14t
    material constitutive equation, 286–287
    material model, 252–253
Johnson–Cook plasticity (JCP) model,
    40–42, 302
Johnson–Holmquist material model, 44

K

KDP crystal. *See* Potassium dihydrogen
    phosphate (KDP) crystal,
    modeling

Kernel functions, 187–188
Kikuchi patterns, 158

L

Lagrange multipliers, sets, 182, 193, 199
Lagrangian-based method, 32
Lagrangian continuum
    boundary conditions, 8
    mass balance, 7–8
    momentum equation, 6–7
    thermal balance, 7
Lagrangian *versus* Eulerian meshes,
    29–30, 29f
The Lennard–Jones (LJ) potential,
    144–145, 144f
Lindgren and coworkers' model, 291
Liquid nitrogen, 213
LJ (the Lennard–Jones) potential,
    144–145, 144f
LS-DYNA software, 36–37, 39, 45
Lumped mass models, 108–109

M

Machined surface alteration, 308–311,
    309f, 310f
Machining-based inverse methods,
    298–299
Machining forces, 306–307, 307f
Machining modeling, material constitutive
    equations
    damage modeling, 296–297
    plasticity model
        microstructure-based models,
            291–295
        phenomenological, 284–288
        physical-based models, 289–291
Machining parameters and boundary
    conditions, 54–57, 56t
    model interface cohesive
        elements, 57t
    tool–workpiece couple, 54, 54f
Machining process and the machine tool
    (MPMT), 108–137
    dynamic performance analysis,
        integrated method
        state space model, establishment,
            110–112, 111f
        theoretical basis, 112–115
    FEM, 131–134, 133f
    overview, 108–109
Maekawa's model, 288

Material removal rate (MRR), 181–182
  current and pulse-off time on, 204f
  current and pulse-on time on, 204f
  estimation
    concurrent, 202f
    MATE in, 201f
    testing, 203t
  normalized, 196, 198
  pulse-on time and pulse-off time on, 204f
MD simulation. *See* Molecular dynamics (MD) simulation
Mean absolute training errors (MATE), 192, 201f
Mechanical micromachining processes, 246
  challenges, 247–250
    MUCT, 248–250
  FEM, 245–273
    microcutting of Ti6Al4V, 251–267
    overview, 245–247
    size effect in, 247–248
Mechanical problem
  boundary conditions, 8
  stress update algorithm, 16–17
Mechanical threshold stress (MTS) model, 289
Mechanistic modeling, 26
MEMS (micro electro mechanical system), 246
Merchant's shear plane model, 27
Mesh-free methods, 32
Meshless local Petrov–Galerkin (MLPG) method, 33
MET (microengineering technologies), 246
Metaheuristic techniques, 194
Metal cutting, 26–28
  deformation zones, 26, 27f
  FEM, 30–32
  process, 2
  SPH, 36–45
Micro electro mechanical system (MEMS), 246
Microengineering technologies (MET), 246
Microfabrication techniques, 246
Micromachining process, 268
Microstructure-based models, 291–292
  empirical model, 292
  physical-based model, 294–295
  semi-empirical model, 292–294
Microsystem technologies (MST), 246
Mie–Grüneisen EOS, 39
Minimum uncut chip thickness (MUCT), 248–250

MLPG (meshless local Petrov–Galerkin) method, 33
Molecular dynamics (MD) simulation, 142
  classical, 143–146
  nanomachining, 141–172
  potentials of, 143
MPMT. *See* Machining process and the machine tool (MPMT)
MRR. *See* Material removal rate (MRR)
MST (microsystem technologies), 246
MTS (mechanical threshold stress) model, 289
MUCT (minimum uncut chip thickness), 248–250

N

Nano-Indenter XP, 117
Nanomachining, large-scale MD simulations, 141–172
  atomistic simulation, 146–158
    contact area determination, 156–157
    dynamically identifying removed matter, 153–156, 156f
    external constraints, boundary conditions, and simulation procedure, 151–152
    preparing model, 146–151, 151f
    removing heat, 152–153
    topography, evaluating workpiece, 157–158
  classical MD simulations, 143–146
  grinding polycrystalline ferrite, 162–172, 163f, 165f, 167f
  overview, 141–143
  system visualization, 158–162, 160f
    atomic displacement, 160–161, 161f
    grain orientation, 158–160
    temperature, 162
Nonlinear SVM regression model, 185f
Normalization coefficient, 216
Numerical discretization/particle approximation, 33–35

O

OFHC (oxygen-free high conductivity copper), 255
Open visualization tool (OVITO), 158–159
Orientation analysis, 159
Orthogonal (2D) cutting models, 31–32, 31f

Orthogonal cutting, simulation
  chip formation process, 62–64
  clearance angle effect, 68–70, 69f
  cutting forces, prediction, 64–66, 65f
  cutting speed effect, 73–74, 73f
  depth of cut a_p effect, 71–73, 72f
  fiber orientations effect, interlaminar
      delamination, 74–76, 74f, 75f
  subsurface damage induced, prediction,
      65f, 66–67, 66f
  tool edge radius effect, 70, 70f
  tool rake angle effect, 67–68, 67f, 68f
Orthogonal parallelepiped workpieces,
      215, 225
OVITO (open visualization tool), 158–159
Oxygen-free high conductivity copper
      (OFHC), 255

P

Particle finite element method (PFEM),
      1–22
  chip formation, 19f
  constitutive model, 12–14
  cutting and feed forces, 20–21, 20f, 20t
  finite element discretization, 10–12
  Lagrangian continuum, 5–8
      boundary conditions, 8
      mass balance, 7–8
      momentum equation, 6–7
      thermal balance, 7
  material response, 21–22
  metal cutting processes, numerical
      simulation, 4–5
      internal variables, data transfer, 5
  orthogonal cutting, 2D plane strain, 19f
  remeshing steps, 6f
  in solid mechanics, 4
  stress update algorithm, 14–17
  variational formulation, 8–10
      mass conservation equation, 9
      momentum equations, 8–9
      thermal balance equation, 9–10
Particle swarm optimization (PSO), 181
Parting line, 31
Pause phase, 87, 89
PDZ (primary deformation zone), 26
PFEM. *See* Particle finite element method
      (PFEM)
Phenomenological plasticity models, 284
  JC's constitutive equation, 286–287
  Maekawa's model, 288

power-law models, 284–286
  ZA model and modified versions,
      287–288
Physical-based models, 289, 294–295
  BCJ material model, 289–291
  Lindgren and coworkers' model, 291
  MTS model, 289
PIM (point interpolation method), 33
Plasma channel, 84–85, 85f
Plasticity model, 284–295
  microstructure-based models,
      291–292
      empirical model, 292
      physical-based model, 294–295
      semi-empirical model, 292–294
  phenomenological plasticity
      models, 284
      JC's constitutive equation, 286–287
      Maekawa's model, 288
      power-law models, 284–286
      ZA model and modified versions,
          287–288
  physical-based models, 289
      BCJ material model, 289–291
      Lindgren and coworkers' model, 291
      MTS model, 289
Point interpolation method (PIM), 33
Polycrystalline ferritic workpiece
      model, 146
Polyhedral template matching (PTM)
      algorithm, 158–159
Potassium dihydrogen phosphate (KDP)
      crystal, modeling, 116–119
      nano-indentation experiment,
          116–117, 117f
      topography requirements, 116f
      Young's modulus and micro-hardness,
          118, 118t
Power-law models, 284–286
Power spectral density (PSD), 148
Primary deformation zone (PDZ), 26
Primary failure, 62
Process parameters, EDM, 100, 101t
  constant parameters, 100–101
  output parameters, 102
  random parameters, 101–102
Progressive mesh, 104, 104f
PSD (power spectral density), 148
PSO (particle swarm optimization), 181
PTM (polyhedral template matching)
      algorithm, 158–159
Pulse phase, 87

Q

Quadratic loss function, 186
Quasi-static approach, 53
Quasi-static conditions, 309–310, 310f
Quasi-static two-temperature method, 153
Quenched steels, 213

R

Response surface methodology (RSM), 299
Robust loss function, 186

S

Scanning electron microscope (SEM)
    images, 43, 85, 85f
Secondary deformation zone (SDZ), 26
Secondary fracture, 62
Semi-empirical model, 292–294
SEM (scanning electron microscope)
    images, 43
Severe plastic deformation (SPD), 26
SFTC DEFORM-2D software, 256
Shape functions, 29
Shear stress, 152, 171
SHPB (Split-Hopkinson pressure bar)
    test, 42
Simulation mesh, 93–94, 93f
Single-shear plane model, 26–27
Slip force calculation, 217–218
Smoothed particle hydrodynamics
    (SPH), 25
    advantages and limitations, 35–36
    versus FEM, 32, 33f
    metal cutting, 36–45
    numerical discretization/particle
        approximation, 33–35
    overview, 32–33
    solution procedure, 35
Smoothing/support domain, 33, 34f
SPD (severe plastic deformation), 26
Special cutting energy, 218
Specific cutting energy, 248
SPH. *See* Smoothed particle
    hydrodynamics (SPH)
Split-Hopkinson pressure bar (SHPB) test,
    42, 283, 297–298
Spread-range (SR) ratio, 193–194, 202f
Spring element group, 131, 131f
Standard stereographic triangle (SST), 159
State space model, 110–112, 111f
    based on FEM, 123–131

air spindle, 125f, 130–131
    fly-cutting machine tool
        configuration, 123–124, 124f
    stiffness equivalence principle,
        127–130, 130f
Sticking–sliding model, 255
Stress update algorithm
    discretized equations, transient
        solution, 14
    mechanical problem, 16–17
    thermal problem, 17
    thermo-elastoplasticity model, 14
Support vector machine (SVM), 181,
    184–189
    internal parameters, 192t
    unified learning, 198t
    nonlinear regression model, 185f
System visualization, tomographs, 158, 160f
    atomic displacement, 160–161, 161f
    grain orientation, 158–160
    temperature, 162

T

Taylor test, 283, 297
TDZ (tertiary deformation zone), 26
Teaching factor (TF), 189
Teaching-learning-based optimization
    (TLBO), 181, 189
    EDM, 180–206
    modified, 200t
    multiobjective, 189–199
Temperature distribution, 21, 22f
Temperature transfer equation and
        equivalent temperature concept,
        94–96, 95f, 96f
Termination criterion, 193
Tertiary deformation zone (TDZ), 26
TF (teaching factor), 189
Thermal problem
    boundary conditions, 8
    stress update algorithm, 17
Thermo-elastoplasticity model, 12–14
    elastic response, 12
    flow rule, 13
    implicit integration scheme, 15
    JC constitutive model, 13–14, 14t
    yield condition, 13
Thermomechanical material properties,
    226–228, 228f, 228t
Thermostatting approach, 153
Three-dimensional (3D) mesomechanic
    model, 53

Thrust force signals, 306, 307f
Time-dependent analysis, 159
Time integration algorithm, 145
Time step, 225
Titanium alloy Ti–6Al–4V, 13
TLBO. *See* Teaching-learning-based
optimization (TLBO)
Tomographs
exemplary atomic displacement, 161f
substrate, 160f, 164f
system visualization, 158–162, 160f
Tool Craft A25 EDM machine, 183
Tool edge radius effect, 70, 70f
Tool rake angle effect, 67–68, 67f, 68f
Topography, evaluating workpiece,
157–158
Triangular element, 127, 127f
Two-dimensional (2D) periodicity, 146, 148

U

Ultrasonic-assisted grinding (UAG), 40
Ultrasonic-assisted scratching (UAS), 40
Undeformed chip thickness, scenarios,
249, 249f
Unified learning system development,
EDM, 184

multiobjective TLBO, 189–199
modifications and marching
procedure, 190–199
SVM, 184–189
testing, 199–206

V

Verlet algorithm, 145–146
von Mises stress field, 21, 22f
von Misses–Huber yield criterion, 13
Voronoi cells, 147–148
Voronoi construction, 146–147, 147f
VUMAT subroutine, 52–53, 56

W

Wear particles, 166, 168–169
Weight-combining method, 195
Workpiece geometry, 225
Workpiece material, 213, 225t

Z

Zener–Hollomon parameter, 292–293
Zerilli–Armstrong (ZA) model and
modified versions, 287–288